Space and Society

The Space and Society series explores a broad range of topics in astronomy and the space sciences from the perspectives of the social sciences, humanities, and the arts. As humankind gains an increasingly sophisticated understanding of the structure and evolution of the universe, critical issues arise about the societal implications of this new knowledge. Similarly, as we conduct ever more ambitious missions into space, questions arise about the meaning and significance of our exploration of the solar system and beyond. These and related issues are addressed in books published in this series. Our authors and contributors include scholars from disciplines including but not limited to anthropology, architecture, art, environmental studies, ethics, history, law, literature, philosophy, psychology, religious studies, and sociology. To foster a constructive dialogue between these researchers and the scientists and engineers who seek to understand and explore humankind's cosmic context, the Space and Society series publishes work that is relevant to those engaged in astronomy and the space sciences, while also being of interest to scholars from the author's primary discipline. For example, a book on the anthropology of space exploration in this series benefits individuals and organizations responsible for space missions, while also providing insights of interest to anthropologists. The monographs and edited volumes in the series are academic works that target interdisciplinary professional or scholarly audiences. Space enthusiasts with basic background knowledge will also find works accessible to them.

More information about this series at http://www.springer.com/series/11929

Konrad Szocik
Editor

Human Enhancements for Space Missions

Lunar, Martian, and Future Missions to the Outer Planets

Springer

Editor
Konrad Szocik
Department of Social Sciences
University of Information Technology
and Management in Rzeszow
Rzeszów, Poland

ISSN 2199-3882 ISSN 2199-3890 (electronic)
Space and Society
ISBN 978-3-030-42035-2 ISBN 978-3-030-42036-9 (eBook)
https://doi.org/10.1007/978-3-030-42036-9

This Springer imprint is published by the registered company Springer Nature Switzerland AG
The registered company address is: Gewerbestrasse 11, 6330 Cham, Switzerland

Introduction

Human enhancement (HE) as such is an idea which is broadly discussed mostly by philosophers and ethicists. In our book, we decided to discuss together two controversial issues: HE and human space missions. Although both issues are controversial, each of them has their own advocates and opponents, respectively. Accordingly, in both cases, strong and clear rationale and justification are challenging. Let us briefly explain what we mean by HE, and what meanings of that term are applied in our volume.

HE in a broad sense includes all ways of improving, "enhancing," betterment of human natural, average physical and psychological condition. In that broad sense, every intentional action taken by humans which aims at improving their capacities and functions is considered as an enhancement. Basic and the most commonly used methods of such enhancement include nutrition, physical exercises, or education. However, such kind of enhancement is not interesting from the philosophical and ethical point of view as long as is not controversial in ethical and moral terms. Someone could argue that there are some kinds of nutrition, exercise, or education which in some specific forms and under some specific conditions may cause ethical issues and, as such, are an interesting object for ethical considerations. However, humans in general agree that our species should eat appropriate food, have some kind of physical activities—mostly when required in specific social and professional groups, and take some minimal level of education. All these activities improve us and get something more than our basic biological equipment may provide. In our book, we do not discuss such kinds of enhancement which are usually broadly accepted. We arc interested in another kind of enhancement, enhancement in a narrow sense. HE in a narrow sense includes such kinds of enhancement which are controversial in an ethical and moral sense. That descriptive definition is required here because there is a couple of activities and actions which cause ethical controversies. HE is controversial when: (a) a particular activity is only one of many ways which lead us to achieve a particular target; (b) when an activity interferes with human biology; (c) when an activity is aimed at trivial purposes; (d) when an activity gives an advantage over others which is considered as a lack of just and fair competition. There is a couple of criteria, and that list is far from being comprehensive. However, when we analyze particular methods of

HE which are usually considered as controversial, they often include at least one of the mentioned criteria. For instance, a hypothetical person in the future who will have modified genes which will provide an advantage over others in some of cognitive capacities, may be accused of behaving in an ethically controversial way because of (a), (c), and (d)—(b) in this case is a context-dependent (it may be reversible—and as such less controversial or uncontroversial, or irreversible—and as such more controversial). However, an analogical genetic modification aimed at preventing disease may lose any ethical controversy. This is why the idea of human enhancement in a broad sense is difficult to be defined in a clear and precise way. Humans, in general, care for means which are used to achieve a particular target. For this reason, an undergraduate who studies hard in a traditional way by reading books is usually considered as a better undergraduate than his hypothetical colleague who would achieve the same target by some kind of medical enhancement (let us assume for the sake of argument that such pill or appropriate genetic modification would be possible in some point in the future). The same could be said about other fields of human life and activity such as physical performance or moral virtues. People value effort and self-sacrifice. However, one could ask if it is not better if humanity could change their moral intuitions and to asses both ways of enhancement in the same way, or to treat both of them as neutral in moral terms. It is worth keeping in mind such factors as saving time or cost/benefit ratio. If pharmacological or genetic cognitive enhancement meets those both criteria better than traditional education or physical exercise do, those factors are strong reasons to change our moral intuitions and cognitive biases.

People also value purposes of HE. HE aimed at trivial purposes is usually criticized. But the same kind of procedure aimed at preventing diseases or possibly research purposes may be treated as acceptable and uncontroversial. However, even in this case, the issue is far from being unequivocal. Not every HE in a narrow sense aimed at medical purposes is considered as morally acceptable. There are at work other criteria such as safety and efficacy. But even if those criteria are met, other ones play a role such as being invasive or non-invasive, reversible or irreversible, heritable or non-heritable, or based on a requirement of informed consent, or not.

Having in mind all of the mentioned criteria and contexts, we decided to move the current discussion on HE on Earth to the context of human space missions. This fact makes our book interesting both for readers interested in the idea of HE and for readers interested in human space missions. The former may treat our book as a kind of thought experiment which considers the idea of HE in a particular context. The space environment is here a case study. In this volume, we argue that space is a specific environment which can change the moral and ethical status of HE when compared with Earth. The latter, who are focused more on space policy and the idea of human spaceflights, may find in our book a source of inspiration and a new framework for their current thoughts. Here we show that one of important but missing contexts in a debate and scholarship of future long-term crewed space missions is an idea of HE, considered as a kind of preventive countermeasure to hazardous factors in space. The idea of HE when applied to the context of space missions does not introduce only the abovementioned ethical issues. That idea is important for scientists, mostly medical and biological, who may study human resistance and performance during long-term human space missions. That idea should be of high value for engineers who study and work on appropriate countermeasures. Finally, political and social scientists together with economists—obviously joined by medics, biologists, engineers and many others—may discuss the idea of HE in relation to the broader issue of rationale for human spaceflights. Dependently not only on a public but also political assessment of importance of human long-term spaceflights, HE may be considered as more or less justified. There is a couple of possibilities—also considered in our book but that topic is still open for further considerations—of hypothetical correlations between HE and rationale for human spaceflights. There are good reasons to assume that when rationale to send humans to space is strong, analogically rationale to enhance astronauts becomes strong as well—still assuming that currently possessed countermeasures to hazardous factors in space are not alternative to HE. However, one could argue why we should not apply HE when its rationale is weak or, at least, unclear and controversial. Another one could argue that humans should always be enhanced independently on the rationale for space mission. However, when the rationale for human spaceflight is relatively weak, an alternative for such mission is a decision about cancelation or postponing that mission until some point in the future when alternatives to HE will become available. The issue of rationale is getting more complicated when the two kinds of targets for human spaceflights—one realistic and occurring today and in the near future, and another one more futuristic but possible in the future—are considered, namely research purposes and the idea of space refuge. Research purposes, mostly astrobiological study of Mars and other space bodies, is one of the main targets of human space missions. If we assume that HE is a reasonable option, it is worth to assess how important is astrobiological research to justify HE. While people who share so-called permissive position to moral issues are prone to accept HE for this and other purposes, conservative supporters of crewed astrobiological research may treat that idea as controversial and, consequently, decide to stop the human space program aimed at the astrobiological research when HE would become obligatory. The issue is changing dependently on the status of space mission. When a mission is exclusivist, rationale for HE seems to be stronger. However, when the idea of inclusivist mission is considered, there are good reasons against the idea of HE. An obligatory HE would be a kind of policy which works against the idea of inclusivism. Such scenario would also change a moral and ethical framework for the idea of HE for space. While we may be prone to accept some kind of HE for a small number of astronauts, HE considered as obligatory policy in a project of mass colonization of space, mostly when motivated by looking for safe space refuge, could go against our moral intuitions.

This volume is divided into two parts. Part I combines conceptual issues with medical and biological. Part II is focused on philosophical and ethical issues which appear on the intersection of philosophy with science and technology. As we want to show in our volume, medical and biological study of human resistance and performance in space, and looking for effective countermeasures shall meet, faster or later, philosophical, mainly ethical issues. It is hard to avoid scientific study of human biology in space without ethical assessment of the acceptability of some of procedures and modifications. This is why part I combines mentioned biological and medical studies with conceptual and sometimes methodological and philosophical considerations. Part II, while dominated by philosophical and ethical approach, is rooted in the scientific background and refers to future prospects of humanity. Scientists skeptical to philosophy should read that second part as a kind of thought experiment but also as an example of futures studies. The value of this book underlined mostly by the second philosophical part lies in the fact that we want to underline long-term consequences of not only space missions as such, but mostly of our decisions which we will make now and in the near future. We are going to show how strictly connected are different issues which, at least apparently, seem to be separated and far from each other.

Part I of the book includes ten contributions. In Chap. 1, Margaret Boone Rappaport and Christopher J. Corbally offer an excellent introduction to the main issue of our volume—human enhancement for spaceflights. They point out that humans still evolve by biological and cultural factors and, as such, artificial enhancement is not necessary as far from biological modifications which happen in an unintentional way, as we can assume. They argue that human enhancement for spaceflights should be considered as one of the possible countermeasures, and specificity of space environment changes its ethical status from controversial to a mere alternative.

In Chap. 2, Arvin M. Gouw discusses a/the possible utility of CRISPR gene editing for astronauts' protection against negative effects of space radiation and microgravity. He considers the possibility of CRISPR-induced hibernation, while hibernation effectively limits the negative effects of space radiation. Another hypothetical application of CRISPR discussed by Gouw is coping with muscle atrophy during space travels. Finally, he shows that therapy/enhancement distinction often fails when applied to space environment, and many of hypothetical modifications considered as an enhancement on Earth get the status of therapeutical interventions when applied to astronauts. To make CRISPR gene editing for spaceflights as much uncontroversial as possible, he recommends limiting CRISPR only to somatic cells, to guarantee its reversibility and to make it inducible only to purposes of space missions.

Evie Kendal, in Chap. 3, develops Gouw's idea to discuss human enhancement for space in terms of prophylaxis. She considers an idea that enhancement in space environment may work as an extension of preventive medicine. Kendal introduces a useful conceptual framework which may be used to assess rationale for particular human enhancement in terms of ethical acceptability and risk/benefit analysis. Her criteria include such ideas as being morally permissible, effects of not being enhanced, and the efficacy of non-biological alternatives, reversibility or irreversibility.

Steven Abood, in Chap. 4, discusses four conceptual approaches to the definition of biological species. He shows how a definitional and conceptual approach may determine our way of thinking about human species. Abood points out that dependently on our way of thinking about humans as a species, we can treat differently more and less radical and invasive human enhancements for space.

Consequently, we may be prone to accept the same enhancements as modifications which do not change our species specificity, or—within another conceptual framework—we may be prone to argue against enhancements which would be perceived as challenging our species identity.

In Chap. 5, Martin Braddock discusses physical limits to human space exploration such as time journey, microgravity, space radiation, and the challenge of isolation in space. Then, he considers possible countermeasures based on human enhancements including gene editing, tissue engineering, and regenerative medicine including 3D bioprinting, exoskeleton, and brain–computer interface. At the end, he considers an idea of human avatar who—if equipped in general artificial intelligence—would be able to extend our presence in remote locations without physical presence of human astronauts.

In Chap. 6, Mark Shelhamer discusses challenges for human vestibular and visual systems which appear during spaceflights, and possible countermeasures. He considers possible enhancements such as vestibular prosthesis, corneal replacement membrane, or retinal implant. He concludes that having in mind risks for human health, some ethical questions are difficult to avoid such as the question about rationale for such missions, or the question about a need to wait for progress in conventional technology which will make possible coping with hazardous factors in space without human enhancement.

In Chap. 7, Andrew B. Newberg and David B. Yaden offer a new conceptual approach to the issue of human enhancement for space. They discuss the overview effect experienced by astronauts in space in terms of possible enhancement. As such, the overview effect is considered as a good candidate to be a personal, psychological enhancement.

Gonzalo Munévar, in Chap. 8, discusses possible applications for space of two kinds of enhancements such as implants and genetic interventions, mostly by CRISPR. He points out that such radical human enhancements for space do not meet two minimal criteria such as safety and efficacy. Munévar argues against those enhancements and suggests that mission planners should focus their attention on non-invasive countermeasures to negative effects of microgravity and space radiation. They include thicker walls of spacecraft using, among others, supplies and polyethylene, or artificial gravity. The key ethical point of Munévar is based on an assumption that supporters of human enhancements for space use a false dilemma which states that human enhancement offers the unique effective countermeasures to hazardous space factors.

Simon P. Worden, in Chap. 9, discusses the idea of so-called ultimate human enhancements. The ultimate human enhancements include nanoprobes which—if sent to exoplanets, for instance, and equipped in biomechanical systems—could imply evolution of new living forms. Such forms could also involve humans if mentioned biomechanical systems would include human genetic code. The idea discussed by Worden is worth deeper consideration. Because there are good reasons to assume that humans will not approach exoplanets in the foreseeable future, uncrewed missions realized by nanoprobes may be likely the unique chance to approach those remote locations in space, mostly when humanity is aimed at implementing new terrestrial life.

In Chap. 10, Riccardo Campa presents a detailed review and analysis of various techniques currently applied in anti-aging medicine. He considers anti-aging therapies in terms of possible means of human enhancement which could be used to regenerate and rejuvenate humans in space missions.

Part II includes nine contributions. In Chap. 11, Ziba Norman and Michael J. Reiss discuss possible advantages of somatic gene editing for long-term human spaceflights. They point out that there are good reasons to treat somatic gene editing in terms of treatment when it is aimed at preventing disease and hazards in space. As such, gene editing of somatic cells should not be treated in terms of moral controversy. They address the risk of formation of new human species separated from human population on Earth. Such speciation, biologically possible under some conditions, would be challenging for humanity and lead to a highly undesirable situation in which one human species—terrestrial or cosmic—would be perceived as better than another.

In Chap. 12, from the viewpoint of virtue ethics, Koji Tachibana examines the ethical validity of so-called drastic genetic enhancement of the human body in space for human survival as a species. He argues that such an enhancement might enable humankind to colonize other planets and accordingly save humanity but could cause various ethical issues. One is the identity problem that an applied gene manipulation would not only change human figure and appearance but also harm the identity as humankind. Another more critical ethical issue is possible exploitation of transitional generations, who would lose their dignity as human beings because they are manipulated and exploited as the means for producing the later generations. He discusses that such a genetic enhancement in space is ethically unacceptable even if it is the only way to save humanity. He then argues the moral possibility of accepting existential risks and human extinction by examining three sorts of voluntary death, namely euthanasia, martyrdom, and Socratic death. Tachibana concludes, from the viewpoint of virtue ethics, the supremacy of human survival as a species can be canceled if the Socratic notion of human virtue is taken into account as one of the ethical criteria for human space explorations and exploitations.

Tony Milligan and Shin-Ichiro Inaba, in Chap. 13, discuss the idea of life extension by human enhancement for the purposes of long-term human spaceflights. They find some benefits of life extension such as the very long time of space expeditions aimed at exploring the Solar System. They consider three objections to the idea of life extension such as the risk of overpopulation, unfair distribution, and a risk of loss of life's meaning.

James S. J. Schwartz in Chap. 14, discusses the idea of a space settlement and spaceflights open for everyone. He starts from an important conceptual remark when he recommends using term "modification" which, in contrast to alternatives such as treatment and enhancement, is a value-neutral term. He offers two arguments against human enhancement for space. One of them is an argument by analogy with disability. Unenhanced people may be excluded from any space program, and they are compared to disabled people on Earth. Schwartz points out that our decision on possible human enhancement for space is a combined result of our free choice and decision to invest funds in a particular area. Instead of funding research useful for future human enhancement, we are free to increase our effort in progress in space habitats which would change the specificity of future missions and possibly make the idea of human enhancement irrelevant. Schwartz argues that humanity is not determined today to choose some particular way of space settlement and to exclude all other alternatives. His main point is to shape an inclusivist space community, and the idea of human enhancement is one of main obstacles to it. His second objection to human enhancement is based on the idea that human space travels and space settlement should be taken for granted as freely accessible for everyone.

In Chap. 15, Kelly C. Smith and Caleb Hylkema assume that space settlement is a reasonable project for humanity due to possible existential risks on Earth, and human enhancement including gene editing should be considered as a serious option. They defend a pragmatic approach to space settlement, in contrast to an idealistic account. They discuss challenges and weak assumptions of those objections to human enhancement for spaceflights which are based on the idea of obligatory informed consent. They argue that genetic enhancement may be a kind of our duty when it will work as a necessary procedure to increase chances for survival and healthy life for space settlers. They also reject a social-inequality argument and show that while such an argument may be reasonable in relation to the terrestrial population, this is not the case of exclusivist community of space settlers. They argue with an idealistic vision of space settlement offered by Schwartz in Chap. 14 of this volume. They point out that at least in the beginning of space colonization, it will be hard—because of pragmatic constraints—to organize the future space settlement in a fully inclusive way.

Ted Peters, in Chap. 16, discusses the idea of genetic enhancement of human moral virtues for the purposes of establishing a morally desirable space community. However, he presents his skepticism to that project. He builds his critical arguments mostly on the virtue ethics which shows that a human morality including moral virtues is only partially determined by human genome. According to Peters, altruism evolves by our free consent and commitment, not by genes alone.

In Chap. 17, Konrad Szocik offers arguments which support the idea of radical human enhancement for space missions. He argues that besides two criteria such as safety and efficacy, there are no other strong objections against human enhancement. He emphasizes that humans already evolve, and a permanent change over time is a typical feature of humans as an individual and as a species. He discusses a particular form of enhancement known as a biomedical moral enhancement. Szocik considers rationale for possible application of moral bioenhancement of future space astronauts and space settlers, but he enumerates possible challenges and objections related to the specificity of human morality.

Milan M. Ćirković, in Chap. 18, criticizes an anthropocentric approach which goes against the idea of human enhancement and space colonization. He points out that both human enhancement and space settlement are reasonable and effective ways in a long-term perspective to mitigate global catastrophic and existential risks on Earth. Ćirković underlines a critical importance of definitional and conceptual issues as well as harmful effects of wrong assumptions which may cause not only logical but also real social and political effects when they affect our thoughts and decisions in such important fields like coping with the effects of climate change or human future in space.

Lluis Oviedo, in Chap. 19, discusses possible functions of religion and religiosity as systems of human enhancement. He analyzes such possible applications of religion as psychological functions, coping with stress, improvement of mental wellbeing, but also possible social and cooperative skills. Oviedo identifies four areas of human mental and psychological life which could be enhanced during long-term space missions, such as a sense of meaning, a psychological ability to cope with failures, temperance and prudence, and such social skills like empathy, compassion, and collaborative attitude. He points out that while those capacities are important for successful space missions, they may be hard to implement by technical means or crew training program. He considers religion and religiosity as possible useful enhancement system for purposes of space missions which, as he shows, may be challenging to the specific nature of religion aimed at theological purposes.

This volume shows different ways of thinking about such an issue as human enhancement for space. Most of the authors discuss enhancements which are aimed at therapeutical and preventive issues which seem more or less evident in the light of hazardous factors in space. Some of them consider more radical enhancements which could easily lead to the evolution of new human sub-species. There is also a difference in ethical approaches and moral intuitions shared by particular contributors. Some of them are prone to accept even radical forms of human enhancements, while others argue against any enhancement. This book has, among many others, two purposes. One of them is to underline the importance of the idea of human enhancement for space missions. This idea is rarely discussed in scientific papers and books in the context of space missions. However, as we think, human space missions may be one of the first case studies in which different forms of human enhancements may be seriously discussed and possibly applied. Another purpose of this book is to connect current discussion in space policy and space philosophy, mostly in relation to the rationale for human space missions, with the idea of human enhancement. While this book is not focused on the issue of the rationale for human spaceflight, it is worth considering both ideas together. Last but not least, we expect that our studies and considerations may be used and applied to many different cases related to the issue of human enhancement on Earth.

Rzeszow, Poland
March 2020

Konrad Szocik

Contents

Part I
Human Enhancements: Biological and Medical Perspectives. Conceptual Issues

Chapter 1
Normalizing the Paradigm on Human Enhancements for Spaceflight

Margaret Boone Rappaport and Christopher J. Corbally

Abstract The authors describe the context and need for, and three methods to achieve a paradigm normalization in space research and space programming for the coming changes to the human species wrought by human enhancements. They explore the relevance of the longstanding nature of human enhancements, which reach back into prehistory, and new future capacities that will render humans able to withstand the rigors of microgravity, the monotony of spaceflight, and the extremes of off-world environments. The authors draw together a wide range of transformative changes to the human genome, human strength, physiology, and cognition that are now perceived as distinct, into a single set of options that must be tested, evaluated, and vetted by both space medicine boards and the public. The backflow of technology from space enterprises to the earthbound human species will make many beneficial changes to our species and the next ones on the human line. These changes have already begun.

1.1 The Metamorphosis that Needed a Paradigm Shift

The wide range of readings in this book on human enhancements for spaceflight provides details of human needs and human futures that will bring on "The Metamorphosis" (Kissinger et al. 2019) of our species on Earth and elsewhere in the solar system. That word, metamorphosis, envisions a human form that exists now, as if protected in a chrysalis, changing in form and function until it springs forth as an apparently new being. Internally, the butterfly, for example, resembles the caterpillar in many physiological workings, but much has changed. A new form of locomotion

M. B. Rappaport (✉)
The Human Sentience Project, Tucson, Arizona, USA
e-mail: msbrappaport@aol.com

C. J. Corbally
Vatican Observatory Research Group, Department of Astronomy,
University of Arizona, Tucson, Arizona, USA
e-mail: corbally@as.arizona.edu

K. Szocik (ed.), *Human Enhancements for Space Missions*, Space and Society,
https://doi.org/10.1007/978-3-030-42036-9_1

has emerged—flight—along with new ways to gather food and new ways to be beautiful. When we work our way through the variety and the invasiveness of the changes surveyed in this book, we conclude that the word, metamorphosis, does not overstate the situation much. There will still be an individual who belongs to the species *Homo sapiens* at the core of all the enhancements we describe, but we should remember to ask ourselves: When do the changes render us a different species? If this happens, is it unexpected or even unfortunate, or do we see ourselves now for what we are: One, in a long line of evolving species in the genus Homo that have been changing since the genus emerged around 2.8 million years ago (Villmoare et al. 2015).

It is useful to recall that the butterfly is not a new species, but a new stage of the same species. That difference is worth remembering. What looks so different and acts so differently may not be so intrinsically different. The resulting version may simply have had "a little tune-up," and so, like we understand with our motorized vehicles—maintenance! The truth must lie somewhere in between "metamorphosis," which sounds so grand, and "tune-up," which sounds so mundane. Hopefully, the result will not scare our families too much, or promise us too much. It is very true that what some of us, the ones who travel in space, will desperately need in the decades to come are ways to survive the rigors of spaceflight. When a few have done so, more will follow, in ever increasing numbers. We will populate the solar system, make no mistake, as the Beatles sang, "with a little help from my friends," or our friends, our human enhancements. They should be looked upon favorably, at first, so at least we can balance the risks and benefits.

1.2 The "Human" in Human Enhancements

Human enhancements are not new, but common to our species, and always have been. They lie along a variety of continua that humans are about to experience even more, from external to invasive, from changes to our hands to changes in the brain networks and genetics that guide them. Human enhancements are necessary and supportive of the enormous challenges of space exploration and space entrepreneurship that will consume our species on Earth and elsewhere for quite a number of centuries. To turn away and decide that human enhancements cannot be managed is to relinquish an important responsibility. They must be managed, and they will be managed, either by us or by our replacements: a new species, an A.I., or a visitor from another system.

Humans with enhancements remain humans. Veterans with functional arm and hand prostheses remain humans. Children born without hearing who receive brain implants so they can "hear" remain humans. Individuals receiving whole-face transplants because of mauling remain humans. They have all received the equivalent of human enhancements. Would we be so callous as to call them "cyborgs"? No, unless that word came to signify something good and worthwhile, which it may. Looking backward for just a moment, instead of forward, like most of this volume, we see clearly that we might feel some kinship with *Homo erectus*, a species who came before us, with a full upright stride but a less developed brain. The species was almost us, but not quite us. Our current form will one day be in the same predicament: The less advanced one will compare itself to the more advanced cyborg. This must be a cooperative meeting because we cannot devolve into social strata defined by enhancements. We are better than that.

At the present, the concept of "cyborg" fails to capture the hopeful and positive aspects of all the genetic, physiological, orthopedic, sensory, cognitive, and psychiatric enhancements on the horizon in response to human spacefaring, other social movements, and technological improvements. Human enhancements will not only change space crew, but humans back on Earth. The backflow of technology will take place in a steady stream, and humans on Earth will benefit. However, many people now picture the human cyborg who goes to space as a brute, outfitted from head to toe with servomechanisms, superhuman implants, in addition to a CRISPR'd genome to be able to survive and not break all his human bones the minute he lands on Mars (Szocik et al. 2019a). This clod appears to be stronger but less careful, sensitive, and refined. That image will likely change because the designers of the cyborgs and the A.I.s will be ourselves, and humans will hesitate to destroy all the beauty they see in the mirror, in favor of a cartoon hero. Will cyborg versions of ourselves be bigger and stronger, and look healthier? Surely, they will, but vitamin supplements accomplished that long ago for the rest of us, and the standards of health and beauty changed. We must not be too fearful of the enhanced physical features and mental capacities of humans yet to come. The developments will likely occur over many years, along many scales, according to personal and program need, type, and affordability.

Many of the positive aspects of human enhancements which we support run contrary to some religious and philosophical positions, especially some types of "naturalism," which view the naturalness of the human body and brain as something to be protected. For individuals who espouse a certain type of naturalism, they see human enhancements as "unnatural," even "defiling," and therefore "ungodly" (*cf* Lamont 1947; Chakroff et al. 2013). These religious and philosophical commitments run deep and should be respected. However, our view is that the general thrust of a strict approach of metaphysical naturalism is to ignore the continuous process of evolutionary change that the human line came from and the continuing changes in body and consciousness that are ongoing. Indeed, there are population forces that may *naturally* "improve" the human genome in the coming centuries and rid it of some of the deleterious genes now retained. It has been suggested that natural selection, which is particularly strong in large populations, may provide a kind of "tune-up" for the human species in the future, while numbers of our species grow into the many billions (Harris 2015, 84). We suggest that it makes increasingly less sense to distinguish between these population forces and human enhancements introduced by humans themselves.

We believe that it is important to ask: Who among us is to say that our species has yet reached perfection? The more we learn about our genetic evolution, the more we realize how flawed it is. It created a species that believes in perfection, as well as spirits and an afterlife, but the genetics at the foundation are a patchwork quilt of modifications that emerged largely by happenstance, until now, when we have finally become involved in our own evolution (Banathy 2003; Anderson 2010; Rappaport and Corbally 2020). Usually, the human traits that emerged in our evolution were beneficial because they were honed by the process of natural selection to be even more adaptive. However, much of the human genome which remains is that due to genetic drift in small groups of early humans, and some of those genes lead to modern illnesses and traits that we withstand in spite of the havoc they wreak, like mental illness (O'Bleness et al. 2012; Varki and Altheide 2005; Rappaport and Corbally 2020). Therefore, our species, *Homo sapiens*, will likely change, through continued evolution, some self-directed, and through human enhancements. It is of paramount importance that humans develop a positive attitude toward the changes to come, so that they can be adequately vetted on the necessary space medicine and ethics boards and in quiet conversations with families and doctors. At the present, there are not many humans who could withstand the harshness of space travel, but there will be more in the future.

1.3 Human Enhancements for Space and Off-World Missions

Human enhancements for spaceflight will always have a mandate that many others do not have: They enable humans to tolerate off-world environments and realize the desire to roam, conquer, and settle new territories for the rest of us. It is a species requirement so ingrained, so much a part of us biologically, it motivated members of the genus Homo who came before us. Long before our species evolved 300–400,000 years ago (Hublin et al. 2017), *Homo erectus* left Africa to colonize Eurasia all the way to the Far East, beginning about one million years ago.

Other forces at work in the development of human enhancements, like war, disability, congenital defects, aging, pandemics, environmental degradation, and the emergence of synthetic biology, nanotechnology, and robotic surgery will all be important, but a leading force will be human space travel because the species will find it such a difficult environment without help. It is the classic *sine qua non*: Without enhancements, human spacefaring will not occur, at least not much, not soon, and not for long periods and distances. From early indications based on neurological testing, the existing human species is just barely, sufficiently neuroplastic to be able to withstand the rigors of space with the current propulsion systems (van Ombergen et al. 2017; Demertzi et al. 2016; Rappaport et al. 2020). A sentient bird could not do it; a sentient packrat could not do it. The sentient wasp would be ill equipped, as would the sentient elephant, but the sentient human can do it, gifted as the species is by corporal flexibility, pervasive adaptability, variation in phenotypic development and adult expression, culture, intelligence, wanderlust, and what we have come to call "verve"—nerve, even recklessness, and a stubborn tendency to try new things—all, we believe, at least partially an inheritance from the ancient apes from which our line descended. Great apes can sometimes behave with blatant demonstrativeness, even fearlessness. The remnant species of great apes from the once large and varied population in Africa and Eurasia remain big, strong, and smart today. We will carry many of their genes to space with us.

Human space travel will likely lead the way for enhancements because of available funding. Sadly, war, pestilence, and the environment have often failed to move many to do much. That will change in the future, when one of these threats gains a structured rationale, a momentum, and money, in that order. Spacefaring is ahead of them all, although that could change, too, if humankind turns its back on the skies and refuses to go to space because it is too difficult and too expensive. That is always possible, but we feel, unlikely. Again, the human propensity to explore, conquer, and settle goes far back into our prehistory. Even if some future political entities eschew spacefaring, others will persevere. Once we left Earth behind, however briefly, humans were destined to do it again and again.

As a species, we evolved in a warm, varied, oxygen-fed place that allowed us to scavenge for what our brains needed most—meat, to feed an organ that still uses more energy than any other. In contrast, space is cold, with no atmosphere, no moisture, no gravity, and no manner to gain sustenance. Even worse, it provides nothing for the non-scientist to do. There will be scientists and physicians on early space missions, but most of the crew will be engineers. What will they do to keep themselves occupied on long, protracted journeys? Induced torpor may help, and eventually full hibernation like the Arctic ground squirrels (Rappaport et al. 2020).

The only other option is to provide all that an earthly environment provides, and more, too. It must provide something to help non-scientists counteract the monotony, confinement, and likely ennui. The latter response will not be altogether insane, because it could be interpreted as a realistic reaction to the circumstances of a lengthy spaceflight. This problem brings us squarely into confrontation with the potential use of psychiatric medications and enhancements that allow humans to tolerate their new environment, one to which they have never had to adapt before, for such long periods of time. Recent analog studies of "extreme teams" that "solve complex problems outside of traditional performance environments and have significant consequences associated with failure" show that affect does change over time and becomes more homogenized. Although there are reported conflicts in all studies, little is known about the management of "team affect" (Bell et al. 2019). Indirectly, of course, these human problems will encourage research on propulsion systems that can deliver us to the asteroids and outer planets more quickly. We note that nanotechnology may be one of the first modalities considered for psychiatric treatment, and we would include it in the broad category human enhancements (*cf* Fond et al. 2013).

1.4 The Dependence of Spacefaring on Human Enhancements

We should recognize the fair warning in front of us: Without human enhancements, space ventures will fail (*cf* Szocik et al. 2019b). We simply do not have the technologies to provide adequate environmental control, not yet and not for a long time. Eventually, humans will construct massive off-world environments that are essentially ships-as-small-planets. They are centuries in the future. Now, the changes the human body will endure have just barely begun to be tested, and for relatively short periods of time, in space. That research gives hints of the dangers to come and the need to manage them very carefully. Human enhancements will at first facilitate human spacefaring but then eventually enable it. Indeed, many types of future human enhancements for spaceflight are today's "treatments," so they will have a good, existing database on them to use in future human enhancement evaluations.

At this early point in the history of human spaceflight, our focus is on musculoskeletal and sensory enhancements, neurological pre-conditioning and treatment programs to counter changes in the brain, and early gene editing techniques to counter, for example, osteoporosis brought on by microgravity (Szocik et al. 2019a). In the future, nanotechnologies will play a greater role, and there will be new types of awareness and new modes of thinking that emerge out of the changes made to our human bodies. Many of these changes will *resemble* the types of changes anticipated by the transhumanist philosophical movement (*cf* Bostrom 2003). However, we emphasize that they may look the same, but not be the same, because the origins lie in different realms. The transhumanist movement emphasizes change in human consciousness that is sought through psychoactive drugs and religious beliefs and practices, sometimes with the aid of physical excess. The human changes envisioned by the transhumanist movement have been bundled, by some, with very straightforward enhancements that may help keep space crew alive. We feel that bundling is faulty.

Our goal is to *un*-bundle transhumanist enhancements and human enhancements for spacefaring and separate the feasible and the practical that are based only on science, from transhumanist goals that are both philosophical and religious. In human space programs, the enhancements will be based on both science and priorities set by the programs and societies that support them. Psychological, cognitive, and consciousness-related phenomena such as altered sensations and perceptions, hallucinations, deep meditative states, as well as hyper-excited trance states may emerge to comprehend life, space, and the cosmos differently and therefore achieve a more comprehensive understanding. However, use of these types of altered states must have a practical purpose in space programs, and experimentation with altered states must have a scientific rationale. Therefore, in devising our normalized paradigm, we include the use of altered states, but only to save lives or to further a specific space program and its mission. At least we agree with one "consequentialist argument" that the risks and benefits of "neuroenhancements" need to be "broken down in detail" (Heinrichs and Stake 2018).

At present, we humans are unable to conceive exactly what will emerge from the physical and cognitive changes that enhancements will support. However, we must believe that they will be wonderful. We two authors, one an anthropologist and the other a Catholic priest (who is also an astronomer), believe our future selves will be better. Why? Partly because we are both optimistic personalities, and partly because the course of social evolution is toward more equitable and peaceful societies, but mainly because the changes that come will be based on the goals, values, and goodness of the human species as it exists now. We know right from wrong, and we will pass that type of evaluation and decision-making ability on to the ones who come later, in their genomes, or if not, we will instill it or install it in future species of the genus Homo, be they organic or artificial. Human enhancements have an extraordinary capacity to go awry, to create "monsters," and to cause dissent and disagreement among those who do not receive them. However, this means that books like this volume are needed even more, in order to begin to plan the evaluation process and vet our future selves.

1.5 Normalizing the Paradigm Means Including All Types of Human Enhancements

Given the massive social and cultural forces at work and the substantial conflict that human enhancements for spaceflight may cause, we offer for consideration a normalization of the paradigm for research on, and program implementation of, human enhancements. We emphasize that ours is a *scientific* paradigm that should be relevant to the needs of space agencies and space entrepreneurs worldwide. The paradigm shift we recommend alters the assumptions, concepts, values, and practices that constitute the current paradigm and draws together a wide range of technologies from many different fields, including nanotechnology, genomics, rehabilitation medicine, as well as the medicines of disabilities and aging, and treatment types that are now so routine that few people think of them as "human enhancements".

For example, lenses are routinely inserted after cataract surgery, and depending on cost, these can restore vision almost completely in some cases. There are some who analyze the existing forms of rehabilitation and restoration and emphasize that the change post-cataract surgery does not make the human "enhanced," except... perhaps a small amount, like a change from 20/40 to 20/20 vision. Is that not an important change for the expert in optics, the woman who embroiders, or the dermatologist who scans one's skin for lesions? At this important, but still less than transformative level of change, we must ask about the ways that decisions are being made concerning "how different" a human enhancement must make a human, in order for it to be called an "enhancement".

Granted, when it comes to facial plastic surgery to reverse aging, a "little nip or tuck" may mean quite a lot in terms of attitude and even potential income, but its social impact is modest. A change in attitude is also wrought by facial plastic surgery to correct a child's cleft palette, a procedure that can completely alter the life of the child receiving this surgery. Children immediately lose the social stigma of such a deformity. Therefore, again, we must consider how new and how different an enhancement must be, and if perhaps a re-conceptualization of a continuum of differences might help human societies incorporate the idea of human enhancements into their view of humanity and re-think their view of human beings' rightful option to change, especially to protect themselves in life-threatening circumstances.

Humans have been changing genetically, culturally, and socio-politically since before the species' globular-shaped skull stabilized around 150,000 years ago (Bruner and Pearson 2013). The human species continued to change afterward and to this day. In the future, human enhancements will be part of that change, whether specific governments, religious, or advocacy groups approve or not. In many real senses, the cat is out of the bag, and human enhancements have a firm foothold in cultures and space programs worldwide, from a modest to a bold and experimental level.

Answers to questions about an acceptable degree of change for a human enhancement slip and slide according to the respondent, the type of enhancement, and its purpose. Even more important, the degree of difference will change over historical time, while new and better enhancements are introduced and people grow accustomed to the idea of new types of human upgrades. Specific enhancements may become more openly obvious in cases where the enhancement has achieved a kind of vogue, and less openly obvious with increasing miniaturization and sequestration of an individual's personal enhancement history in private medical records. Careers and businesses should also be mentioned that exist in special venues, such as sports, arts, and music. Their views toward enhancements may vary according to their particular characteristics. For example, there has long been a prohibition against metabolic enhancements for athletes, but saxophone players are allowed to smoke marijuana before a performance. The decisions that societies have made are detailed, complicated, and sometimes arbitrary. Where they must not be arbitrary is in a space program, because too much is at stake.

Our view is that a "natural" versus "unnatural" distinction holds less and less credence, and efforts to combat "unnatural" processes such as genetically modified organisms, like crops, are not going to work. There are few non-GMO crops left because of crossbreeding between wild and cultivated parts of the land. A type of exquisite pretentiousness attends many people's efforts to fight "unnatural" changes, and we recommend a more practical approach that views humans and all other living plants and animals as continuously changing, in both their genetic composition and their inherently variable, phenotypic expressions. We cannot afford to be too fastidious when trying to understand the effects on human beings of perhaps the most "unnatural" environment of all—space. It has an inflexibility to it that will require space program designers to be as clever as possible with whatever tools they have at hand, in order to keep crew alive. If that means feeding rations of genetically modified and nutritionally upgraded food to the crew, then that is what we will need to provide. Greenhouses can fail, as can waste recycling systems. Redundant systems are needed, and backed up, themselves, by human enhancements that prolong human life in the most dangerous of circumstances for long periods of time.

1.6 Three Approaches to Normalize the Paradigm on Human Enhancements, from Common Sense, Statistics, and Mathematics

1.6.1 The Normalized Paradigm Places Enhancements in a Standard State

To this point, we have mainly used an informal definition of "to normalize" that most closely approximates "to return to a standard state". We have reached out conceptually and identified many types of human enhancements that can be classified along a set of scales from invasive to non-invasive, hidden-to-open, serious-to-frivolous, physical to consciousness-related, life-saving to potentially life-threatening, inexpensive to priceless, as well as other dimensions to be defined in the future. An increasingly wide range of surgeries, mechanical and tissue implants, medications, and consciousness-altering drugs are all included. We must consider all of them because each one might save human lives in space or in off-world environments like space stations, Earth's moon, Mars, or in the microgravity of asteroids. Whether any specific enhancement is advisable for a particular mission would be determined by a space medicine review board, but at least the previous understandings of human enhancements as odd, unusual, extreme, or "unfair" (because other humans do not receive them) will have been debated and the risks and benefits of potentially life-saving measures adequately vetted. Our sense is that the public debate has well started and will continue.

1.6.2 The Normalized Paradigm Conceives of Naturally Occurring Modifications that Occur Along a Normal Bell Curve

A second meaning is captured in Fig. 1.1, by a bell curve along the horizontal axis. It derives from the field of statistics. It implies that data can be collected and analyzed on human enhancements according to scales of specifications that are relevant to any particular space mission. This perspective sees human enhancements as a set of real modifications to human biology, brain, and consciousness.

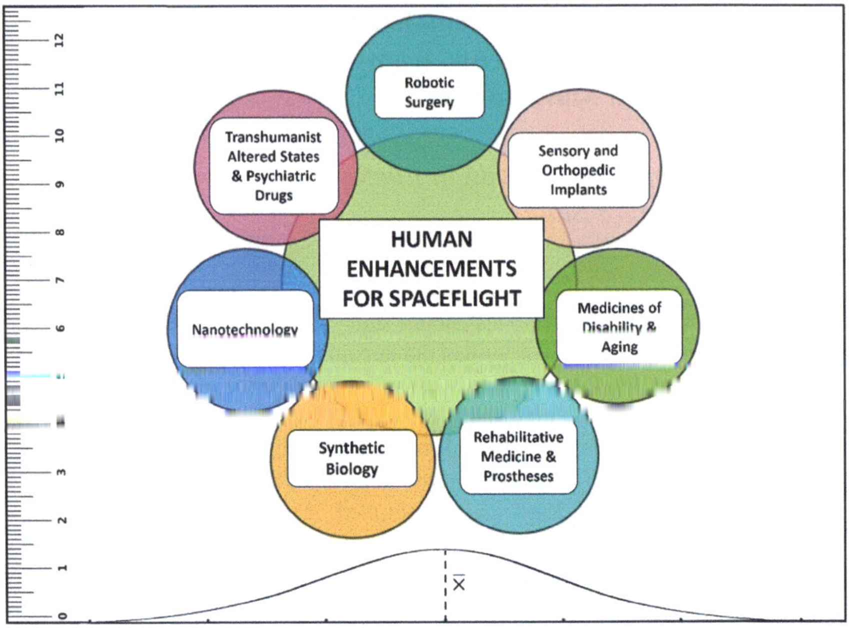

Fig. 1.1 Expanded evaluation research paradigm for comparative assessment of human enhancements

1.6.3 The Normalized Paradigm Transforms Variables Related to Enhancements and Enables Comparative Evaluations

Equally important is a third definition of "to normalize" that is borrowed from mathematics and means to reduce variables to a standard metric or scale. This is captured in Fig. 1.1 by a ruler on the vertical axis. Indeed, it is an operation like this that allows the comparison of the risk–benefit ratios of different drugs, types of neurological damage resulting from spaceflight, or, as here, human enhancements. It will be enormously useful to begin to analyze human enhancements along the same scales and vet them along the same dimensions. Human spacefaring will consider a very wide range of enhancements, some of which we cannot anticipate because the sequelae of long-term spaceflight are still largely unknown. What will be used to combat them? We do not yet know. After risk–benefit analyses, which operationalize the values of the program and the society that supports the mission, a cost–benefit analysis will then be appropriate. An organized approach to these analyses will be essential because of the number and variety of human enhancements that require vetting before use.

1.6.4 Evaluation and Decision-Making for Mission Success

Broader societal complications will filter into the decision-making process for human space missions, which is a less complicated context than Earth. When it comes to human spaceflight, the priority is on survival with an ability to complete tasks that lead toward mission success. This priority for human enhancements in the inhospitable environment of space helps to clarify many of the medical decisions to be made. Surviving space and microgravity are not situations for the prideful or vain, be they crew or program managers back on Earth. Unforeseen situations will arise in which an available, stored enhancement is needed to save a life in the case of injury, incapacity, or the need for metabolic stasis until return. In other cases, the new field of synthetic biology may allow the 3D "printing" of prostheses both old and new and, those used on Earth and eventually, only imagined in space or on an asteroid, in the case of an emergency (*cf* Snyder et al. 2019). Robotic surgeries may allow printed and stored prostheses to be inserted in spaceflight.

In light of the extreme nature of the space environment, much of the philosophical discussion on whether human enhancements are "good" or "bad," and especially, questions on whether an enhancement affects the "germline" or just an individual (e.g., Rüther and Heinrichs 2019)—these seemingly important issues lose meaning in relation to mission crew in the context of space. For example, the germline argument against human enhancements pales in comparison with known effects of simply leaving Earth: Radiation dose is excessive in spaceflight, on the International Space Station, and it will be even greater on a mission to Mars. In a sense, Mars crew will have already decided that their genomes will be changed by their acceptance of their mission. The question for each person is whether they then want to have children. Radiation remediation is a serious issue for all human spacefaring and will be for the foreseeable future. Some philosophical considerations, while interesting to the dilettante, become irrelevant.

We recommend that space program managers draw in and include genetic techniques like CRISPR and place them in the same analytical framework as psychiatric medications that proponents of the transhumanism movement might have favored. That may seem like a radical re-organization of the changes humans have proposed making to their minds and bodies. It may seem like "comparing apples and oranges," and in many cases, it will be similar. Nevertheless, with concepts borrowed from mathematics, and made practical with the use of good evaluation methodologies, different types of human enhancements can come under the same umbrella. For those that pass rigorous assessments, they can help to save the lives of exploratory space crew and the hardy souls who excavate the first greenhouses on Mars and the first mines on asteroids containing precious metals needed by industries on Earth.

1.6.5 Continual Re-evaluation of Enhancement Needs

It will be essential to repeatedly assess whether humans need a specific human enhancement in space. For example, there are early signs that the human body can adapt, to an extent, to some aspects of weightlessness, over a period of time and on repeated exposure. Furthermore, the concept of "need" may transform over the following centuries, especially while space-based businesses come to reap handsome profits. In those circumstances, individual crew may choose to risk their health and well-being in order to make financial gain. While it may be possible for Occupational Safety and Health Administrations to operate on Earth, it may be much more difficult to regulate businesses off-world, where the "nationality" of certain locations may be undetermined. Who has jurisdiction? Similarly, who is present to supervise and evaluate excessive use of human enhancements or the dangers of non-use?

1.7 Enhancement or Extinction, that May Be the Question

Homo sapiens is only one in a series of species that emerged along branching lines from the late Miocene apes, and survived. All the other examples of our genus Homo have gone extinct. An important question to ask is whether human enhancements pave the way toward improvement of our species or toward extinction. As a group of competing and cooperating societies now on Earth, we need to ask this question soon, because the technologies are fast developing to create artificial intelligences, cyborgs, and simply human space crew who need an enhancement to survive.

It is also useful to ask if human enhancements will be necessary to save our species in a competition with a new being, an artificial one this time, who will be physically stronger, think faster, and perhaps exterminate us (Rappaport and Corbally 2020). While unimaginable in many ways, it is possible that this artificial intelligence may not value our survival—in fact, value anything—and we as a species may need the enhancements that we describe to save our own lives on Earth.

Be forewarned: There is a metamorphosis coming, but also a competition, if not with our own manufactured beings, then with someone else's. Humans may need enhancements, even more transformative ones, to protect the futures of our species or our descendant species. There is value in our consideration of the risks and benefits that human enhancements imply. They offer us both physical and mental capacities that our species and earlier species on our line have not yet had. In a sense, human enhancements allow us to skip forward in our own evolution by supplementing our natural gifts.

Human enhancements are all inherently risky, because they outfit the human species with organs, capacities, strengths, and perhaps even perspectives that evolution has not yet provided, either through natural selection, genetic drift, or other population mechanisms. They are purposefully installed, not anticipated patiently for millennia. In devising human enhancements, we take our own human evolution "by the horns" and make it do what we want it to do, and what we need our species to be in order to tolerate this dangerous new environment we are about to enter *en masse*—space.

1.8 Evaluation of a Full, Normalized Set of Human Enhancements

There are many who will call the incorporation of enhancements into human development and evolution, foolhardy. There will be strong countercurrents to any and all introductions of new technologies that give humans additional capacities, instead favoring a more natural approach. There will be public debate on the distribution of these enhancements, and there will probably be eras in which they cease to be technologically developed. However, it is important to be very clear: They are already in development. They have always been, since the origins of our self-aware species. In the early hunter's impatience and desire to throw his spear farther, and to devise a spear thrower to extend the reach of his own arm, he protects and feeds himself and his social group. Today, our toolkits follow a long line of creative embellishments to the human form and consciousness. Do we imagine that the spear thrower, itself, affected only the hunter's ability to throw a spear? It affected much more, including how the hunter felt about himself and his abilities, and how secure his family was.

In order to vet them properly, we take all possible enhancements as a group, along normalized scales of evaluation, and as a "normal" group of human innovations—normal, because humans thought of them for constructive purposes. If human enhancements are developed whose purpose is solely to harm or destroy anyone, then we do not consider them within this same paradigm and they are better conceived as "weapons." We make this distinction carefully, knowing full well that all pharmaceuticals and modifications to the human body can sometimes have adverse effects. We exclude the modifications whose sole purpose is destruction or defilement of humans, animals, or the environments in which they thrive.

Human enhancements will be examined, discussed, accepted, or rejected. They will not simply disappear, and they will eventually render some humans seemingly "superhuman", perhaps in the same way that the hunter with his spear thrower was first viewed as above and beyond his former self. If we have the data and we have the right methods to assess the data on human enhancements, we can begin the long process of deciding whether they should be implemented. Space crew must be considered initially in this process, because without human enhancements, they will not complete their dangerous missions. We humans can send robot explorers in their place for a while, but this will not satisfy our species for long. We will go. We will explore, as far as our current technology will allow us.

References

Anderson, M. L. (2010). Neural reuse: A fundamental organizational principle of the brain. *Behavioral and Brain Sciences, 33,* 245–313.

Banathy, B. H. (2003). Self-guided/conscious evolution. *Systems Research and Behavioral Science, 20*(4), 309–321.

Bell, S. T., Brown, S. G., & Mitchell, T. (2019). What we know about team dynamics for long-distance space missions: A systematic review of analog research. *Frontiers in Psychology, 10,* 811.

Bostrom, N. (2003). Human genetic enhancements: A transhumanist perspective. *Journal of Value Inquiry, 37,* 493–506.

Bruner, E., & Pearson, O. (2013). Neurocranial evolution in modern humans: The case of Jebel Irhoud 1. *Anthropological Sciences, 121,* 31–41.

Chakroff, A., Dungan, J. A., & Young, L. (2013). Harming ourselves and defiling others: What determines a moral domain? *PLoS ONE, 8*(9), e74434.

Demertzi, A., Van Ombergen, A., Tomilovskaya, E., et al. (2016). Cortical reorganization in an astronaut's brain after long-duration spaceflight. *Brain Structure and Function, 221,* 2873–2876.

Fond, G., Macgregor, A., & Miot, S. (2013). Nanopsychiatry–the potential role of nanotechnologies in the future of psychiatry: A systematic review. *European Neuropsychopharmacology, 23*(9), 1067–1071.

Harris, E. E. (2015). *Ancestors in our genome: The new science of human evolution.* Oxford, England: Oxford University Press.

Heinrichs, J.-H., & Stake, M. (2018). Enhancement: Consequentialist arguments. *ZEMO, 1,* 321–342.

Hublin, J.-J., Ben-Ncer, A., Bailey, S. E., Freidline, S. E., Neubauer, S., Skinner, M. M. Bergmann, I., et al. (2017). New fossils from Jebel Irhoud, Morocco and the pan-African origin of *Homo sapiens. Nature, 546,* 289–292.

Kissinger, H. A., Schmidt, E., & Huttenlocher, D. (2019). The metamorphosis. *The Atlantic.*

Lamont, C. (1947). Naturalism and the appreciation of nature. *The Journal of Philosophy, 44*(22), 597–608.

O'Bleness, M., Searles, V. B., Varki, A., Gagneux, P., & Sikela, J. M. (2012). Evolution of genetic and genomic features unique to the human lineage. *Nature Reviews Genetics, 13,* 853–866.

Rappaport, M. B., & Corbally, C. (2020). *The emergence of religion in human evolution.* Abingdon-on-Thames, UK: Routledge.

Rappaport, M. B., Szocik, K., & Corbally, C. (2020). Neuroplasticity as a foundation for human enhancements in space. *Acta Astronautica, 175,* 438–446.

Rüther, M., & Heinrichs, J.-H. (2019). Human enhancement: Deontological arguments. *ZEMO, 2,* 161–178.

Snyder, J. E., Walsh, D., Carr, P. A., & Rothschild, L. J. (2019). A makerspace for life support systems in space. *Trends in Biotechnology, S0167–7799*(19)30110–30116.

Szocik, K., Campa, R., Rappaport, M. B., & Corbally, C. (2019a). Changing the paradigm on human enhancements: The special case of modifications to counter bone loss for manned mars missions. *Space Policy, 48,* 68–75.

Szocik, K., Norman, Z., & Reiss, M. J. (2019b). Ethical challenges in human space missions: A space refuge, scientific value, and human gene editing for space. *Science and Engineering Ethics,* 1–19.

Van Ombergen, A., Demertzi, A., Tomilovskaya, E., et al. (2017). The effect of spaceflight and microgravity on the human brain. *Journal of Neurology, 264,* S18–S22.

Varki, A., & Altheide, T. K. (2005). Comparing the human and chimpanzee genomes: Searching for needles in a haystack. *Genome Research, 15,* 1746–1758.

Villmoare, B., Kimbel, W. H., Seyoum, C., Campisano, C. J., DiMaggio, E. N., Rowan, J., et al. (2015). Early *Homo* at 2.8 Ma from Ledi-Geraru, Afar, Ethiopia. *Science, 347,* 1352–1355.

Chapter 2
CRISPR Challenges and Opportunities for Space Travel

Arvin M. Gouw

Abstract CRISPR gene technology has been scientifically exciting because of its great potential to alter any DNA. There are scientific risks with CRISPR gene editing, but risk assessment is not impossible and is currently in progress. Space travel presents a unique set of challenges and opportunities for CRISPR gene editing. Biomedically, space travel is a major health hazard for astronauts. Research on the genetic and molecular mechanisms behind these biomedical challenges presents opportunities for CRISPR to serve as a biomedical intervention for astronauts. This article gives several examples where such mechanisms have been reported. However, the guidelines regarding CRISPR applications are still very generic and difficult to implement in specific cases. The focus of this paper is on the therapy/enhancement distinction, which many guidelines mention, to help us discern when CRISPR applications are justifiable. I propose a reworking of the definition of normal in order to be able to distinguish therapy from enhancement, where the former is deemed more appropriate for CRISPR applications than the latter. Space travel is a unique and interesting scenario to test the applications of the therapy/enhancement distinction as guidelines for CRISPR gene editing.

2.1 What is CRISPR?

What does CRISPR stand for? CRISPR actually stands for Clustered Regularly Interspaced Short Palindromic Repeats. That does not mean much for those who do not work closely with CRISPR. CRISPR in many people's minds conjures the image of designer babies. Though CRISPR does have the potential to edit the human genome, CRISPR was discovered as part of a bacterial immune system against viruses or bacteriophages (Barrangou et al. 2007; Haurwitz et al. 2010). When viruses insert their genetic material into bacteria, bacteria respond by incorporating some of that viral genetic material into their genomes. This allows bacteria to cleave viral DNA

A. M. Gouw (✉)
Harvard Divinity School Center for Science, Religion, and Culture,
Stanford University School of Medicine, Stanford, USA
e-mail: arvgouw@stanford.edu

K. Szocik (ed.), *Human Enhancements for Space Missions*, Space and Society,
https://doi.org/10.1007/978-3-030-42036-9_2

the next time around using their Cas enzyme(Terns and Terns 2011). Since this CRISPR/Cas system is conserved in virtually every species, scientists have been able to tweak the CRISPR/Cas system such that the Cas enzyme will cleave only specific genes as specified by the scientist(Lo et al. 2013). Scientists do that by inserting what is called single guide RNA (sgRNA)(Doudna and Charpentier 2014).

Like with all genetic engineering methods in the past, CRISPR has its risks. One of the most concerning risks is being off-target, or missing the target gene. Though sgRNAs are designed to target only a single gene, it is always a concern that they could hit any of the other 20,00–30,000 genes in the human genome. Scientists keep improving the CRISPR/Cas system to allow for better accuracy in gene editing (Tan et al. 2020; Kim et al. 2020). Though this is a serious technical problem in the application of CRISPR technology, it is not impossible to overcome.

Though off-target effects may sound very concerning, on-target effects may be more dangerous. The reason is because we do not always know the effects of deleting a single gene (or multiple genes for that matter) in a complex network of genes. These are called on-target effects, or unintended effects (Lee and Kim 2019). There are current examples of this, such as the elimination of sickle-cell disease leading to an increased risk of contracting malaria (Bosley et al. 2015). It is impossible to know every possible outcome of genetic modification, thus only research and time will tell us more about these on-target unintended consequences. However, the difficult question is always, how much experimentation is needed before something is ready to be implemented? This is a challenge to proponents of the Precautionary Principle.

2.2 Biomedical Challenges and Possible Solutions in Space Travel

There are multiple challenges and possible solutions for astronauts in space. I would like to highlight four common biomedical challenges that astronauts have to face: radiation and carcinogenesis, immune system dysregulation, bone and muscle atrophy, and neurocognitive impairment. Currently, most solutions to these problems involve the use of physical protective barriers, exercise, and drugs. Unfortunately, absorption, distribution, metabolism, and excretion (ADME) profile of drugs change in space (Eyal and Derendorf 2019; Berman and Eyal 2019). Given the fact that drug mechanisms work differently in space than on earth, this gives room for genetic engineering as a possible solution. The differences in pharmacokinetic and pharmacodynamic profiles of drugs in space necessitate experimentation of pharmacokinetic and pharmacodynamic (PK/PD) studies in space, such as drug testing on 3D-bioprinted cell cultures (Shao et al. 2019; Creff et al. 2019) under simulated microgravity (Hammond et al. 2016; Ma et al. 2019).

Genetic modification may bypass this problem if somehow biological adaptation will attune itself to the extent of the challenge in space. In other words, biological adaptations are designed to be modulated in response to the degree of the stimulus.

For example, in the case of stress, we know that chronic stress induces a different set of responses compared to acute stress. Thus, genetic modification in specific pathways will automatically adapt to the kind and degree of the challenge presented to astronauts in space. This genetic modulation may prove to be a more reliable means of adaptation versus drug mechanisms that are unpredictable in space. In the following section, I will discuss the four major biomedical issues that astronauts have to face and explore the genetic mechanisms behind them to assess where CRISPR gene editing may be helpful.

2.2.1 Radiation and Carcinogenesis

There are several types of radiation encountered in space, but all are commonly referred to as particle nuclei of high energy and charge (HZE) (Guo et al. 2015). High radiation causes carcinogenesis or death. Radiation increases innate immune activity, which induces inflammation and can cause tumorigenesis. This can be seen in the Hiroshima atomic bomb survivors (Hayashi et al. 2012). There are of course multiple pathways by which radiation induces carcinogenesis. HZE ions have been implicated in breast cancer, lung cancer, liver cancer, and acute myeloid leukemia, among others (Bielefeldt-Ohmann et al. 2012; Rivina and Schiestl 2013). Mechanistically, radiation could directly amplify cancer-causing genes, known as oncogenes, such as MYC. Breast epithelial cells exposed to radiation are shown to gain MYC amplification and initiation of breast cancer (Wade et al. 2015). The situation is of course worse when a carrier of a mutant copy of a tumor suppressor gene is predisposed to HZE. Genetic mouse models carrying an *Apc* mutant gene develop higher grade intestinal cancer upon HZE exposure (Datta et al. 2013). Microarray signatures of irradiated mouse lungs show similar gene signatures to those of early stage lung cancer patients (Delgado et al. 2014). However, this does not mean that it would be simple to genetically design a way to protect cells from radiation, because the key genes need to be identified (Barcellos-Hoff et al. 2015).

In dealing with radiation, unfortunately shielding is not very effective against cosmic radiation, particularly because of the duration of exposure during space travel (Durante 2014). Similarly, radioprotective drugs are currently not advanced enough to be effective in counteracting cosmic radiation (Kennedy 2014). In addition to the use of physical barriers to reduce space radiation, scientists have been studying hibernation as a possible solution to provide protection against radiation. Hibernation is state in which metabolic activity is reduced (Heldmaier et al. 2004; Cerri et al. 2016). Hibernation has been positively selected by many animals, presumably to save energy and avoid predators, because hibernation has been found in proto-mammals (150 million years ago). Hibernation may have been a useful defense mechanism against predators such as dinosaurs during cold climates (Cerri et al. 2016). This is indeed consistent with the surprising recent finding that dinosaurs did not hibernate, even though they lived in cold temperatures and required a large food intake (Woodward et al. 2011). Hibernation gives animals higher resilience to stress. For example, lethal hemorrhagic shock in a non-hibernator was shown to be non-lethal in hibernators (Bogren et al. 2014). Similarly, hibernating cells are able to resist stress better than normal cells (Talaei et al. 2011). Some of the protective effects are due to dopamine, serotonin, and the H_2S pathways (Dugbartey et al. 2015a, b).

Radiation can be considered as a stress stimulus. Hibernation has been shown to be radioprotective not only due to cell cycle arrest as previously thought, but also through the hypoxia-inducible factor (HIF) pathway (Cheng et al. 2015; Fu et al. 2015) or hypothermia (Lisowska et al. 2014). Moreover, the radioprotective effects of hibernation have been known since the 1950s from other animal studies (Smith and Grenan 1951; Mraz and Praslicka 1961; Barr and Musacchia 1969). More specifically, squirrels have been known to be resistant to lethal doses of radiation during hibernation (Musacchia and Barr 1968, Jaroslow et al. 1969).

Let us look into both natural and synthetic causes of hibernation. When it comes to natural causes of hibernation, the lowest body temperature ever recorded for a living human was 13.7 °C with circulatory arrest after an accident, in Anna Elisabeth Johansson Bagenholm (Gilbert et al. 2000). Even more dramatic was the resuscitation of the clinically dead 13-month-old Erika Nordby at 16 °C in a −24 °C outside temperature (Cerri et al. 2016). If there is any hope for CRISPR-induced hibernation, we need to find synthetic cases of hibernation. It is possible to induce hibernation synthetically without low temperatures using 5′-AMP, or H_2S (Fu et al. 2006; Zhang et al. 2006; Blackstone et al. 2005). This synthetic modulation usually fails in non-hibernating animals (Haouzi et al. 2008), except in specific cases in certain brain areas (Cerri et al. 2013; Tupone et al. 2013a, b). Therefore, it is theoretically possible, though still very challenging, that CRISPR modification in key brain areas such as the Raphe Pallidus could induce hibernation. It is even more promising now that some have suggested a genetic component to hibernation (Grabek et al. 2019; Lane et al. 2011; Hadj-Moussa et al. 2019), which provides the possibility for CRISPR modulation of these “genes of the undead” (Hadj-Moussa et al. 2019) to induce or suppress hibernation (Fernandez-Tornero 2018; Wu et al. 2002).

However, the pathways that are induced by hibernation are shared by cancers to promote their survival. Metabolically, it has been known that reduced metabolism due to hypoxia provides protection against radiation (Lyman and Fawcett 1954, Patterson et al. 1957). It is thus tempting to genetically activate the hypoxia pathway for radioresistance; however, it is also now known that cancers often activate the hypoxia-inducible factor (HIF) pathway (Stine et al. 2015; Le et al. 2014). In turn, cancer genes activate hypoxia metabolic programs to support proliferation (Gouw et al. 2019, 2016). In fact, radiation-resistant cancers often activate the HIF pathway (Yue et al. 2019; Wang et al. 2019).

2.2.2 *Dysregulated Immune System*

In addition to radiation exposure, microgravity has been found to alter the immune system. It activates certain immune cells while inactivating others. More specifically, immune suppression of T lymphocytes has been recorded (Dang et al. 2014; Hughes-Fulford 2011; Maccarrone et al. 2003; Battista et al. 2012; Martinez et al. 2015). However, the innate immune system has been reported to be activated. Microgravity modulates IL-6 (Ma et al. 2014; Muid et al. 2013; Smith 2018) and causes immune system dysregulation (Van Walleghem et al. 2017; Kita et al. 2004; Crucian and Sams 2009), which also depends on the duration of exposure to microgravity (Kita et al. 2004; Crucian et al. 2008). Though the activation of the immune system is generally considered to be beneficial in fighting pathogenesis, ranging from bacteria to cancer (Mittal et al. 2014), many have argued that there are two general pathways by which hyperactivation of the immune system promotes carcinogenesis (Barcellos-Hoff 1998b). First, the intrinsic pathway involves genetic mutations, which activate the innate immune system, leading to chronic inflammation. For example, Ras in cancer cells activates the inflammatory IL-8, and Bcl2 induces necrosis, which in turn activates the innate immune system through toll-like receptors (TLRs) (Mantovani et al. 2008; Sparmann and Bar-Sagi 2004). Second, the extrinsic pathway involves a dysregulated immune system that is unable to resolve an infection, leading to chronic inflammation. Such chronic inflammation leads to DNA damage in cells, which in turn leads to tumorigenesis (Guerra et al. 2007; Farber et al. 1990). In both pathways, a dysregulated innate immune system is pro-tumorigenic.

However, it is not only microgravity in space that could alter the immune system. Radiation exposure can activate macrophages, which in turn cause tissue damage due to constant inflammation. Such inflammation promotes tumorigenesis (Mukherjee et al. 2014; Coates et al. 2008). In general, irradiated cells release free radicals that activate innate immune cells, including the polarization of macrophages from M1 to M2 (Coates et al. 2008), to release TNFa and reactive oxygen and nitrogen species, which cause chromosomal instability in hematopoietic cells (Lorimore et al. 2008). Such chronic inflammation predisposes the tissue to be tumorigenic (Barcellos-Hoff 1998a, b; Wright and Coates 2006). This pro-tumorigenic inflammation model is also consistent with the finding that many age-associated diseases and aging have activated the innate immune system, with higher serum levels of IL-6 (Sarkar and Fisher 2006).

2.2.3 *Osteoporosis and Muscular Atrophy*

One of the first identified medical problems in astronauts was bone loss and muscle loss. Numerous studies have been done to identify various mechanisms behind bone loss. Exercise and pharmacological intervention have been used as the main treatments to reduce bone loss in space (Cavanagh et al. 2005; Diao et al. 2018; Kast et al. 2017). Bisphosphonates have been evaluated in bed-rest studies that emulate microgravity conditions (Grigoriev et al. 1998; Endo and Matsumoto 2008, 2012). Studies of quadriplegics and paraplegics treated with anti-resorptive drugs have also shown some promising results (Shapiro et al. 2007). Studying these drug mechanisms will allow us to identify key genetic and molecular pathways behind bone loss and bone recovery. The identification of key pathways behind bone loss is the first stepping stone to CRISPR engineering to battle bone loss.

The Wnt/beta catenin pathway involving low-density lipoprotein receptor-related proteins (LRP) 5 and 6 (Li et al. 2005; Bao et al. 2012) has been implicated in the regulation of osteoblastogenesis (Hong et al. 2019; Chen et al. 2019). This gives hope that genetic modulation of specific genes in the Wnt/beta catenin pathway can reverse or prevent bone loss in space (De Santis et al. 2018). Other proposals for CRISPR genetic engineering to battle bone loss have also been reported (Carmeliet et al. 1998; Yuan et al. 2019, Xix Congresso Nazionale S.I.C.O.O.P. Societa' Italiana Chirurghi Ortopedici Dell'Ospedalita' Privata et al. 2019).

Microgravity causes not only bone loss, but also loss of muscle mass. Since one of the first applications of CRISPR was to knockout the myostatin gene to prevent muscle loss, it is plausible it could be used to inhibit the myostatin pathway to prevent muscular atrophy during space flight. Myostatin knockout has doubled muscle mass in multiple animals (Zhang et al. 2019; Wang et al. 2018; Wei et al. 2016).

2.2.4 Neurocognitive and Psychological Deficits

Two other problems that astronauts face during spaceflight are altered blood flow, which cause neurocognitive and psychological deficits. Microgravity affects the brain's blood circulation, causing hypokinesia as well as circadian and cardiopulmonary problems (Norsk 2014). Prolonged isolation has also been reported to induce an altered cardiovascular system (Gunga et al. 1996a, b).

Virtual reality environments have been created to simulate space conditions for astronaut training (Aoki et al. 2007; Cater and Huffman 1995; Harm et al. 2007). Numerous psychoactive drugs have also been prescribed for astronauts to battle depression. Given the neurological nature of these psychological problems, genetic modification by CRISPR will be much more challenging, because the gap between psychology and genetics is much wider than the gap between genetics and cancer (Gouw 2018b).

2.3 What Are the Guidelines Regarding CRISPR?

Given the prospects for CRISPR gene editing to overcome the aforementioned challenges, I would like to briefly focus on the concerns raised by the European Academies Science Advisory Council (EASAC), and two American institutions: The American College of Medical Genetics and Genomics (ACMG) and the National Academies of Science, Engineering, and Medicine.

First, the EASAC published a report called "Genome editing: scientific opportunities, public interests, and policy options in the European Union." Second, ACMG published a report: "Genome editing in clinical genetics: points to consider." (Acmg Board of Directors 2017). Third, the National Academies of Science, Engineering, and Medicine also raised concerns in their comprehensive report on CRISPR. All three documents raise the same scientific concerns: off-target effects, on-target effects, epigenetic effects, and chimerism. Epigenetic effects mean factors beyond genetics that alter the intended genetic modification. For example, it is possible that insertion of a radiation resistance gene is epigenetically turned off, rendering that genetic modification useless. Chimerism means that there is always the possibility that CRISPR modification may not take place in all the cells that we intended to change. An organism or an organ will then have a chimera of the original unmodified genetics, as well as the newly modified genetics. Beyond these four scientific concerns, some social justice concerns were also raised. In short, the main concern is access to CRISPR technology (Gouw 2019). The concern is that only the rich will be able to afford genetic engineering, the GenRich, while most people will not be able to afford genetic engineering (the GenPoor) (Fears and Ter Meulen 2017). Thus for our scenario, does it mean only the GenRich will be able to migrate to Mars in the future? Overall, these documents agree that further research on the risks of genome editing is needed, and that discussions that promote public engagement to combat social inequality and justice are not to be overlooked (de Lecuona et al. 2017).

For our purposes, I would like to highlight the fact that there is an overall agreement from the three aforementioned documents in promoting CRISPR only for therapy but not enhancement. However, none of those documents define clearly what those terms mean: therapy, enhancement. I and others have argued that this therapy/enhancement distinction is problematic for several reasons (Gouw 2018a). The therapy/enhancement distinction implies that there is a universal "norm" that everyone can agree on. However, there is never a definition of what "normal" means. If we were to adopt the definition of "healthy" from the World Health Organization" to distinguish between therapy and enhancement, then everyone will fall short of being "healthy". This is because WHO's definition encompasses "a state of complete physical, mental and social well-being and not merely the absence of disease or infirmity" (Nobile 2014). Similar to "healthy", other words are equally problematic: "impairment", "normal state of health," "natural", "normal", "native" (Lustig et al. 2008). Thus, before we can proceed to utilizing the terms therapy and enhancement, we first need to consider what "natural" or "normal" means to be able to distinguish therapy from enhancement (Gouw 2018a, 2019).

2.4 Revisiting the Therapy Versus Enhancement Distinction

In the context of genomic engineering, enhancement is defined relative to being above the "normal well-being" of a person. Others have noticed that the distinction between therapy and enhancement is problematic and propose to not use this criterion at all. When it comes to space exploration, biomedical interventions that would normally have been considered as enhancement would be considered a medical necessity for astronauts (Szocik 2020; Szocik and Braddock 2019). For example, being genetically modified to double muscle mass would be considered to be an enhancement for an athlete, but it would be considered to be a therapy for an astronaut due to microgravity. Thus, the same standard paradigm can be used to show how this therapy/enhancement distinction can be useful depending on specific situations, as several of the other contributors to this volume also pointed out (Szocik et al. 2019).

Despite the absence of an absolute static standard of nature, it is not impossible to use the therapy/enhancement distinction to help our ethical discussions. But this can only happen if we avoid defining "normal" or "natural" following classical Aristotelian definitions of "natural" as the Roman Catholic Church often does. I propose to reframe the definition of normal following a combination of two criteria: ability to be alive and being close to the statistical average of the population. The first criterion makes sense because etymologically, the study of diseases, known as pathology, is the study of pathos, or suffering (Gottweis 2005). This can be seen as a minimal criterion for the need for therapy. The second criterion of observing the statistical average of the population functions more as a limiting criterion. Anything above the statistical norm will fall under enhancement.

Some may argue that the second criterion will fall into relativism, because one can easily justify a new population subset to derive the average, and suddenly enhancement becomes therapy. Though this second criterion does allow for such flexibility, I consider it as a major strength of this definition. An error of sampling or defining a population is an error of methodology, but not an error of the working of my proposed definitions of normal, therapy, and enhancement.

When applied to the context of space travel, the first criterion points out that since normal well-being involves the need to be alive, biomedical interventions to allow astronauts to survive space travel would be considered as therapy and not enhancement. Second, though spaceflight biomedical interventions are often considered as enhancement on earth, the notion of average in the second criterion allows us to redefine the population selection for the measurement of average. In other words, if we are applying therapeutic interventions only to astronauts, then the statistical average would be taken from only astronauts as its sample size. This statistical view of normal along with its two criteria allows us to differentiate therapy from enhancement for any given circumstances throughout space and time.

In other words, I believe the therapy versus enhancement recommendation from the multiple scientific guidelines can still work even for genomic engineering for astronauts, if it considers the criteria of survival and population sampling as proposed above. While the first criterion serves to identify the "floor" for applying therapy, the second criterion serves as the "ceiling" for therapy and "second floor" for enhancement.

2.5 Concluding Remarks

In terms of future directions, assuming that I have overcome opposition of biomedical intervention of astronauts as enhancement instead of therapy, there are obviously many other considerations before moving forward with CRISPR-ing our astronauts should the technology become possible. First, a cost–benefit analysis of various treatments should be done. Depending on the circumstances, genetics could be either the first line or the last line of intervention (Szocik et al. 2019). Second, since biomedical interventions, especially CRISPR gene editing, are done to the astronaut's body, the astronaut's preference must be taken into consideration. Third, we must consider what happens to astronauts before, during, and after the CRISPR modification. Pre-flight modification may be needed for studying and assessing the success of the modification. However, pre-flight modification may lead to enhancement on earth. Similarly, upon return from space travel, such modifications may persist and astronauts will gain enhancements. Following that line of reasoning, we must consider whether it is a somatic modification or germline modification. Germline modification would be passed down to future generations. One possible solution to this would be freezing gametes before receiving germline genetic modification, but does this also mean astronauts have to be sterilized upon their return?

I would propose several suggestions that would make CRISPR modification less controversial for space travel. All three suggestions are based on the engineering of transgenic mouse models that are in current use for medical research. First, CRISPR modifications should be done in a tissue-specific manner. In other words, the modification should be done only to a single organ or a certain body area instead of on the whole body. Sometimes people refer to this as somatic modification instead of germline modification. Second, the modification should be designed to be reversible. This has several obvious advantages. Reversibility will solve the problem of enhancement of astronaut before and after the flight. Moreover, reversibility allows us to reverse off-target or unintended consequences of the modification. Third, the modification should be inducible, preferably inducible only upon space travel circumstances only. This has the potential of overcoming enhancement upon return to earth, because the genetic modifications would not be induced or activated under normal earth conditions.

Though we are still far from where we can apply CRISPR gene editing to astronauts, and it will be awhile before long-duration space travel will take place, it is important to assess the opportunities and challenges of using CRISPR technology for space travel. Moreover, space travel can be used as a very unique and interesting scenario to test the currently evolving ethical paradigms for CRISPR research and applications.

References

Acmg Board of Directors. (2017). Genome editing in clinical genetics: Points to consider-a statement of the American College of Medical Genetics and Genomics. *Genet Med, 19*(7), 723724. https://doi.org/10.1038/gim.2016.195.

Aoki, H., Oman, C. M., & Natapoff, A. (2007). Virtual-reality-based 3D navigation training for emergency egress from spacecraft. *Aviation, Space and Environmental Medicine, 78*(8), 774–783.

Bao, J., Zheng, J. J., & Wu, D. (2012). The structural basis of DKK-mediated inhibition of Wnt/LRP signaling. *Science Signal, 5*(224), 22. https://doi.org/10.1126/scisignal.2003028.

Barcellos-Hoff, M. H. (1998a). How do tissues respond to damage at the cellular level? The role of cytokines in irradiated tissues. *Radiation Research, 150*(5 Suppl), S109–S120.

Barcellos-Hoff, M. H. (1998b). The potential influence of radiation induced microenvironments in neoplastic progression. *Journal of Mammary Gland Biology and Neoplasia, 3*(2), 165–175. https://doi.org/10.1023/a:1018704806635.

Barcellos-Hoff, M. H., Blakely, E. A., Burma, S., Fornace, A. J., Jr., Gerson, S., Hlatky, L., et al. (2015). Concepts and challenges in cancer risk prediction for the space radiation environment. *Life Science and Space Reserach (Amst), 6*, 92–103. https://doi.org/10.1016/j.lssr.2015.07.006.

Barr, R. E., & Musacchia, X. J. (1969). The effect of body temperature and postirradiation cold exposure on the radiation response of the hibernator Citellus tridecemlineatus. *Radiation Research, 38*(2), 437–448.

Barrangou, R., Fremaux, C., Deveau, H., Richards, M., Boyaval, P., Moineau, S., et al. (2007). CRISPR provides acquired resistance against viruses in prokaryotes. *Science, 315*(5819), 1709–1712. https://doi.org/10.1126/science.1138140.

Battista, N., Meloni, M. A., Bari, M., Mastrangelo, N., Galleri, G., Rapino, C., et al. (2012). 5-Lipoxygenase-dependent apoptosis of human lymphocytes in the international space station: Data from the ROALD experiment. *The FASEB Journal, 26*(5), 1791–1798. https://doi.org/10.1096/fj.11-199406.

Berman, E., & Eyal, S. (2019). Drug interactions in space: A cause for concern? *Pharmaceutical Research, 36*(8), 114. https://doi.org/10.1007/s11095-019-2649-9.

Bielefeldt-Ohmann, H., Genik, P. C., Fallgren, C. M., Ullrich, R. L., & Weil, M. M. (2012). Animal studies of charged particle-induced carcinogenesis. *Health Physics, 103*(5), 568–576. https://doi.org/10.1097/HP.0b013e318265a257.

Blackstone, E., Morrison, M., & Roth, M. B. (2005). H_2S induces a suspended animation-like state in mice. *Science, 308*(5721), 518. https://doi.org/10.1126/science.1108581.

Bogren, L. K., Murphy, C. J., Johnston, E. L., Sinha, N., Serkova, N. J., & Drew, K. L. (2014). 1H-NMR metabolomic biomarkers of poor outcome after hemorrhagic shock are absent in hibernators. *PLoS ONE, 9*(9), e107493. https://doi.org/10.1371/journal.pone.0107493.

Bosley, K. S., Botchan, M., Bredenoord, A. L., Carroll, D., Charo, R. A., Charpentier, E., et al. (2015). CRISPR germline engineering–the community speaks. *Nature Biotechnology, 33*(5), 478–486. https://doi.org/10.1038/nbt.3227.

Carmeliet, G., Nys, G., Stockmans, I., & Bouillon, R. (1998). Gene expression related to the differentiation of osteoblastic cells is altered by microgravity. *Bone, 22*(5 Suppl), 139S-143S. https://doi.org/10.1016/s8756-3282(98)00007-6.

Cater, J. P., & Huffman, S. D. (1995). Use of the remote access virtual environment network (RAVEN) for coordinated IVA-EVA astronaut training and evaluation. *Presence (Camb), 4*(2), 103–109. https://doi.org/10.1162/pres.1995.4.2.103.

Cavanagh, P. R., Licata, A. A., & Rice, A. J. (2005). Exercise and pharmacological countermeasures for bone loss during long-duration space flight. *Gravitational and Space Biology Bulletin, 18*(2), 39–58.

Cerri, M., Mastrotto, M., Tupone, D., Martelli, D., Luppi, M., Perez, E., et al. (2013). The inhibition of neurons in the central nervous pathways for thermoregulatory cold defense induces a suspended animation state in the rat. *Journal of Neuroscience, 33*(7), 2984–2993. https://doi.org/10.1523/JNEUROSCI.3596-12.2013.

Cerri, M., Tinganelli, W., Negrini, M., Helm, A., Scifoni, E., Tommasino, F., et al. (2016). Hibernation for space travel: Impact on radioprotection. *Life Science and Space Research (Amst), 11*, 1–9. https://doi.org/10.1016/j.lssr.2016.09.001.

Chen, Y., Aiken, A., Saw, S., Weiss, A., Fang, H., & Khokha, R. (2019). TIMP loss activates metalloproteinase-TNFalpha-DKK1 axis to compromise Wnt signaling and bone mass. *Journal of Bone and Mineral Research, 34*(1), 182–194. https://doi.org/10.1002/jbmr.3585.

Cheng, L., Lisowska, H., Sollazzo, A., Wegierek-Ciuk, A., Stepien, K., Kuszewski, T., et al. (2015). Modulation of radiation-induced cytogenetic damage in human peripheral blood lymphocytes by hypothermia. *Mutation Research, Genetic Toxicology and Environmental Mutagenesis, 793*, 96–100. https://doi.org/10.1016/j.mrgentox.2015.06.007.

Coates, P. J., Rundle, J. K., Lorimore, S. A., & Wright, E. G. (2008). Indirect macrophage responses to ionizing radiation: Implications for genotype-dependent bystander signaling. *Cancer Research, 68*(2), 450–456. https://doi.org/10.1158/0008-5472.CAN-07-3050.

Creff, J., Courson, R., Mangeat, T., Foncy, J., Souleille, S., Thibault, C., et al. (2019). Fabrication of 3D scaffolds reproducing intestinal epithelium topography by high-resolution 3D stereolithography. *Biomaterials, 221*, 119404. https://doi.org/10.1016/j.biomaterials.2019.119404.

Crucian, B. E., Stowe, R. P., Pierson, D. L., & Sams, C. F. (2008). Immune system dysregulation following short- vs long-duration spaceflight. *Aviation, Space and Environmental Medicine, 79*(9), 835–843. https://doi.org/10.3357/asem.2276.2008.

Crucian, B., & Sams, C. (2009). Immune system dysregulation during spaceflight: Clinical risk for exploration-class missions. *Journal of Leukocyte Biology, 86*(5), 1017–1018. https://doi.org/10.1189/jlb.0709500.

Dang, B., Yang, Y., Zhang, E., Li, W., Mi, X., Meng, Y., et al. (2014). Simulated microgravity increases heavy ion radiation-induced apoptosis in human B lymphoblasts. *Life Sciences, 97*(2), 123–128. https://doi.org/10.1016/j.lfs.2013.12.008.

Datta, K., Suman, S., Kallakury, B. V., & Fornace, A. J., Jr. (2013). Heavy ion radiation exposure triggered higher intestinal tumor frequency and greater beta-catenin activation than gamma radiation in APC(Min/+) mice. *PLoS ONE, 8*(3), e59295. https://doi.org/10.1371/journal.pone.0059295.

de Lecuona, I., Casado, M., Marfany, G., Lopez Baroni, M., & Escarrabill, M. (2017). Gene editing in humans: Towards a global and inclusive debate for responsible research. *Yale Journal of Biology and Medicine, 90*(4), 673–681.

Delgado, O., Batten, K. G., Richardson, J. A., Xie, X. J., Gazdar, A. F., Kaisani, A. A., Girard, L., Behrens, C., et al. (2014). Radiation-enhanced lung cancer progression in a transgenic mouse model of lung cancer is predictive of outcomes in human lung and breast cancer. *Clinical Cancer Research, 20*(6), 16101622. https://doi.org/10.1158/1078-0432.CCR-13-2589.

De Santis, M., Di Matteo, B., Chisari, E., Cincinelli, G., Angele, P., Lattermann, C., et al. (2018). The role of Wnt pathway in the pathogenesis of OA and its potential therapeutic implications in the field of regenerative medicine. *BioMed Research International, 2018*, 7402947. https://doi.org/10.1155/2018/7402947.

Diao, Y., Chen, B., Wei, L., & Wang, Z. (2018). Polyphenols (S3) Isolated from cone scales of *Pinus koraiensis* alleviate decreased bone formation in rat under simulated microgravity. *Science Report, 8*(1), 12719. https://doi.org/10.1038/s41598-018-30992-8.

Doudna, J. A., & Charpentier, E. (2014). Genome editing. The new frontier of genome engineering with CRISPR-Cas9. *Science, 346*(6213), 1258096. https://doi.org/10.1126/science.1258096.

Dugbartey, G. J., Bouma, H. R., Strijkstra, A. M., Boerema, A. S., & Henning, R. H. (2015). Induction of a torpor-like state by 5'-AMP does not depend on H_2S production. *PLoS ONE, 10*(8), e0136113. https://doi.org/10.1371/journal.pone.0136113.

Dugbartey, G. J., Talaei, F., Houwertjes, M. C., Goris, M., Epema, A. H., Bouma, H. R., & Henning, R. H. (2015). Dopamine treatment attenuates acute kidney injury in a rat model of deep hypothermia and rewarming—The role of renal H_2S-producing enzymes. *European Journal of Pharmacology, 769*, 225–233. https://doi.org/10.1016/j.ejphar.2015.11.022.

Durante, M. (2014). Space radiation protection: Destination Mars. *Life Science and Space Research (Amst), 1,* 2–9. https://doi.org/10.1016/j.lssr.2014.01.002.

Endo, I., & Matsumoto, T. (2008). Bisphosphonate and mechanical stress on bone. *Clin Calcium, 18*(9), 13211326. CliCa080913211326.

Endo, I., & Matsumoto, T. (2012). Space flight/bedrest immobilization and bone. Bisphosphonate and the loss of bone mineral due to space flight or prolonged bed rest. *Clin Calcium, 22*(12), 18631870. CliCa121218631870.

Eyal, S., & Derendorf, H. (2019). Medications in space. In search of a pharmacologist's guide to the galaxy. *Pharmaceutical Research, 36*(10), 148. https://doi.org/10.1007/s11095-019-2679-3.

Farber, J. L., Kyle, M. E., & Coleman, J. B. (1990). Mechanisms of cell injury by activated oxygen species. *Laboratory Investigation, 62*(6), 670–679.

Fears, R., & Ter Meulen, V. (2017). How should the applications of genome editing be assessed and regulated? *Elife 6*. https://doi.org/10.7554/eLife.26295.

Fernandez-Tornero, C. (2018). RNA polymerase I activation and hibernation. Unique mecha- [illegible] *[illegible]* 9(1), [illegible]. https://doi.org/10.1080/21541264.2017. [illegible].

Fu, Z., Chen, D., Cheng, H., & Wang, F. (2015). Hypoxia-inducible factor-1alpha protects cervical carcinoma cells from apoptosis induced by radiation via modulation of vascular endothelial growth factor and p53 under hypoxia. *Medical Science Monitor, 21,* 318–325. https://doi.org/10.12659/MSM.893265.

Fu, Z. L., Feng, Y. B., Xu, H. X., Zhang, X. P., Shi, C. Z., & Gu, X. (2006). Role of norepinephrine in development of short-term myocardial hibernation. *Acta Pharmacologica Sinica, 27*(2), 158–164. https://doi.org/10.1111/j.1745-7254.2006.00245.x.

Gilbert, M., Busund, R., Skagseth, A., Nilsen, P. A., & Solbo, J. P. (2000). Resuscitation from accidental hypothermia of 13.7 degrees C with circulatory arrest. *Lancet, 355*(9201), 375–376. https://doi.org/10.1016/S0140-6736(00)01021-7.

Gottweis, H. (2005). Regulating genomics in the 21st century: from logos to pathos? *Trends in Biotechnology, 23*(3), 118–121. https://doi.org/10.1016/j.tibtech.2005.01.002.

Gouw, A. (2019). The CRISPR advent of lulu and nana. *Theology and Science, 17*(1), 9–12.

Gouw, A. M., Margulis, K., Liu, N. S., Raman, S. J., Mancuso, A., Toal, G. G., Tong, L. (2019) The MYC oncogene cooperates with sterol-regulated element-binding protein to regulate Lipogenesis essential for neoplastic growth. *Cell Metabolism, 30*(3), 556–572. https://doi.org/10.1016/j.cmet.2019.07.012.

Gouw, A. (2018a). Challenging the Therapy/Enhancement Distinction in CRISPR Gene Editing. In D. Boonin (Ed.), *The palgrave handbook of philosophy and public policy* (pp. 493–508). New York: Palgrave Macmillan.

Gouw, A. (2018b). Genetic virtue program: An unfeasible neo-pelagian theodicy? *Theology and Science, 16*(3), 273–278.

Gouw, A. M., Toal, G. G., & Felsher, D. W. (2016). Metabolic vulnerabilities of MYC-induced cancer. *Oncotarget, 7*(21), 29879–29880. https://doi.org/10.18632/oncotarget.7223.

Grabek, K. R., Cooke, T. F., Epperson, L. E., Spees, K. K., Cabral, G. F., Sutton, S. C., et al. (2019). Genetic variation drives seasonal onset of hibernation in the 13-lined ground squirrel. *Communications and Biology, 2,* 478. https://doi.org/10.1038/s42003-019-0719-5.

Grigoriev, A., Morukov, B., Stupakov, G., & Bobrovnik, E. (1998). Influence of bisphosphonates on calcium metabolism and bone tissue during simulation of the physiological effects of microgravity. *Journal of Gravitational and Physiology, 5*(1), P69-70.

Guerra, C., Schuhmacher, A. J., Canamero, M., Grippo, P. J., Verdaguer, L., Perez-Gallego, L., et al. (2007). Chronic pancreatitis is essential for induction of pancreatic ductal adenocarcinoma by K-Ras oncogenes in adult mice. *Cancer Cell, 11*(3), 291–302. https://doi.org/10.1016/j.ccr.2007.01.012.

Gunga, H. C., Kirsch, K. A., Rocker, L., Maillet, A., & Gharib, C. (1996a). Body weight and body composition during sixty days of isolation. *Advances in Space Biology and Medicine, 5,* 39–53. https://doi.org/10.1016/s1569-2574(08)60052-0.

Gunga, H. C., Kirsch, K., & Rocker, L. (1996b). Erythropoietin in blood volume regulation under real and simulated micro-g conditions. *Journal of Gravitational Physiology, 3*(2), 1–4.

Guo, J., Zeitlin, C., Wimmer-Schweingruber, R. F., Hassler, D. M., Ehresmann, B., Kohler, J., et al. (2015). MSL-RAD radiation environment measurements. *Radiation Protection Dosimetry, 166*(1–4), 290–294. https://doi.org/10.1093/rpd/ncv297.

Hadj-Moussa, H., Watts, A. J., & Storey, K. B. (2019). Genes of the undead: hibernation and death display different gene profiles. *FEBS Letters, 593*(5), 527–532. https://doi.org/10.1002/1873-3468.13338.

Hammond, T., Allen, P., & Birdsall, H. (2016). Is there a space-based technology solution to problems with preclinical drug toxicity testing? *Pharmaceutical Research, 33*(7), 1545–1551. https://doi.org/10.1007/s11095-016-1942-0.

Haouzi, P., Notet, V., Chenuel, B., Chalon, B., Sponne, I., Ogier, V., & Bihain, B. (2008). H_2S induced hypometabolism in mice is missing in sedated sheep. *Respiratory Physiology and Neurobiology, 160*(1), 109–115. https://doi.org/10.1016/j.resp.2007.09.001.

Harm, D. L., Taylor, L. C., & Bloomberg, J. J. (2007). Adaptive responses in eye-head-hand coordination following exposures to a virtual environment as a possible space flight analog. *Journal of Gravitational and Physiology, 14*(1), P83–P84.

Haurwitz, R. E., Jinek, M., Wiedenheft, B., Zhou, K., & Doudna, J. A. (2010). Sequence- and structure-specific RNA processing by a CRISPR endonuclease. *Science, 329*(5997), 1355–1358. https://doi.org/10.1126/science.1192272.

Hayashi, T., Morishita, Y., Khattree, R., Misumi, M., Sasaki, K., Hayashi, I., et al. (2012). Evaluation of systemic markers of inflammation in atomic-bomb survivors with special reference to radiation and age effects. *The FASEB Journal, 26*(11), 4765–4773. https://doi.org/10.1096/fj.12-215228.

Heldmaier, G., Ortmann, S., & Elvert, R. (2004). Natural hypometabolism during hibernation and daily torpor in mammals. *Respiratory Physiology and Neurobiology, 141*(3), 317–329. https://doi.org/10.1016/j.resp.2004.03.014.

Hong, G., He, X., Shen, Y., Chen, X., Yang, F., Yang, P., et al. (2019). Chrysosplenetin promotes osteoblastogenesis of bone marrow stromal cells via Wnt/beta-catenin pathway and enhances osteogenesis in estrogen deficiency-induced bone loss. *Stem Cell Research and Therapy, 10*(1), 277. https://doi.org/10.1186/s13287-019-1375-x.

Hughes-Fulford, M. (2011). To infinity … and beyond! Human spaceflight and life science. *The FASEB Journal, 25*(9), 2858–2864. https://doi.org/10.1096/fj.11-0902ufm.

Jaroslow, B. N., Robbins, M., Stalling, S. A. (1969). Radiosensitivity of intestinal crypt cells in ground squirrels during arousal from hibernation. ANL-7635. *ANL Report*, 39–41.

Kast, J., Yu, Y., Seubert, C. N., Wotring, V. E., & Derendorf, H. (2017). Drugs in space: Pharmacokinetics and pharmacodynamics in astronauts. *European Journal of Pharmaceutical Sciences, 109S,* S2–S8. https://doi.org/10.1016/j.ejps.2017.05.025.

Kennedy, A. R. (2014). Biological effects of space radiation and development of effective countermeasures. *Life Science and Space Research (Amst), 1,* 10–43. https://doi.org/10.1016/j.lssr.2014.02.004.

Kim, H. K., Lee, S., Kim, Y., Park, J., Min, S., Choi, J. W., et al. (2020). High-throughput analysis of the activities of xCas9, SpCas9-NG and SpCas9 at matched and mismatched target sequences in human cells. *Nature Biomedical Engineering, 4*(1), 111–124. https://doi.org/10.1038/s41551-019-0505-1.

Kita, M., Yamamoto, T., Imanishi, J., & Fuse, A. (2004). Influence of gravity changes induced by parabolic flight on cytokine production in mouse spleen. *Journal of Gravitational and Physiology, 11*(2), P67–P68.

Lane, J. E., Kruuk, L. E., Charmantier, A., Murie, J. O., Coltman, D. W., Buoro, M., et al. (2011). A quantitative genetic analysis of hibernation emergence date in a wild population of Columbian ground squirrels. *Journal of Evolutionary Biology, 24*(9), 1949–1959. https://doi.org/10.1111/j.1420-9101.2011.02334.x.

Le, A., Stine, Z. E., Nguyen, C., Afzal, J., Sun, P., Hamaker, M., et al. (2014). Tumorigenicity of hypoxic respiring cancer cells revealed by a hypoxia-cell cycle dual reporter. *Proceedings of the*

National Academy of Science of the United States of America, 111(34), 12486–12491. https://doi.org/10.1073/pnas.1402012111.

Lee, M., & Kim, H. (2019). Therapeutic application of the CRISPR system: Current issues and new prospects. *Human Genetics, 138*(6), 563–590. https://doi.org/10.1007/s00439-019-02028-2.

Li, X., Zhang, Y., Kang, H., Liu, W., Liu, P., Zhang, J., et al. (2005). Sclerostin binds to LRP5/6 and antagonizes canonical Wnt signaling. *Journal of Biological Chemistry, 280*(20), 19883–19887. https://doi.org/10.1074/jbc.M413274200

Lisowska, H., Brehwens, K., Zolzer, F., Wegierek-Ciuk, A., Czub, J., Lankoff, A., et al. (2014). Effect of hypothermia on radiation-induced micronuclei and delay of cell cycle progression in TK6 cells. *International Journal of Radiation Biology, 90*(4), 318–324. https://doi.org/10.3109/09553002.2014.887233.

Lo, T. W., Pickle, C. S., Lin, S., Ralston, E. J., Gurling, M., Schartner, C. M., et al. (2013). Precise and heritable genome editing in evolutionarily diverse nematodes using TALENs and CRISPR/Cas9 to engineer insertions and deletions. *Genetics, 195*(2), 331–348. https://doi.org/10.1534/genetics.113.155382.

Lorimore, S. A., Chrystal, J. A., Robinson, J. I., Coates, P. J., & Wright, E. G. (2008). Chromosomal instability in unirradiated hemaopoietic cells induced by macrophages exposed in vivo to ionizing radiation. *Cancer Research, 68*(19), 8122–8126. https://doi.org/10.1158/0008-5472.CAN-08-0698.

Lustig, B. A, Brody, B. A., & McKenny, G. P. (2008). Altering nature. 2 vols, *Philosophy and medicine*. United States: Springer.

Lyman, C. P., & Fawcett, D. W. (1954). The effect of hibernation on the growth of sarcoma in the hamster. *Cancer Research, 14*(1), 25–28.

Ma, X., Dewan, S., Liu, J., Tang, M., Miller, K. L., Yu, C., et al. (2019). 3D printed micro-scale force gauge arrays to improve human cardiac tissue maturation and enable high throughput drug testing. *Acta Biomaterialia, 95,* 319–327. https://doi.org/10.1016/j.actbio.2018.12.026.

Ma, X., Pietsch, J., Wehland, M., Schulz, H., Saar, K., Hubner, N., et al. (2014). Differential gene expression profile and altered cytokine secretion of thyroid cancer cells in space. *The FASEB Journal, 28*(2), 813–835. https://doi.org/10.1096/fj.13-243287.

Maccarrone, M., Battista, N., Meloni, M., Bari, M., Galleri, G., Pippia, P., et al. (2003). Creating conditions similar to those that occur during exposure of cells to microgravity induces apoptosis in human lymphocytes by 5-lipoxygenase-mediated mitochondrial uncoupling and cytochrome c release. *Journal of Leukocyte Biology, 73*(4), 472–481. https://doi.org/10.1189/jlb.0602295.

Mantovani, A., Allavena, P., Sica, A., & Balkwill, F. (2008). Cancer-related inflammation. *Nature, 454*(7203), 436–444. https://doi.org/10.1038/nature07205.

Martinez, E. M., Yoshida, M. C., Candelario, T. L., & Hughes-Fulford, M. (2015). Spaceflight and simulated microgravity cause a significant reduction of key gene expression in early T-cell activation. *American Journal of Physiology: Regulatory, Integrative and Comparative Physiology, 308*(6), R480–R488. https://doi.org/10.1152/ajpregu.00449.2014.

Mittal, D., Gubin, M. M., Schreiber, R. D., & Smyth, M. J. (2014). New insights into cancer immunoediting and its three component phases–elimination, equilibrium and escape. *Current Opinion in Immunology, 27,* 16–25. https://doi.org/10.1016/j.coi.2014.01.004.

Mraz, L., & Praslicka, M. (1961). Influence of hypothermia on the survival of mice after large doses of x-radiation. *Nature, 189,* 677–678. https://doi.org/10.1038/189677a0.

Muid, S., Froemming, G. R., Ali, A. M., & Nawawi, H. (2013). Interleukin-6 and intercellular cell adhesion molecule-1 expression remains elevated in revived live endothelial cells following spaceflight. *Malaysian Journal of Pathology, 35*(2), 165–176.

Mukherjee, D., Coates, P. J., Lorimore, S. A., & Wright, E. G. (2014). Responses to ionizing radiation mediated by inflammatory mechanisms. *The Journal of Pathology, 232*(3), 289–299. https://doi.org/10.1002/path.4299

Musacchia, X. J., & Barr, R. E. (1968). Survival of whole-body-irradiated hibernating and active ground squirrels; Citellus tridecemlineatus. *Radiation Research, 33*(2), 348–356.

Nobile, M. (2014). The WHO definition of health: A critical reading. *Medicine and Law, 33*(2), 33–40.

Norsk, P. (2014). Blood pressure regulation IV: Adaptive responses to weightlessness. *European Journal of Applied Physiology, 114*(3), 481–497. https://doi.org/10.1007/s00421-013-2797-2.

Patterson, W. B., Lyman, C. P., & Patterson, H. R. (1957). Growth of human tumors in hibernating hamsters. *Proceedings of the Society for Experimental Biology and Medicine, 96*(1), 94–97. https://doi.org/10.3181/00379727-96-23402.

Rivina, L., & Schiestl, R. (2013). Mouse models of radiation-induced cancers. *Advances in Genetics, 84,* 83–122. https://doi.org/10.1016/B978-0-12-407703-4.00003-7.

Sarkar, D., & Fisher, P. B. (2006). Molecular mechanisms of aging-associated inflammation. *Cancer Letters, 236*(1), 13–23. https://doi.org/10.1016/j.canlet.2005.04.009.

Shao, J., Ma, J., Lin, L., Wang, B., Jansen, J. A., Walboomers, X. F., et al. (2019). Three-dimensional printing of drug-loaded scaffolds for antibacterial and analgesic applications. *Tissue Engineering Part C Methods, 25*(4), 222–231. https://doi.org/10.1089/ten.TEC.2018.0293.

Shapiro, J., Smith, B., Beck, T., Ballard, P., Dapthary, M., BrintzenhofeSzoc, K., & Caminis, J. (2007). Treatment with zoledronic acid ameliorates negative geometric changes in the proximal femur following acute spinal cord injury. *Calcified Tissue International, 80*(5), 316–322. https://doi.org/10.1007/s00223-007-9012-6.

Smith, F., & Grenan, M. M. (1951). Effect of hibernation upon survival time following whole-body irradiation in the marmot (Marmota monax). *Science, 113*(2946), 686–688. https://doi.org/10.1126/science.113.2946.686.

Smith, J. K. (2018). IL-6 and the dysregulation of immune, bone, muscle, and metabolic homeostasis during spaceflight. *NPJ Microgravity, 4,* 24. https://doi.org/10.1038/s41526-018-0057-9.

Sparmann, A., & Bar-Sagi, D. (2004). Ras-induced interleukin-8 expression plays a critical role in tumor growth and angiogenesis. *Cancer Cell, 6*(5), 447–458. https://doi.org/10.1016/j.ccr.2004.09.028.

Stine, Z. E., Walton, Z. E., Altman, B. J., Hsieh, A. L., & Dang, C. V. (2015). MYC, metabolism, and cancer. *Cancer Discovery, 5*(10), 1024–1039. https://doi.org/10.1158/2159-8290.CD-15-0507.

Szocik, K. (2020). Is human enhancement in space a moral duty? Missions to mars, advanced AI and genome editing in space. *Cambridge Quarterly of Healthcare Ethics, 29*(1), 122–130. https://doi.org/10.1017/S0963180119000859.

Szocik, K., & Braddock, M. (2019). Why human enhancement is necessary for successful human deep-space missions. *New Bioethics, 25*(4), 295–317. https://doi.org/10.1080/20502877.2019.1667559.

Szocik, K., Norman, Z., & Reiss, M. J. (2019). Ethical challenges in human space missions: A space refuge, scientific value, and human gene editing for space. *Science and Engineering Ethics.* https://doi.org/10.1007/s11948-019-00131-1.

Talaei, F., Bouma, H. R., Van der Graaf, A. C., Strijkstra, A. M., Schmidt, M., & Henning, R. H. (2011). Serotonin and dopamine protect from hypothermia/rewarming damage through the CBS/H_2S pathway. *PLoS ONE, 6*(7), e22568. https://doi.org/10.1371/journal.pone.0022568.

Tan, J., Zhang, F., Karcher, D., & Bock, R. (2020). Expanding the genome-targeting scope and the site selectivity of high-precision base editors. *Nature Communications, 11*(1), 629. https://doi.org/10.1038/s41467-020-14465-z.

Terns, M. P., & Terns, R. M. (2011). CRISPR-based adaptive immune systems. *Current Opinion in Microbiology, 14*(3), 321–327. https://doi.org/10.1016/j.mib.2011.03.005.

Tupone, D., Madden, C. J., & Morrison, S. F. (2013a). Central activation of the A1 adenosine receptor (A1AR) induces a hypothermic, torpor-like state in the rat. *Journal of Neuroscience, 33*(36), 14512–14525. https://doi.org/10.1523/JNEUROSCI.1980-13.2013.

Tupone, D., Madden, C. J., & Morrison, S. F. (2013b). Highlights in basic autonomic neurosciences: central adenosine A1 receptor—The key to a hypometabolic state and therapeutic hypothermia? *Auton Neurosci, 176*(1–2), 1–2. https://doi.org/10.1016/j.autneu.2013.02.004.

Van Walleghem, M., Tabury, K., Fernandez-Gonzalo, R., Janssen, A., Buchheim, J. I., Chouker, A., et al. (2017). Gravity-related immunological changes in human whole blood cultured under simulated microgravity using an in vitro cytokine release assay. *Journal of Interferon and Cytokine Research, 37*(12), 531–540. https://doi.org/10.1089/jir.2017.0065.

Wade, M. A., Sunter, N. J., Fordham, S. E., Long, A., Masic, D., Russell, L. J., et al. (2015). c-MYC is a radiosensitive locus in human breast cells. *Oncogene, 34*(38), 4985–4994. https://doi.org/10.1038/onc.2014.427.

Wang, G., Xiao, L., Wang, F., Yang, J., Yang, L., Zhao, Y., & Jin, W. (2019). Hypoxia inducible factor-1alpha/B-cell lymphoma 2 signaling impacts radiosensitivity of H1299 non-small cell lung cancer cells in a normoxic environment. *Radiation and Environmental Biophysics, 58*(3), 439–448. https://doi.org/10.1007/s00411-019-00802-4.

Wang, X., Niu, Y., Zhou, J., Zhu, H., Ma, B., Yu, H., et al. (2018). CRISPR/Cas9-mediated MSTN disruption and heritable mutagenesis in goats causes increased body mass. *Animal Genetics, 49*(1), 43–51. https://doi.org/10.1111/age.12626.

Wei, Y., Chen, Y., Qiu, Y., Zhao, H., Liu, G., Zhang, Y., et al. (2016). Prevention of muscle wasting by CRISPR/Cas9-mediated disruption of myostatin in vivo. *Molecular Therapy, 24*(11), 1889–1891. [illegible]

[illegible], H. N., Rich, T. H., Chinsamy, A., & Vickers-Rich, P. (2011). Growth dynamics of Australia's polar dinosaurs. *PLoS ONE, 6*(8), e23339. https://doi.org/10.1371/journal.pone.0023339.

Wright, E. G., & Coates, P. J. (2006). Untargeted effects of ionizing radiation: Implications for radiation pathology. *Mutation Research, 597*(1–2), 119–132. https://doi.org/10.1016/j.mrfmmm.2005.03.035.

Wu, Q., Sugimoto, K., Moriyama, K., Adachi, Y., Nakayama, A., & Mori, J. K. (2002). Cloning of hibernation-related genes of bullfrog (Rana catesbeiana) by cDNA subtraction. *Comparative Biochemistry and Physiology Part B: Biochemistry and Molecular Biology, 133*(1), 85–94. https://doi.org/10.1016/s1096-4959(02)00117-3.

Xix Congresso Nazionale S.I.C.O.O.P. Societa' Italiana Chirurghi Ortopedici Dell'Ospedalita' Privata, Aicale, A. R., Tarantino, D., Maccauro, G., Peretti, G. M., & Maffulli, N. (2019). Genetics in orthopaedic practice. *Journal of Biological Regulators and Homeostatic Agents, 33*(2 Suppl. 1):103117.

Yuan, J., Tickner, J., Mullin, B. H., Zhao, J., Zeng, Z., Morahan, G., & Xu, J. (2019). Advanced genetic approaches in discovery and characterization of genes involved with osteoporosis in mouse and human. *Front Genet, 10,* 288. https://doi.org/10.3389/fgene.2019.00288.

Yue, X., Lan, F., & Xia, T. (2019). Hypoxic glioma cell-secreted exosomal miR-301a activates Wnt/beta-catenin signaling and promotes radiation resistance by targeting TCEAL7. *Molecular Therapy, 27*(11), 1939–1949. https://doi.org/10.1016/j.ymthe.2019.07.011.

Zhang, J., Kaasik, K., Blackburn, M. R., & Lee, C. C. (2006). Constant darkness is a circadian metabolic signal in mammals. *Nature, 439*(7074), 340–343. https://doi.org/10.1038/nature04368.

Zhang, T., Lu, Y., Song, S., Lu, R., Zhou, M., He, Z., Yuan, T., Yan, K., & Cheng, Y. (2019). 'Double-muscling' and pelvic tilt phenomena in rabbits with the cystine-knot motif deficiency of myostatin on exon 3. *Biosci Reports, 39*(5). https://doi.org/10.1042/BSR20190207.

Chapter 3
Biological Modification as Prophylaxis: How Extreme Environments Challenge the Treatment/Enhancement Divide

Evie Kendal

Abstract This chapter proposes that extreme environments shift the boundary between treatment and enhancement, such that biological modifications intended to protect space travellers' health can be considered prophylactic measures, rather than human enhancements. In considering which modifications might reasonably fall within this category, the following criteria will be explored: severity of predicted damage without biological intervention; the nature and acceptability of the proposed intervention; the efficacy of any non-biological alternatives; possible reversibility of the intervention; short- and long-term impacts; and incidental risks and benefits beyond the intended purpose. This discussion will also take into account arguments about the proper scope of medicine and ethical implications of transhumanism.

3.1 Introduction

Space is a hostile environment for humans, and much of aerospace medicine is focused on minimising the negative effects of space travel on the human body. Astronauts face a number of significant health risks, including exposure to radiation and microgravity—factors that have been associated with increased lifetime cancer risk, loss of organ function, visual acuity, bone and muscle density, and a variety of genetic and cognitive changes (Blaber et al. 2010; Szocik et al. 2019; Mann et al. 2019a). The isolation and stress of prolonged space travel are also concerning for astronauts' psychological wellbeing (Marušič et al. 2014).

Due to the inherent pressures of operating in extreme environments, remote and Antarctic medicine are useful analogues for aerospace medicine. However, this chapter will argue that many of the differences in medical practice in such environments provide what might otherwise be considered a *lower* standard of care for patients and workers—albeit due to reasonable practical constraints—rather than

E. Kendal (✉)
Swinburne University of Technology, Hawthorn, Victoria, Australia
e-mail: ekendal@swin.edu.au

K. Szocik (ed.), *Human Enhancements for Space Missions*, Space and Society,
https://doi.org/10.1007/978-3-030-42036-9_3

developing more effective interventions targeted to these conditions. Translated to the space environment, such an approach might support prophylactic appendectomies for space doctors, for example, but not biological or genetic manipulation to improve organ function.

This chapter proposes that extreme environments shift the boundary between treatment and enhancement, such that biological modifications intended to protect against bone loss in microgravity or increase human resilience against radiation, etc., can be considered prophylactic measures, rather than human enhancements. In considering which modifications might reasonably fall within this category, the following criteria will be explored: severity of predicted damage without biological intervention, the nature and acceptability of the proposed intervention; the efficacy of any non biological alternatives; possible reversibility of the intervention, short and long-term impacts, and incidental risks and benefits beyond the intended purpose. This discussion will also take into account arguments about the proper scope of medicine and ethical implications of transhumanism.

By establishing what makes space exploration and aerospace medicine special, this chapter aims to demonstrate that targeted modifications in human biology that might be deemed inappropriate enhancements on Earth can represent necessary protective measures for space travellers. As such, adopting a more nuanced approach to the treatment/enhancement divide might help improve astronaut health and safety, while avoiding concerns about misuse on Earth.

3.2 The Threats of Space Travel and Current Mitigation Strategies

Space travel poses a number of significant threats to human health. Exposure to cosmic radiation, for example, has been shown to increase cancer risk, with solar flares representing a particular concern. As Koike et al. (2005) note, when able to be predicted, those flares expected to eject a lot of high-energy protons (termed solar particle events) are considered sufficiently hazardous to contraindicate space flight during the relevant period. However, due to the necessary time commitments, a crewed mission to Mars or beyond would be unable to avoid exposure in the same manner (Koike et al. 2005). At present, the only mitigation strategy for the crew of such a mission would rely on structural and personal protective materials. Radiation is also implicated in fertility loss, vision impairment and neurological damage, thereby justifying why people on Earth who work in careers involving increased exposure to radiation are required to wear protective equipment (e.g. a radiographer's lead apron) and employers are ethically and legally required to reduce risk by limiting exposure. Szocik and Wójtowicz (2019) claim such strategies are unlikely to be practical for a long-term space mission though, where the average dose of radiation is approximately 100 times that of Earth and typically cannot be limited. Nevertheless, the ethical obligation to protect space industry employees from occupational health and safety risks remains.

Space travel is also associated with musculoskeletal changes, including bone demineralisation and muscle atrophy (Uri and Haven 2005; Blaber et al. 2010). The effects of microgravity are known to increase astronauts' risk of developing osteoporosis, through bone density loss and reduced bone formation, while muscular damage is associated with the development of orthostatic intolerance, including dizziness, fainting and hypotension on standing and coordination issues, including ataxia (Blaber et al. 2010). At present, these risks are predominantly managed through diet and exercise regimes, but returned astronauts are still expected to experience significant weakness and discomfort when reintroduced to Earth gravity conditions. As bone and muscle loss increases according to the duration of exposure to microgravity, longer-term missions are expected to lead to significant cumulative damage to space travellers. The impact of altered gravitational conditions on posture has also been associated with increased risk of renal stone formation, cardiovascular changes and the development of oedema (Uri and Haven 2005; Blaber et al. 2010).

Neuro-cognitive issues have also been identified in astronauts, in addition to psychological distress associated with prolonged isolation from loved ones and confinement in close quarters with co-workers (Mann et al. 2019b). The stress and high stakes of operating in an extreme environment mean difficulties concentrating and sleeping are particularly dangerous for space crews, while a lack of privacy, entertainment and psychological support has been associated with mental health concerns in these and other analogous conditions, such as Antarctic expeditions and confined environments (Salamon et al. 2018). While astronauts are vigorously evaluated for psychological resilience and mental health status, studies show social conditions of isolation and environmental exposure to microgravity are associated with increased levels of stress and anxiety and decreased cognitive performance (Marušič et al. 2014). Recommendations to redress these stressors include pharmacological interventions, virtual reality entertainment and regular mental health screenings by aeromedical examiners (Salamon et al. 2018; Scarpa and Sventek 2017). Space adaptation syndrome refers to the collection of central nervous system effects seen in astronauts, including sensory changes, altered perception, synaptic loss, neurochemical imbalances, memory loss, dizziness, vertigo, nausea and vomiting (Koike et al. 2005; Blaber et al. 2010). These symptoms directly compromise what Marušič et al. (2014) term the "human advantage" of crewed missions—the superior cognitive skills of a human for decision-making compared with a machine. When added to the psychological symptoms associated with circadian rhythm disruption, sleep deprivation and persistent high stress, the need for more effective methods to protect astronaut wellbeing is apparent.

The above stressors have also been implicated in immune suppression among space travellers. Endocrine dysregulation and viral reactivation (e.g. of latent herpes virus) have been seen in astronauts, placing them at higher risk of infectious diseases both during space travel and for a period of time on return to Earth (Mann et al. 2019b). Immune function changes are both associated with, and compounded by, the other negative health outcomes listed here; however, few mitigation strategies have been developed to counteract the negative health impacts of space travel on the immune system.

A final area of interest here is the differences in standards of care for emergency situations occurring in extreme environments. The classic example is the use of prophylactic appendectomies for space and Antarctic mission doctors, due to the risks posed by acute appendicitis where limited emergency medical resources are available (Australian Antarctic Division 2015). While this intervention—the removal of a healthy organ that might never actually become diseased—would be deemed inappropriate in most other contexts, the unique health threats of working in a remote, isolated environment are considered justified grounds to engage this practice. Similarly, other medical and dental emergencies may require treatment using equipment, expertise or methods that would be considered substandard on Earth. Alternatives would require building spacecraft and stations with medical facilities of similar sophistication as what would be available to patients on Earth, an undertaking that would likely be prohibitive both in terms of practicality and expense. The use of telemedicine and robot-assisted surgeries are already helping space travellers, but for longer-term, deep-space missions, lag time will limit the potential of these options.

More effective ways to manage the health risks associated with space travel might include biologically modifying astronauts such that their bodies, and particularly skin, are more radiation resistant, and bone and muscle are protected from demineralisation and atrophy. Cognitive, psychological and immunological issues could be managed through a number of proposed neurological or genetic modifications, which could also have an impact on metabolic efficiency, energy levels and sleep needs (Szocik et al. 2019). Biological modifications aimed at optimising organ system function and preventing damage and disease could also circumvent the need for some emergency interventions. While it is beyond the scope of this chapter to consider the scientific possibilities of these so-called "human enhancements", there are a number of ethical issues that warrant attention when considering what risk/benefit thresholds might justify such interventions. This chapter will consider these biological modifications from a bioethical perspective, focusing on the interests of both space travellers and human societies on Earth.

3.3 Ethical Concerns with Biological Modification

As demonstrated above, there are various potential benefits to improving human resilience against the unique health threats of operating in the space environment. However, when such improvements rely on modification of the human body, a number of ethical concerns arise. Bioconservatives—both religious and secular—appeal to ideas of unnaturalness to oppose biological modification and/or human enhancement. From the religious perspective, this can be due to a belief it is immoral or hubristic to interfere with divine design, or simply counterproductive, as humans have been created in such a way as naturally promotes flourishing (Burdett and Lorrimar 2019). In the latter case, since human enhancements do not align with the natural order, which was supposedly created with human flourishing in mind, the logic follows that they must ultimately diminish wellbeing. From the secular perspective, human enhancement might be considered to undermine ideals of individuality and personal freedom, by circumscribing which human characteristics are desirable and seeking to homogenise members of a social group—in this case, astronauts—to exhibit these qualities. In his book *The Case Against Perfection*, bioethicist Michael Sandel argues that such goals may threaten the core of what makes us human—a uniqueness that includes our achievements and flaws (Sandel 2007). Similarly, Jürgen Habermas refers to this "self-instrumentalisation and self-optimisation" of human nature as being dependent on a number of assumptions about what makes a "good life" that may not be valid or hold across cultures (Habermas 2018, 20). Even if modifications were restricted to those aimed at avoiding disease, he notes consensus regarding what might constitute sufficient justification for biological intervention would be limited to extreme cases.

The bioconservative argument against biological modification goes beyond safety concerns to focus on whether it is morally appropriate for humans to attempt to control their own nature. A related concern that is prominent in the literature is the potential for human enhancements to become the norm in society, thereby pressuring citizens to submit to invasive modifications in order to meet increasingly demanding standards (Ruggiu 2018). This argument often appeals to the idea that natural limitations can be liberating, as well as showing respect for the diversity of human gifts and aptitudes (Sandel 2007). Others are concerned that allowing some to access biological modifications for socially desirable traits will undermine equality, with Francis Fukuyama (2002) claiming that the basis of liberal democracy is our shared human nature. Once some members of the population—presumably the more socio-economically advantaged—can escape some of the vulnerabilities of being human, this solidarity may be lost. Furthermore, allowing some biological modifications to be done may serve as a justification for more and more significant interventions into human biology. This classic "slippery slope" argument suggests that if we allow any human enhancements, we might end up in a situation where there are no limits placed on the kinds of enhancements people might demand.

Szocik and Wójtowicz (2019) claim that the philosophical literature focused on human enhancement has heretofore neglected to engage with the complexities of space travel. When transposing the concerns listed above to the space industry context, it is possible to speculate why—many of these opposing arguments lose at least some weight when considered in isolation of the systems we typically associate with human societies. Astronauts as a population are not representative of the general population, and pressures to engage in enhancement technologies, while certainly of potential concern within this small sample, are unlikely to influence the vast majority of people. Further, the kinds of enhancements an astronaut might want may have limited applicability to non-spacefaring humans. Space is already an unnatural environment for humans to be found, so appeals to "naturalness" may serve only as an argument to prevent space travel entirely, rather than any biological modifications required to make such travel safer. Potential justice concerns that only the wealthy will be able to access these technologies, thereby introducing a class system of enhancement "haves" and "have nots", is also going to have limited influence, as astronauts are already a highly specialised group.

While broad concerns about the impact of biological modification of astronauts on the wider human population cannot be completely disregarded, the scale of potential impact and the relative ease with which extreme environment eligibility criteria could be placed on access to these technologies goes some way to alleviating these concerns. However, the ethical issues specific to space industry employees do warrant closer examination. Even imagining a scenario in which the total number of people susceptible to coercion is small, it is still important to ensure there are safeguards in place to protect astronauts' autonomy when selecting whether to engage in biological modification practices. This is particularly the case when considering the competitive nature of the industry. As Le Dévédec (2019) notes.

> [illegible] possibility of altering and optimising biological norms in themselves, human enhancement practices are far from participating in workers' emancipation. On the contrary, they foster greater adaptability in workers to more intense working conditions and they intensify the internalisation of the neoliberal norms of self-surpassment and performance (4).

He notes that the use of technoscientific and pharmaceutical enhancements is already commonplace in high-pressure workplace environments, such as the military, or in careers where the limitations of the body are a particular impediment, such as elite sports (Le Dévédec 2019). Given that space and other extreme environments are similarly stressful and physically demanding, the drive to pursue certain modifications is likely to follow suit. Psychostimulant use, for example, is increasingly seen among workers in careers where focus is required under conditions of sleep deprivation, such as truck drivers or surgeons (Le Dévédec 2019). As noted in the previous section, operating in space involves the same demands, under potentially even more risky conditions. And just as one truck driver might feel compelled to use amphetamines to work through the night, especially if their peers are doing likewise, so might an astronaut feel undue pressure to submit to new technologies aimed at optimising their physical and psychological functioning on mission. From a bioethical perspective, this suggests that several thresholds have to be met to justify the use of biological modification in astronauts, including both cost/benefit analyses and stringent informed consent procedures. This chapter will now consider these issues in turn.

3.4 Biological Modification as Prophylaxis and the Case for "Somaforming"

One method of justifying biological modification for astronauts is to consider the risks of not enhancing their physical and mental resilience to the space environment, especially in the context of long-term space missions. Considering the insufficiency of current mitigation strategies against the unique health threats outlined earlier in this chapter, alongside the sustained exposure to these occupational risks that a long-term mission would entail, there is a strong case to be made that there is an ethical obligation to protect astronaut wellbeing through a variety of measures. At a minimum, these include developing more effective protective gear, and potentially more significant interventions, including biological modification. Importantly, due to the increased risks associated with space as an extreme environment, these interventions can be classified as prophylactic measures, rather than human enhancements. This is partially due to the fact that when employed in their intended environment, far from conferring superhuman abilities, the proposed modifications would merely provide some protection against unusually severe health risks. Or, as Szocik et al. (2019) state, for biologically modified space travellers, "acquired advantages just might barely counteract comparable threats" (73). As with all preventive medicine, modifications to counteract the effects of microgravity, for example, would only be made available to this at-risk population, thereby minimising the risk of influencing medical practice on Earth. Such a modification would be analogous to providing influenza vaccinations to prevent disease—a benefit is only present if the recipient is at some risk of exposure to the virus. Receiving the vaccination leads to an "unnatural" advantage—acquired immunity—but it is of no use in the absence of the threat itself.

Another way to approach the justification of biological modification is to consider them as directed evolution. As Szocik and Wójtowicz (2019) note:

> ...the space environment is not a natural environment of human evolutionary adaptedness. ... We argue that, because of this lack of adaptedness to live in space, the treatment-enhancement distinction fails. We consider human enhancement in space as a non-trivial undertaking which – if ever applicable – will be a medical treatment-like procedure aimed at increasing the chances for human survival in space (5-6).

As current compensatory measures are inadequate to the challenge of protecting human health in this extreme environment, biological modifications that make humans better suited for life and work in space can be viewed as a way of levelling the playing field between humans living on Earth and those living in space or on other planets. While the former benefit from natural evolution adapted to Earth's conditions, the latter require interventions to achieve a similar level of safety. If recasting modifications in this way, the usual separation between treatment and enhancement becomes less obvious, and the case that "enhancements" are simply an extension of preventive medicine becomes more compelling. This also goes some way towards addressing objections that the proper scope of medicine, and particularly state-funded health care, should be restoring "species-typical normal functioning", rather than providing enhancements for some citizens (Daniels 2000). After all, it is species-typical to be able to withstand background radiation, breathe normally in the atmosphere and retain bone and muscle density through everyday activity. Biological modifications that return a human form to this condition while in space could be seen less as enhancing the body and more as treating it to cope with its unique environmental stressors.

That the treatment-enhancement divide has received so much attention in bioethics scholarship also speaks to concerns about transhumanism more generally. While it must be accepted that standard medical therapy today would have been considered dramatic human enhancement in the past—antibiotics, vaccinations, artificial respiration[1]—there still seems to be a belief that there is a line that can be drawn between "fair" restorative therapy and potentially "unfair" human enhancement. Cohen (2014) claims this can be partially attributed to the fact physicians and nurses traditionally "viewed themselves as the curers of disease and caretakers of the diseased, not as part of the profession aimed at improving wellness more generally" (650). This also explains some of the difficulty preventive medicine has gaining traction, despite the field having a pronounced impact on population health at low expense. The lack of a specific disease and cure to point to often undermines support for health promotion interventions. In a similar vein, while most would not object to research focused on improving treatments for the cancers arising from cosmic radiation, there is still scepticism regarding the appropriateness of modifying the human body to prevent the damage accumulating in the first place.

In her fiction novel, *To Be Taught If Fortunate* (2019), Becky Chambers proposes a term that avoids the transhumanist label and captures the idea of adapting a human body to life in space or on another planet: somaforming. The following excerpts paint a good picture for what this might look like, taken from the perspective of the novel's protagonist:

> In the early decades of human spaceflight, six months in low-Earth orbit – a mere two hundred miles up – were enough to raise your overall cancer risk a few notches. The further you head into interplanetary space… the worse the exposure becomes.
>
> Human spaceflight was stalled for decades because of this, crippled by the technological nut that could not be cracked: how do you keep humans alive in space during the length of time it takes to reach other planets? We beat our heads against the drafting table, trying to build tools that could do what our anatomy could not. We wrapped our brains around algorithms, trying to create artificial intelligence that could venture to other worlds for us. But our machines were inadequate, and our software never woke up. …To properly survey a place, you need boots on the ground. You need human intuition. You need eyes that can tell when something that looks like a rock might be *more* than a rock.
>
> It ended up being far easier, once the science matured, to engineer our bodies instead.
>
> We do not change much – nothing that would make us unrecognisable, nothing that would push us beyond the realm of our humanity and nothing that changes how I think or act or perceive. Only a small number of genetic supplementations are actually possible, and none of them are permanent. … Hence, the enzyme patch: a synthetic skin-like delivery system gives our bodies that little bit extra we need to survive on different worlds (13-14).

This fictional account of why retaining the human element of space exploration is beneficial, aligns with Marušič et al.'s (2014) claims about the superior decision-making capacity of human travellers over machines. As Chambers' character notes, a human will be able to use their intuition to observe conditions in a way a machine will not. In terms of ethical considerations, the character continues:

> We astronauts are not superheroes, nor shape-shifters. We are as human as you.
>
> I am as biased as can be, but I believe somaforming is the most ethical option when it comes to setting foot off Earth. I am an observer, not a conqueror. I have no interest in changing other worlds to suit me. I choose the lighter touch: changing myself to suit them (14).

This also speaks to concerns about how terraforming other worlds for human settlements might impact the natural evolution of other biosignatures. For astronauts, by this account it is plausible to consider biological modification as simply being "treated" to safely live in space.

3.5 Ethical Prioritisation of Biological Modification for Space Travel

The final consideration of this chapter is how to prioritise and ethically evaluate different potential modifications for space travellers to maximise benefits while minimising foreseeable harms. The argument for why better health protective measures are required if we intend to continue human space exploration has already been made, so this section will focus more on how to select appropriate interventions while avoiding negative ramifications, including for Earth populations. The proposal here is to establish a point-based system that can classify modifications as morally impermissible or permissible, on the basis of cost–benefit analyses, with the potential for a third category—a moral imperative to engage biological modification to avoid extreme harm.

In the human enhancement literature, there is often a distinction drawn between enhancements that are reversible versus irreversible, with the former typically being considered less ethically challenging. However, Cohen (2014) claims this represents "more of a continuum" than a binary opposition (649), with other authors further differentiating between temporary, long-lasting or permanent interventions. For example, Ruggiu (2018) uses the terms "provisional" to refer to pharmaceutical enhancements and "permanent" to refer to genetic modification, implying the latter warrants greater ethical scrutiny (85). Szocik and Wójtowicz (2019) claim the "most radical" modifications will likely be both irreversible and invasive, and potentially inheritable by offspring (6). This section will only consider interventions that are intended for competent adults able to consent to their own treatment, with further work needed to evaluate the ramifications for affected embryos and fetuses, and issues surrounding unconscious and other non-consenting patients.

The first threshold to reach when attempting to justify any biological modification aimed at improving astronaut wellbeing is whether the intervention itself is morally permissible, e.g. does it introduce unacceptable risks, either to astronauts or the broader community? This might be treated as a binary: yes/no. If this threshold is met, the next most important consideration is the severity of predicted damage without the biological intervention. A high risk of significant damage would contribute several points in favour of modification, whereas the efficacy of any non-biological alternatives to protect astronaut health might be used to offset this score. While Szocik and Wójtowicz (2019) claim it might always be better to enhance astronaut cognition and immunity, even in situations where "non-enhancement countermeasures are available and effective enough", a point system that can reasonably balance the two factors might help elucidate which biological modifications should be prioritised (1). Other issues like the length of time the astronauts will be required to withstand exposure to the unique health threats of the space environment can all be factored into the severity of threat calculation. The invasiveness of the intervention will also need to be weighed against the health benefit conferred, as is typical for cost–benefit analyses in health care. For example, the positive points accruing to a modification that aims to avoid an increased lifetime cancer risk will necessarily be weighed against the "cost" of the intervention itself, such that a very invasive or painful modification would require a very large health benefit to be considered ethically permissible. While there is no consensus in the literature regarding the use of "invasive" as a clinical descriptor, the definition put forward by Cousins et al. (2019) focuses on method of access to the body (e.g. penetration of the skin), the type of instruments used in the procedure and the skill level required for the practitioners involved. These all directly impact the risks involved for the patient such that the standards for informed consent necessarily become more stringent the more invasive the procedure involved.

The next consideration focuses on the short- and long-term impacts of the intervention and whether it is reversible. Here, there may be disagreement regarding which elements should provide a positive weighting in favour of modification and which a negative. For example, a permanent intervention might be judged more favourably than a temporary one, if the person receiving it intends to live in an off-world settlement indefinitely. Having to repeatedly undergo therapy to retain the health advantages required to live safely in the relevant environment might be considered onerous or could increase the likelihood of negative side effects. However, in many circumstances, reversibility would be considered desirable, especially for the majority of candidates who will eventually return to Earth conditions. Finally, incidental risks and benefits beyond the intended purpose of the modification need to be considered, including the potential for social harms on Earth. As stated previously, the sphere of influence for astronaut enhancements may often be limited, but any threat that allowing modifications in this population will lead to greater inequality on Earth must be factored into ethical evaluation of the proposed modifications. This is also the point at which unintended impacts on embryos or fetuses might be considered in the future.

The above characterisation highlights a few obvious questions that will need to be addressed before such a system could be used to classify biological modifications and rank them in terms of research priority. The first is who should decide what is more important between, for example, avoiding a slippery slope on Earth and avoiding prolonged suffering for humans in space? After all, some of the considerations noted in this chapter have incommensurable goals. However, established policies in many healthcare systems around the world already rely on a similar system when justifying health expenditure, with quality-adjusted life years (QALYs) used to compare health states and intervention risk/benefits that would otherwise be incomparable. Using a similar method with a particular focus on collecting the views of astronauts and aeromedical examiners could create an algorithm by which such judgments could be made. Interventions could be classified as morally permissible, impermissible or in the extreme, potentially imperative. After which, the major ethical concerns would be promoting astronaut autonomy and bodily integrity regarding whether to accept or refuse available biological modifications, including ensuring detailed informed consent procedures are followed that provide the best available data on the risks and benefits of the proposed interventions and work to avoid undue pressure or coercion to engage with these for space workers.

3.6 Conclusion

In conclusion, biological modification for astronauts and other space travellers has the potential to mitigate some of the unique health threats of operating in the space environment. As analogies with other extreme environments have shown, in the absence of so-called "enhancement" technologies, standards of care might otherwise be lower for humans in space than those on Earth. Given the lack of evolutionary adaptation to the space environment, some modifications can reasonably be considered to collapse the traditional treatment-enhancement distinction and are unlikely to cause significant ethical problems for societies on Earth, such as a slippery slope to more extreme human enhancements. Finally, it is possible to evaluate and prioritise potential biological modifications on the basis of the preventive health benefits conferred and the efficacy of any non-biological alternatives. The most important considerations moving forward are how best to promote astronaut autonomy, health, and safety, regarding the use of biological modifications for space travel.

References

Australian Antarctic Division. (2015). *Do you need your appendix removed before you go?* Australian Government Department of the Environment and Energy website, available at: https://www.antarctica.gov.au/about-antarctica/people-in-antarctica/health. Last updated 27 Nov 2015, accessed 28 October 2019.

Blaber, E., Marçal, H., & Burns, B. P. (2010). Bioastronautics: The influence of microgravity on astronaut health. *Astrobiology, 10*(5), 463–473.

Burdett, M., & Lorrimar, V. (2019). Creatures bound for glory: Biotechnological enhancement and visions of human flourishing. *Studies in Christian Ethics, 32*(2), 241–253.

Chambers, B. (2019). *To be taught if fortunate.* UK: Hodder and Stoughton.

Cohen, G. (2014). What (If Anything) is wrong with human enhancement? what (If Anything) is right with It? *Tulsa Law Review, 49*(3), 645.

Cousins, S., Blencowe, N. S. & Blazeby, J. M. (2019). What is an invasive procedure? A definition to inform study design, evidence synthesis and research tracking. *BMJ Open 32.* https://doi.org/10.1136/bmjopen-2018-028576.

Daniels, N. (2000). Normal functioning and the treatment-enhancement distinction. *Cambridge Quarterly of Healthcare Ethics, 9*(3), 309–322.

Fukuyama, F. (2002). *Our posthuman future: Consequences of the biotechnology revolution.* New York: Farrar, Straus and Giroux.

Habermas, J. (2018). *The future of human nature.* Cambridge: Polity Press.

Koike, Y., Frey, M. A., Sahiar, F., Dodge, R., & Mohler, S. (2005). Effects of HZE particle on the nigrostriatal dopaminergic system in a future mars mission. *Acta Astronautica, 56,* 367–378.

Le Dévédec, N. (2019). The biopolitical embodiment of work in the era of human enhancement. *Body & Society.* https://doi.org/10.1177/1357034X19876967.

Mann, V., Sundaresan, A., & Chaganti, M. (2019). Cellular changes in the nervous system when exposed to gravitational variation. *Neurology India, 67*(3), 684–691.

Mann, V., Sundaresan, A., Mehta, S. K., Crucian, B., Doursout, M. F., & Devakottai, S. (2019). Effects of microgravity and other space stressors in immunosuppression and viral reactivation with potential nervous system involvement. *Neurology India, 67*(2), S198-203.

Marušič, U., Meeusen, R., Pišot, R., & Kavcic, V. (2014). The brain in micro- and hypergravity: The effects of changing gravity on the brain electrocortical activity. *European Journal of Sport Science, 14*(8), 813–822.

Ruggiu, D. (2018). Implementing a responsible, research and innovation framework for human enhancement according to human rights: The right to bodily integrity and the rise of 'enhanced [illegible] 10(1), [illegible]

[illegible], Horack, J. M., & Newton, E. K. (2018). Application of virtual reality for crew mental health in extended-duration space missions. *Acta Astronautica, 146,* 117–122.

Sandel, M. 2007. *The case against perfection: Ethics in the age of genetic engineering* Cambridge: Cambridge University Press.

Szocik, K., & Wójtowicz, T. (2019). Human enhancement in space missions: From moral controversy to technological duty. *Technology in Society, 59,* 101156.

Szocik, K., Campa, R., Boone Rappaport, M., & Corbally, C. (2019). Changing the paradigm on human enhancements: The special case of modifications to counter bone loss for manned Mars missions. *Space Policy, 48,* 68–75.

Scarpa, P. J., & Sventek, J. C. (2017). Aerospace medical association letter to the editor. *Environmental Health, 16,* 132.

Uri, J. J., & Haven, C. P. (2005). Accomplishments in bioastronautics research aboard international space station. *Acta Astronautica, 56,* 883–889.

Chapter 4
Crossing the Posthuman Rubicon: When Do Enhancements Change Our Definition of Human?

Steven Abood

Abstract To understand what we might become in the age of enhancements and the implications of this transformation, we first must, as the Oracle at Delphi advised, know ourselves. This requires clarifying our understanding of how we define ourselves as *Homo sapiens*. Four conceptual frameworks are discussed: that of the reproductivists, and their traditional biological species concept model, as well as those of the compositionalists, functionalists, and fundamentalists. By clarifying our understanding of the ground we stand upon, we will be better able to decide how we will step into tomorrow.

4.1 Introduction

Caesar stood on the brink. The River Rubicon lays before him, demarcating Cisalpine Gaul to the north and Roman Italy to the south. The Roman Senate had warned Caesar that if he entered Italy by crossing the Rubicon at the head of his troops, he would be declared an enemy of Rome and he and his soldiers would be sentenced to death. To remain in Gaul would mean forfeiting his power to his enemies in Rome. Crossing the river into Italy would mean war, and either oblivion or immortality. Caesar pondered his choice as the shivering men of the 13th Legion gazed at their beloved leader through the falling snow and awaited their fate. Then Caesar knew, and he spoke. "The die is cast," he told them and stepped forward (Freeman 2009; Goldsworthy 2006; Plutarch 1470; Suetonius 121). And so on that January day in 49 BC, Julius Caesar stepped into the river and crossed into eternity. He left the bank a man and emerged on the other side of the Rubicon as something else.

How will we know when we have become something else? How will we know if we've crossed our Rubicon, achieved a speciation event, passed the point of no return? With a looming mission to Mars and novel biological and mechanical enhancement technology including the CRISPR-Cas9 genome editing system, nanotechnology,

S. Abood (✉)
Department of Biological Sciences, Florida International University, 11220 SW 8th Street, Miami, FL 33199, USA
e-mail: saboo001@fiu.edu

K. Szocik (ed.), *Human Enhancements for Space Missions*, Space and Society,
https://doi.org/10.1007/978-3-030-42036-9_4

and brain–computer interfaces, these questions are preeminent. For the first time in human history, we are approaching a juncture where we can choose to leave our terrestrial homes and journey to a new world of differing conditions that will significantly change us through natural selection. To reach and survive this environment, we may also choose to undergo an unnatural selection, a directed evolution, through the use of biological and mechanical enhancements. Independent of any interplanetary mission, enhancement technology is and will continue to change us. But are these changes differences in degree or kind? Where is our river that divides human and posthuman? To know when we're on the other side of that bank, we must understand the ground we're currently standing on and how we define when that ground changes. How we choose to view ourselves will change the outcome of this calculus. Four lenses, those of the reproductivists, compositionalists, functionalists, and fundamentalists, to be defined in the following sections, differentially focus our perspectives on these questions. Let us begin first with how we currently define our human species and whether this captures the essence of who we, or whether our current construct falls short.

4.2 Reproductivists: Traditional Reproductive Notions of Speciation and Their Deficits

The most widely accepted theoretical framework to describe the dividing line which separates humans from all other species, and all other species from each other, is the biological species concept (Mayr 1942). This states that organisms belong to the same species if they can and are likely to interbreed to produce viable and fertile offspring. Speciation, the creation of a new species, occurs when a subgroup of the founder species can no longer produce fertile offspring with members of the founder species. With significant import for a mission to Mars, speciation often occurs when subgroups from an original population become isolated, and different selective pressures operate on each group resulting in reproductive incompatibility. We can label proponents of the biological species concept as "reproductivists" to distinguish this way of defining a species from others we will discuss.

4.2.1 Prezygotic Barriers

Proponents of defining different species based on their inability to produce fertile offspring with one another delineate two major kinds of reproductive barriers based on when these barriers act (de Queiroz 2005). Prezygotic barriers prevent members of different species from producing a zygote, the first cell of an offspring of sexual reproduction formed by the fusion of the male and female gametes of the parents. According to the biological species concept, two species are different simply because they don't mate with one another. It doesn't matter whether two organisms can, in actuality, mate with one another and produce fertile offspring. If mating doesn't occur, the two organisms belong to different species.

Mating may not occur because of habitat, temporal, and behavioral isolation. In habitat isolation (which is also called geographic isolation), two organisms prefer different habitats and are thus unlikely to encounter one another. In temporal isolation, two organisms reproduce at different times of the day or year. Behavioral isolation results when one organism finds another unappealing and unsuitable for sexual intercourse, for any of a multitude of reasons.

Defining organisms as different species simply because they are unlikely to mate is highly problematic. It leads to absurd results. Consider if such a framework was applied to *Homo sapiens* as it is to all other species on Earth by its proponents. Female supermodels and unemployed men would be behaviorally isolated from reproducing and thus defined as separate species. North and South Koreans would be defined as separate species due to their geographic isolation from one another. Taking geographic isolation one significant step further, the second a spacecraft left the terra firma of Earth with no intention of returning, its voyagers would no longer be *H. sapiens*, as their physical separation prevents them from reproducing with Earth's inhabitants. Under the biological species concept, a species could also rapidly diverge, then converge, and then diverge again. All it would take is one individual to cross over from South to North Korea or vice versa, have viable-offspring producing sex, and viola, North and South Koreans go back to being the same species again. The same would occur if the second after takeoff of the nonreturnable spacecraft from Earth, it was discovered that a stowaway from Earth was aboard. If he or she had sex that with crew members resulting in fertile offspring, then again, viola, the voyager species and the Earthbound species become one once again. Should the demarcation of species be so instantaneously malleable? Some measure of the utility of all definitional categories, including what defines us as humans, is the permanence and stability of the category. Additionally, separating organisms into different species simply because they don't prefer to mate lacks definitional specificity. How many organisms who don't prefer to mate does it take? Over what time period? While not without its usefulness as a preliminary measure of evidence to demarcate a species, those who separate species solely on such unstable and unspecific elements are on fragile ground.

Behavioral isolation due to a temporary factor such as infection by a virus that causes a female to find a male malodorous, and thus sexually unappealing, is also fraught with definitional peril as a basis for speciation. In contrast, behavioral isolation due to a behavior significantly emanating from an evolved brain difference would be on much firmer ground.

Prezygotic barriers to reproduction can also include inherent biological factors which prevent two organisms from forming a zygote. The prezygotic barrier of gametic isolation occurs when male and female gametes fail to combine into a zygote. Mechanical isolation is another prezygotic barrier based on inherent biological factors which prevents two organisms from engaging in sexual intercourse due to the shape or size of their bodies.

The prezygotic barriers of gametic and mechanical isolation are firmer ideological constructs upon which to demarcate species than habitat, temporal, or behavioral isolation. When our conception of what a particular species is rests upon a factor or set of factors that is relatively stable and inherent in that species at least for the evolutionary epoch for when the defining is taking place, instead of on outside factors, we remain consistent with how our cognitive modules usually define things. A boat does not become a car simply because it rests on dry land. Likewise, you do not become something significantly different from your sibling, from a species perspective, simply because you find yourself on the other side of a mountain range. We must not confuse the common precursors to evolutionary change with the change itself. That would be putting the definitional cart before the evolutionary horse.

4.2.2 Postzygotic Barriers

Postzygotic barriers prevent a zygote formed from the gametes of organisms of different species from developing into fertile adults. The chromosomal mismatch may be lethal to the embryo. Alternatively, the embryo may reach sexual maturity but the chromosomal mismatch may cause it to be infertile. In either case, the postzygotic barrier leads to the end of that individual's genetic line. Postzygotic barriers present less definitional difficulties in the demarcation of species than the prezygotic barriers of habitat, temporal, or behavioral isolation. This is because postzygotic barriers concern a feature that is inherent and stable in a particular organism. But they present difficulties as well. Are we to define every adult who is infertile as an individual of another species? Why stop there? How about a woman who has reached menopause? Is she a human one day when she still retains her last ovum, and something else the day she loses it? Here, we would be tracking backward to a prezygotic barrier, but is the conceptual difference substantially different? A woman who cannot reproduce and a woman who can produce only infertile offspring are still at the ends of their genetic lines, which seems to be the core concept upon which postzygotic barriers revolve. If this is the dividing line, then a man who fails to attract a woman due to insufficient resources, poor social graces, or foul odor (behavioral inhibitions) would be in the same ideological boat as a man unable to produce fertile offspring due to a physical insult to his reproductive apparatus such as irradiation or exposure to a virus. Again, a definitional dividing line for species is too porous a barrier if it can readily change from one day to the next.

4.2.3 Allopatric Speciation

As we'll delve into more detail shortly, biological or mechanical enhancements could in themselves be considered the basis of a new definitional construct of speciation. But these technological enhancements could also lead to speciation as defined by these more traditional notions of speciation. How might this occur? Allopatric speciation occurs when a group from an ancestral population evolves into a separate species after a period of geographical separation. Natural selection may act upon a geographically separated species leading to a genetic divergence where prezygotic or postzygotic barriers to reproduction develop. If one group's environment selects for a large body size, and the other group's environment selects for a small body size, they may be a mechanical prezygotic barrier to reproduction, for instance. Enhancements necessary to survive a Mars mission or reproduce on a Mars base may in themselves constitute a basis for speciation. The enhancements themselves would unnaturally select for speciation. But even independent of this ideological basis for speciation, the enhancements could lead to speciation under traditional notions of speciation. Enhancements which made travel to Mars possible for some would lead to a geographical separation. The Martian environment could then naturally select for speciation. Likewise, enhancements necessary to survive in a vehicle undertaking a long interstellar journey would also lead to a geographic separation and eventually, a natural selection for speciation. These scenarios predict a two-step process: first, unnatural selection through the directed evolution of enhancement to allow for a geographical separation, and second, natural selection acting on the enhanced population due to the novel geographic environment. The first may be sufficient for speciation, or speciation may require both steps, depending on the definition of speciation that is adopted.

4.2.4 Sympatric Speciation

In a similar fashion, enhancements may result in sympatric speciation. In sympatric speciation, groups from an ancestral population evolve into separate species without any geographical separation. How could speciation occur when the founder species remains in the same general location as the derivative species? Although no physical barrier exists that would restrict gene flow, the use of different habitats or resources can occur. For whatever reason, some members of a group may prefer resources more prevalent in the north of a habitat and others may prefer more resources in the south. For example, due to a genetic mutation in their taste buds, the northern group members may prefer a food source that grows more abundantly in the north because of the soil composition there. If the groups segregate into different geographical areas despite being physically able to intermingle, they will be less likely to interbreed with members of the other geographical area. If different selective forces act on the different geographical regions, speciation can then occur. Experiments on *Drosophila pseudoobscura* (fruit flies) illustrate that segregating populations of the same species in different areas with different food sources can lead to reproductive isolation, a necessary step toward sympatric speciation (Dodd 1989).

On Earth before a space voyage, on a Martian habitat, or on an intergenerational spacecraft, sympatric speciation may occur when those with enhancements willfully segregate from those without enhancements. Segregation may even occur between those with different kinds of enhancements. Some members on a Mars colony may choose to survive the harsh Martian atmosphere through the use of genetic or mechanical enhancements. As humans are predisposed to segregate based on actual and perceived differences, reproductive subpopulations may emerge based on enhancements despite any physical barriers separating populations or biological barriers preventing successful reproduction. Reproductive subpopulations may emerge even with solely mechanical enhancements. For instance, a wearable brain–computer interface or implanted chip which allows communication in a virtual world would unnaturally select for sexual relations among those with access to this technology. Of course, over time, genetic and even epigenetic changes from the use of such mechanical enhancements could lead to a dependence upon them, further segregation, and the increased likelihood of speciation.

4.2.5 *The Question of Subspecies:* **H. sapiens augmentum***?*

The concept of hominid subspecies can be justly called an ideological morass as evidenced by the introduction of the subspecies framework in the 1920s and 1930s, followed by its eradication. But the subspecies concept may have relevance in the coming age of enhancements and therefore may undergo a timely resurrection. In the beginning of the twentieth century, Neanderthals were classified as part of the species *H. sapiens* (White et al. 2003). Differences between anatomically modern humans and Neanderthals resulted in application of a subspecies framework, with the former classified as *H. sapiens sapiens* and the latter classified as *H. sapiens neanderthalensis.* As more data accumulated regarding differences in compositional structure between anatomically modern humans and Neanderthals, taxonomists abandoned this subspecies classification and separated Neanderthals into a separate species, *Homo neanderthalensis.* Recently, the subspecies concept in general has fallen into disfavor among some taxonomists (Futuyma 1986).

But does the abandonment of the subspecies framework, or an analogous theoretical construct, make sense? Without a further deconstruction of the concept of species, whatever we call it, how are we to account for subpopulations that differ significantly in genetic composition, but can still produce fertile offspring? Recent DNA analysis of 1523 people from around the world reveals that early modern humans, Neanderthals, and Denisovans interbred and produced fertile offspring far more often than previously believed (Vernot et al. 2016). But this interbreeding occurred with variable frequencies or success rates among the ancestors of modern human populations. Modern Africans typically lack Neanderthal or Denisovan DNA in their genomes. In contrast, Neanderthal DNA is embedded in the genomes of modern Eurasians. And modern Melanesians contain both Neanderthal and Denisovan DNA in their genomes. Since the crux of the biological species concept is to define species based on which organisms can successfully produce fertile offspring with which, these genomic differences may justify some form of taxonomic subclassification if we hold the biological species concept to have ideological validity. Alternatively, the differences between the amalgamations of ancient DNA in subpopulations may not be significant enough to segregate a population into differing species or subspecies. We are all hybrids. Furthermore, the biological species concept rests on what subpopulations modern organisms can currently interbreed with and produce fertile offspring, not the ability or inability, or likelihood or unlikelihood, of their ancestors doing so in the past.

We've already surveyed numerous problems with the biological species concept. These breaches in the dam of the concept may lead us to question the core of the concept itself. Of all the distinguishing characteristics of *H. sapiens*, is it really logical to define our species based on reproductive ability alone? And couldn't the goalpost of the production of fertile offspring change with novel reproductive technologies? Should two organisms who undergo the technologically assisted reproduction of fertile offspring count as members of the same species if such reproduction wasn't possible or likely without the technology? For instance, reproductive technologies such as the CRISPR-Cas9 system (or other gene editing systems) and in vitro fertilization may be able to modify the gametes of two incompatible organisms so that they form a healthy zygote that could then go on to produce fertile offspring once it reaches sexual maturity. Exowombs, artificial wombs that would allow for an extracorporeal pregnancy, may also allow two organisms from naturally incompatible species to produce fertile offspring. At Juntendo University in Tokyo, Japan, researchers developed an extrauterine fetal incubation system where 14 goat fetuses were placed in artificial amniotic fluid and kept alive for three weeks (Kuwabara et al. 1987; Klass 1996). At the Children's Hospital of Philadelphia, researchers developed another extrauterine system where fetal lambs were placed in artificial amniotic fluid in a mechanical system that provided oxygen and nutrients and removed waste (Partridge et al. 2017). The fetal lambs were placed in a darkened environment reminiscent of the inside of a womb and heard the recorded sounds of their mothers' heart beats. The fetal lambs developed normally for a month. These efforts may contribute to the development of a working exowomb prototype that may push the limits of the possible and test the utility of the biological species concept framework.

Perhaps a central problem with the biological species concept, especially as we enter the age of enhancement, is that it defines human beings in terms of their inability to do something. Consistent with the framework, an organism is a human being and not something else because he or she cannot reproduce and create fertile offspring with other organisms. As discussed, this could change with in vitro fertilization and CRISPR-Cas9 technology. Likewise, for better or worse, this could lead to the defining of different species or subspecies if humans genetically diverge within a Mars population, or between Mars and Earth, and are no longer able to successfully produce fertile humans with one another.

Within this definitional framework of defining a species based on reproductive ability, if enhancements prevent an enhanced population from reproducing with a non-enhanced population and producing fertile offspring, then the enhanced population could rightly be considered something other than *H. sapiens*. At the least, such reproductive barriers may justify the adoption of a subspecies classification of enhanced humans, such as *H. sapiens augmentum*, which roughly translates as wise man who is augmented or enhanced. Alternatively, *H. sapiens supra* could be used, a term that contains an even more pronounced (and perhaps troubling) value judgment of the status conferred by enhancements, translating roughly as wise man who is above or superior to his non-enhanced self.

What might *H. sapiens augmentum* look like in a Martian environment? Various biological and mechanical enhancements may combine to help early settlers survive the novel environmental challenges of Mars. Radiation shielding will be a foremost concern to protect against the higher levels of radiation on Mars (Barker and Gilroy 2017). Imperfect shielding from biological enhancements, mechanical enhancements, wearable devices, and architectural structures may lead to higher mutation rates, which will in turn lead to both higher rates of cancer and beneficial mutations that will expedite natural selection. Settlers may be augmented with nanotechnology to help fight cancer and other medical conditions that the Martian environment makes more likely. If enhancements do not fully arrest natural selection, skin pigmentation may also change to help block the absorption of radiation (Freeman 2017).

As gravity is only 38% that of Earth, and bones must work against the force of gravity to remain strong, enhancements and selection for more durable bone compositions may both contribute to the prevention of dangerous fractures (Hawkey 2005). Given the increased risk for fractures due to the low gravity of Mars, caesarian sections may be favored to prevent against hip fractures during childbirth. A high prevalence of caesarian sections on Mars may select for babies with larger heads, a phenomenon already evidenced on Earth (Mitteroecker et al. 2016). Larger heads will allow for larger brains that may be able to control enhanced biological and mechanical tools more proficiently than those with smaller brains.

Diminished Martian sunlight will also spur both artificial selection through enhancements and natural selection. These processes will act upon the eyes to enhance sight but also upon the biochemical processes to enhance the synthesis of vitamin D. The synthesis of vitamin D in the skin of *H. sapiens* on Earth requires sunlight (Wacker and Holick 2013). A lack of sunlight therefore results in dangerously low levels of vitamin D, resulting in a condition called hypovitaminosis D which can lead to rickets, osteoporosis, increased cancer risk, increased risk of pelvic fractures during childbirth, hyperparathyroidism, and increased risk of seizures in infants born to affected mothers (Stuijt 2009; Douglas 2007; Keim 2007). Vitamin supplementation, especially on a Mars base with limited supplies and manufacturing capabilities, may not completely remedy the problem. In a study of 178 women wearing traditional Muslim garb that blocked sunlight from reaching their skin, 176 of the 178 women (99%) had hypovitaminosis D (Mishal 2001; Saadi et al. 2007). Even after three months of daily vitamin D supplementation, 70% of the women still did not have the minimum required healthy blood plasma levels of vitamin D. A biological enhancement may therefore be necessary for Martian colonists to produce adequate levels of vitamin D despite diminished sunlight. A lack of sunlight also detrimentally affects circadian rhythms, mood, and sleep (Abood 2019). Enhancements may be utilized to help rectify these disruptions.

Reproductive isolation leading to speciation as defined by the biological species concept may occur due to the development of incompatible immune systems between early and new Martian settlers, or between Martian voyagers and those who remain on Earth. The lack of bacterial life on Mars would likely lead to deficient innate immunity systems and sluggish acquired immunity systems for bacteria indigenous to Earth. New settlers carrying such bacteria would therefore pose a serious health risk for early settlers on Mars, leading to reproductive segregation, a prelude for speciation under the biological species concept (Freeman 2017). But what if enhancement technologies, such as nanotechnology that enhances the immune system, allow immune-deficient Martians to reproduce with those carrying bacteria from Earth? In this scenario, under the biological species concept, instead of enhancements leading to speciation, enhancements would prevent speciation by allowing reproduction between these two populations to occur.

Are there other bases apart from the biological species concept to define an enhanced person as another species or subspecies, a *H. sapiens augmentum*? The remainder of this chapter will discuss three definitional possibilities beyond the reproductivist dividing line.

4.3 Compositionalists

Instead of drawing the line in the sand of speciation based on reproductive ability or reproductive behavior as reproductivists do, compositionalists segregate species based on essential features of their physical composition which they believe are significant enough to demarcate them from one another. The history of natural science is replete with distinctions between species made due to skeletal divergences. Could species be defined based on physical differences between enhanced and non-enhanced persons, or between those with different physical enhancements?

Aristotle believed that the essence of a thing determined what it was, although in his conception this essence could be something different than its physical composition (Aristotle 330 B.C.). The Romans translated his phrase "to ti esti" ("the what it is") by coining the term "essential," which they derived from the Latin verb esse, "to be." In the late fourteenth century, essential became anglicized as "essence."

Like our conception of a speciation dividing line based on physical composition, Aristotle's essence contrasted with the function of a thing. A collection of jewelry, for instance, could be defined and divided into different piles through a functionalist lens by the function each piece of jewelry serves (Smith 2012). Necklaces ordain the neck. Bracelets ordain the wrists. Earrings decorate the ears. Alternatively, though a compositionalist lens, the same collection of jewelry could be defined and divided based on each piece's essential composition. Through this lens, gold pieces of jewelry have the essential physical composition of 79 protons (as all things made out of gold do), and silver pieces of jewelry have the essential physical composition of containing 43 protons (as all things made out of silver do). Of course, our understanding about what constitutes the physical essence of a thing will advance as science advances.

Protons were discovered only in 1920, by Ernest Rutherford. But centuries before that, John Locke took the mantle of a compositionalist when he postulated that the essence of gold consisted of the microscopic properties of gold (Locke 1689).

When it comes to defining and dividing living things, a modern compositionalist may take the same approach. While a functionalist would group both birds and planes together based on their function as flying things, a strict compositionalist would separate them due to one being composed of feathers and living tissue, and the other composed of metal and mechanical parts. An extremely permissive compositionalist may group birds and planes together, given that both possess wings. But let's say we replace a bird's wings with mechanically enhanced wings made of metal. Is the bird no longer a bird simply because an enhancement changed its composition? A compositionalist would argue yes, while a functionalist would argue no

Moving from non-human animals to humans, what about our composition defines us as humans? As mentioned, there are the distinctive skeletal and other physical features of humans that researchers have pondered about for decades including the features that make us prodigious endurance hunters, including large skulls which while moving fast stabilize our body like the weights placed on top of a skyscraper shifting in the wind, enlarged semicircular canals of the inner ear which are responsible for balance, the nuchal ligament in the back of our neck that stabilizes us when we're moving fast, low wide shoulders and narrow waists which help a runner's torso twist independently and aid the opposing motion of arms and legs, large gluteus maximuses which contract while running and help prevent our trunk from lurching forward, thicker leg bones and joints to compensate for all our weight on two limbs instead of four, slow twitch muscle fibers in our legs, springy Achilles tendons, arched shock absorbing feet, and short, flexible toes that help propel us forward as they bend (Spoor et al. 1994; Pontzer 2007; Steudel-Numbers 2006; Ruff et al. 1999; Latimer and Lovejoy 1989; Ker et al. 1987; Lieberman 2006).

As evidenced by the preceding description, structure is closely intertwined with function. While Aristotle believed that the essential essence of humans lays in their ability to reason, Boethius believed that it was the ability to engage in moral reasoning, writing that "…a man who loses his goodness ceases to be a man…and turns into a beast" (Boethius 524). Both Aristotle and Boethius could therefore reasonably be categorized as functionalists. But because structure is so closely linked to function, it would not take much to transform them into compositionalists. Since an iron rod blew through railroad foreman Phineas Gage's prefrontal cortex in 1848, an injury he astonishingly survived but that drastically changed his personality, we have understood that many aspects of our personality are localized in different parts of our brains and depend upon their physical compositions (Harlow 1868; Glenn et al. 2011; Ratiu and Talos 2004; Damasio et al. 1994). The amygdala in psychopaths, individuals marked by a profound lack of empathy, and who would fail Boethius's standard of the capacity to morally reason, or at least to act rightly based on moral reasoning, is 18% smaller than non-psychopaths (Yang et al. 2009). Psychopaths would not be who they are if their amygdalas and other brain areas resembled the composition of those with a profound degree of empathy. We can define psychopaths based on their behavior, but this behavior is largely determined by the structural compositions of their brains and genomes. A strict compositionalist with perfect knowledge of the connectome, genome, and epigenome for psychopathy may define a person as *H. sapiens psychopathia* based on their composition alone before any behavioral markers of psychopathy are manifested.

The origin of the organic matter which makes up the composition of a person may matter to strict compositionalists. They may define a person born on Mars and a person born on Earth of different species simply due to the different geographical origins of the organic materials which compose their bodies.

If CRISPR-Cas9 or other gene editing systems significantly change the composition of a subgroup of *H. sapiens*, then a compositionalist may readily deem this subgroup a separate species or subspecies. But even those with mechanical enhancements may eventually be deemed a separate species or subspecies if the use of these enhancements significantly changes brain or other bodily structures and these changes are transmitted generationally through genetic or epigenetic mechanisms. Already, mechanical enhancements such as smartphones and other computational devices have resulted in decreases in gray matter in brain areas such as the anterior cingulate cortex, an area important for attentional control (Wilmer et al. 2017).

Factors such as significance of change, separability/permanence (i.e., can the enhancement be reversed or removed?), dependence (i.e., can the individual thrive or survive without the enhancement? Will the individual's biology change in a way where he or she can no longer thrive or survive without the enhancement?), and generational transference (i.e., is the enhancement or consequences of use of the enhancement passed from one generation to the next) may determine whether a particular enhancement will confer *H. sapiens augmentum* status. Even a complex and highly advantageous enhancement may not be deemed sufficient to confer upon its user a separate species or subspecies status if it does not meet some or all of these criteria. Although an automobile is highly complex and useful, for example, and humans spend so significant a portion of their lives in them that they may be mistaken for biological–mechanical hybrids or cyborgs, these "wearable" mechanical enhancements do not permanently change the composition of their users. Most would agree that a human doesn't cease to be a human simply because he or she is driving a car.

4.4 Functionalists

Closely linked to the compositionalists, but distinct enough to sometimes arrive at different outcomes for the same group of organisms when it comes to the question of speciation, are the functionalists. Functionalists group items or living things based on a function or a collection of functions they deem essential. Returning to our analogy about jewelry, a functionalist would characterize and separate individual pieces of jewelry within a collection based on its function of alternatively ordaining necks, ears, or wrists, while a compositionalist would characterize and separate them based on their compositions of gold or silver.

A strict functionalist may categorize a human with an immune system enhanced by nanotechnology as a separate species or subspecies. Other functionalists may not categorize the enhanced function of an immune system as sufficiently distinctive or important to base a speciation distinction upon.

Often speciation theorists will combine a number of compositional features, a number of functional attributes, or a combination of both, to arrive at a definitional category for *H. sapiens*. This approach attempts to circumvent the problem inherent with basing speciation on a single trait that other species may also possess. A similar problem occurs with a checklist approach for the definition of life, a prerequisite for all current conceptions of humanity. A theorist may observe that most living things move, grow, and utilize energy, but if he bases his definition of life on only one of these attributes a critic may answer that wind moves, crystals grow, and fire uses energy, but we do not consider these to be alive. At least two problems occur with such a checklist approach to speciation. One is that some individual members otherwise deemed to be *H. sapiens* don't satisfy one or more of the attributes utilized to define them. This may be due to age or a physical or cognitive deficit due to genetics or physical injury. Many functionalists would consider bipedalism, the ability to walk on two legs, as an important function of a human being. Yet, a person may not walk due to young or old age, infirmity, a genetic deficit, or a physical injury. One solution is to define humans based on a checklist of representative attributes, without requiring the outlier members of the species to meet every requirement. A second problem with checklists is that non-humans may also possess the various compositional or functional traits that were thought to be possessed only by humans. This is the same problem encountered when attempting to define life through the use of a checklist approach. When Plato defined human beings as featherless bipeds, the rival philosopher Diogenes of Sinope burst into his lecture wielding a plucked chicken and shouted, "Here is Plato's man!" (Laertius 200). Of course, the more lengthy the checklist, the more likely one is able to avoid the second problem of an overinclusive definition where non-humans are considered humans. But as we so often can't have our cake and eat it too, avoiding the second problem through an overly detailed checklist of traits that must be satisfied to be a human puts us at jeopardy for the first, an underinclusive definition where some humans aren't counted as humans although they should be.

An alternative approach is to avoid checklists of required traits and focus on a singular trait that is so fundamentally different in degree or kind from the rest of life as to separate us irrevocably from it. This is the essence of fundamentalism.

4.5 Fundamentalists

There are at least two species of fundamentalism. One postulates that differences in degree of an attribute are so profound that they essentially amount to a difference in kind. The second kind of fundamentalism concerns a difference in perceiving or interacting with the world that is so radically different that a difference in kind categorization seems most appropriate.

Candidates for the first kind of fundamentalism, traits we share with other organisms but that we possess to a much more developed degree, a degree so high that it can be essentially argued that a difference in kind emerges, are aplenty. Consciousness, free will, reasoning, and moral reasoning are all suspects and closely related. Rene Descartes wrote that, "animals are mere machines but man stands alone" (Descartes 1637). It remains to be seen whether humans possess a unique ghost in the machine, one that allows them to sever themselves from the supposed causality of everything that has come before. But if so, our higher levels of cognition that appear to allow us to choose a path other than that of instinctual limbic reactivity may underlie this ability, as well as distinguish and define us. There is a lack of compelling evidence that any other hominin produced art, the artifacts of *H. sapiens*' ability to think abstractly. We abstractly reason about others' mental states in a way that no other organism is known to do. This higher capacity to form a theory of other minds is one attribute of ours that allows us to connect with others of our kind to an unprecedented degree, forming a cooperative neural network to imagine tasks of tremendous creativity, and then engineer them into reality. These cooperative endeavors would not be possible without the cultural learning and precise communication enabled by our language abilities that are also developed to a higher degree than any known organism. Unlike all other mammals whose larynx is positioned opposite their first to third vertebrae, the larynx in *H. sapiens* descends around age two so that it is opposite the fourth to seventh cervical vertebrae (Rhodes 2016). This prohibits *H. sapiens* from simultaneously swallowing and breathing, making it more likely we'll choke while eating or drinking, but results in an enlarged pharyngeal chamber, enabling us to engage in articulate speech as no other mammal can. A compositionalist may focus on the positioning of the larynx at a lower position in *H. sapiens* than other mammals as most important; a functionalist may argue that the key is the production of articulate speech, or otherwise advanced communication, no matter how this is achieved structurally, while a fundamentalist may focus in on how profoundly this articulate speech separates *H. sapiens* from the communication abilities of all others. In addition to articulate speech, cooperation is also enhanced by an extremely high degree of prosociality in humans. Comparative studies between humans and chimpanzees show that humans are hypercooperative and will help others to a higher degree (Warneken 2015). Human children are less selective about who they share with, while chimpanzees largely only share with close relatives, reciprocating partners, or potential mates (Silk et al. 2013).

Controversy remains about what traits piggybacked on others and how far one can reduce a trait or series of traits to a foundational trait from which others emerge. Our ability to imagine the thoughts of other minds and communicate with others allows for grand acts of altruism such as coordinated global aid efforts after a natural disaster, but also for terrorist attacks and organized genocides. Both our capacity for good and our capacity for evil have been cited as that which is fundamental to our humanity. However, if these traits are merely derivative, are they truly fundamental or outcomes of something more preeminently fundamental?

From what concoction of traits or modules do the emergent properties of what makes us human derive? How linked are these traits? That is, did a confluence of necessary traits independently arise or was there one prime mover which drew *H. sapiens* away from the caves of other hominins and into the light? Candidates have been proposed, such as the emergence of *NOTCH2NL* genes (Suzuki et al. 2018). Whatever the spark, something caused us to produce superior tools and other creations than all other organisms approximately 80,000 years ago (Jerison 1973). When *H. sapiens* first emerged about 200,000 years ago, we shared the planet with at least four other hominins—Neanderthals, Denisovans, the "hobbit" *Homo floresiensis*, and a mysterious fourth group (Hogenboom 2015). But only we survived.

Charles Darwin famously wrote that, "There is no fundamental difference between man and the higher mammals in their mental faculties" (Darwin 1871). He argued that all differences were "of degree, not of kind." But to postulate that the gulf between humanity's creations and those of all other organisms merely derives from a difference in degree seems myopic. Of course, some of the dispute may simply be over wordplay. In the sense that we're all composed of atoms and living cells, we are all the same and all differ merely by degree. But the arrangement of our genomes and the development of cognitive modules and our ability to create and interact in ways unknown to the rest of life on Earth seem to justify an assignment of a different kind instead of merely a different degree. A threshold of degree can be reached that can trigger an emergent property. A wave contains more properties within it and thus is a different kind of thing than merely the molecules of water in a drinking glass. Through the abundance of degree, it has achieved a difference in kind. The designed world around us is a testament to a property or properties in *H. sapiens* that through an abundance of degree has emerged into a difference in kind. If you want to understand the artist, look at his works. And our works are grand.

The second kind of fundamentalism may be ushered in by enhancements which result in a completely different way of perceiving or interacting with the world. The change would be so radical to the way humans have lived and experienced life that it would automatically result in a difference in kind. An enhancement in function such as the ability to move at greater speeds than before may not satisfy this requirement but moving in a completely different way, such as flying, swimming while able to breathe underwater, or even teleporting, would. Of course, the prerequisite in this case would be that movement must be considered a fundamentally important attribute to the human experience.

What aspects of the human experience could enhancements potentially modify so that a difference in kind would be automatically achieved, justifying for fundamentalists the notion that a brave new species has been created? What fundamental changes would justify a redefinition of us? The following candidates encompass extreme changes that would likely satisfy even the strictest fundamentalist's requirements for speciation. Although they may seem fantastical, it is important to start at the brink to delineate the fundamentalism concept before moving into murky areas and less obvious cases in future works. A second point on the fantastical nature of the following proposed candidates: that which was once dubbed science fiction has often become science fact. Kahlil Gibran wrote that our children, "dwell in the house of tomorrow, which you cannot visit, not even in your dreams" (Gibran 1923). Perhaps Gibran was right, and we cannot clearly see where our future lies. But perhaps we can catch glimpses, especially if we are our own designers. If so, perhaps we should not dismiss these waking dreams so readily merely because they seem fantastic in our house of today.

4.5.1 Immortality

The first emperor of China, Qin Shi Huang Di, emptied the royal coffers to send his advisors far and wide to find the secret of eternal life. One intrepid advisor returned after years of travel with a supposedly magical elixir formulated by the most respected Taoist masters. Unfortunately for the emperor, the majority of this concoction consisted of mercury, a toxic element, and he subsequently died upon drinking it. To add insult to injury, he was buried in a tomb containing rivers of flowing mercury, a microcosm of China's great rivers (Gascoigne 2003).

Qin Shi Huang Di's quest for immorality was an ancient attempt to change a fundamental tenet of our existence: We are impermanent. Just as all living things that have come before us, we will live, and then we will die. Could it be otherwise? There is evidence that death may not be the end of our conscious experience. There are accounts of patients whose brains had apparently ceased all activity later recounting the presence of objects and circumstances in their hospital rooms once they were revived (Parnia 2014). Approximately four percent of the population according to a Gallup survey has experienced a near death experience with common features independent of age, time, place, culture, and religious beliefs (Parnia 2014). But regardless of this great question of whether a kind of after-death immortality exists, can we manufacture our own immortalities? The cross-cultural vampire myths labeled those with human form who were immortal or who lived abnormally prolonged lives as something apart from *H. sapiens* (Barber 1988). It is likely that we too, especially the fundamentalists among us, would label a subgroup of us who achieved grossly extended life spans or immortality through biological or mechanical enhancements as another species or subspecies.

Enhancements that prolong life must address one or more of the three primary causes of death: aging, disease, and physical trauma (Hayflick 2007). Even if these three causes were somehow negated, a person could still expire if he lacked energy input from food scarcity or water deprivation.

The problem of aging is thought to be tied to the Hayflick limit, the number of times a cell divides before it undergoes senescence, the loss of its structural integrity leading to cell death (Wai 2004; Shay and Wright 2000). The Hayflick limit appears to depend on the length of chromosomal telomeres, which decrease with every cell division, and may be an evolutionary compromise between selecting for cancer and selecting for aging (Bremermann 1982). Genetically engineering mice to produce 10 times the normal level of telomerase, an enzyme which protects telomeres from shortening, increased their life spans by 50% (Kirkwood 1977). These and other experiments, as well as observations of long-living species and case studies of humans with oddly long life spans, have led to speculation that life span can be extended. Researchers who analyzed a Greenland shark in a recent study estimated that it was between 272 and 512 years old, which would make it the longest-living vertebrate (Nielsen et al. 2016). Frenchwoman Jeanne Calment, the longest-living human with verified records, lived to the age of 122 years, 164 days, passing away on August 4, 1997.

Apart from arresting aging, enhancements to stave off death, such as medical nanorobotics, must also address life-threatening infections and injuries. The biochemical process of death must be stopped, and then the underlying injury or infection that initiated the process of death must be addressed (Parnia 2014). One strategy to arrest the process of death is suspended animation, where the biochemical processes of death are slowed so that they can be arrested and reversed. Scientists are studying organisms like the wood frog, which appears to utilize cryoprotectants to freeze without damaging cells, as well as cases of unintentional freezing in humans and animals, to gain insight into the possibilities of suspending animation (Costanzo et al. 2015). When scientists thawed 300 prehistoric worms recovered from permafrost above the Arctic Circle, two of them revived and began moving and eating. They were estimated to be 32,000 years old and 41,700 years old, respectively (Weisberger 2018). Suspended animation may be utilized one day for prolonged space travel, minimizing the need for an intergenerational mission and caloric provisions.

The Greeks believed that when Eos, the goddess of the dawn, asked Zeus to make her lover Tithonus immortal, she forgot to ask that he also be granted eternal youth (Hamilton 1942). When old age came upon him, he continued to live but could scarcely move. Life extension must ideally be achieved with an onus placed on quality of life. This may be facilitated through the use of tissue engineering through the use of an organism's own stem cells to prevent rejection, and by taking lessons from organisms like the hydra which has a seemingly indefinite capacity for self-renewal (Boehm et al. 2012; Fujisawa 2003; Martínez 1998; Estep 2010).

Digitally facilitated immortality has also been hypothesized though the use of mind uploading, the transference of memories, personality, and consciousness into another medium (Kurzweil 2005; Martin 1971; Cohan 2013). This could be another carbon-based body, a robotic or synthetic body, or a disembodied medium such as a localized or distributed computer network existing either in stasis or in a virtual world. An "uploaded astronaut," freed from the infinitesimal range of precise environmental conditions and consistent amounts of proper sustenance required for carbon-based life forms, would have tremendous potential for interstellar travel (Prisco 2012). The Blue Brain Project, a Swiss research initiative, aims to create a digital reconstruction of a human brain and has successfully simulated part of a rodent brain (Fildes 2009). Digital immortality may require or be facilitated by an artificial intelligence. One approach would be to utilize this artificial intelligence to create or shape the digital personality of the individual through the construction of a digital avatar based on digital records of a person's behavior while biologically living.

4.5.2 Omnipresence Part 1: Traveling Through Time

A kind of time-traveling omnipresence or omnitemporality could emerge if a disembodied consciousness was transmitted to different times or laid dormant and then reactivated after a great span of time had passed. The following concepts of physics illustrate, for purposes of our thought experiment on how the most radical technological changes could affect our definitional concepts of speciation, how time travel may be at least theoretically possible. Time travel is obviously beyond the reaches of our current technological grasp.

Albert Einstein's theory of gravity unites space and time together as the phenomenon of spacetime (Allday 2019). Spacetime curves in the presence of mass and the theories of special and general relativity suggest that suitable geometries of spacetime or specific types of motion in space might allow time travel into the past and future if these geometries or motions were possible (Thorne 1994). Spacetime allows for the possibility of wormholes, tunnels that could theoretically serve as shortcuts through spacetime that connect otherwise very distant parts of the universe. Travel through wormholes would allow for the bridge of billions of light-years or more, or through different points in time, including travel to the past. Time travel through wormholes would possibly require one opening of the wormhole to be moving at a substantial velocity compared to the other. Many physicists believe that natural wormholes were formed in the Big Bang through which time travel may be possible (Thorne 1994). Many also believe that wormholes are constantly emerging in and out of existence on the quantum scale, perhaps a billion times smaller than an electron (Rodrigo 2011). According to some physicists, it may be possible to capture one of these quantum wormholes, enlarge it to human scale, and use it to travel through time (Rodrigo 2011).

4.5.3 Omnipresence Part 2: Teleporting or Rapidly Traveling Through Space

If space and time are a singular phenomenon called spacetime, the wormholes we discussed in the preceding section would allow for not only travel through time, but travel to vastly distant points in space. Here, we continue our thought experiment on how radical technological change could fundamentally alter definitional categories of speciation.

Fundamental changes that may radically change our conceptions about the possibilities of travel through the universe include advanced propulsion systems. Another proposal to bridge vast gulfs of distance rapidly is through teleportation. Physicist Michio Kaku proposes transmitting the data regarding the composition of a human body—an estimated 2.6 followed by 42 zeroes of bits of data, via X-rays to a satellite network and then to a quantum computer at a distant location to then be reconstituted.

4.5.4 Omnipresence Part 3: Distributed Consciousness

Instead of sending you or an entire record of you via teleportation, the essence of you, that is, your consciousness, personality dispositions, and memories, could be transmitted. Quantum entanglement and quantum teleportation could be involved in such a feat, although we have yet to utilize such phenomenon to transport anywhere near the amount of data required to encapsulate the essence of a human brain. Quantum entanglement occurs when two particles are somehow entangled, so that an action performed by one affects the other, no matter how far apart in the universe the two are Schrödinger (1935, 1936). When the state of the first particle is measured, the second particle somehow knows that it should be in a mirror image state. This information about state seems to travel instantaneously, without a speed of light limit. When this happens, the quantumly entangled particles undergo quantum teleportation. Einstein called it "spooky action at a distance." Chinese scientists have recently quantumly teleported a packet of information from Tibet to a satellite in orbit, 870 miles away (Emspak 2017).

Another approach to a distributed consciousness would be to multiply your consciousness by creating virtual copies of yourself. Mind uploading an actual replica of the connections of a person's brain, or their general personality proclivities and generalized memories based on a lifetime of Internet use and habits, and combining these templates with an artificial intelligence, would theoretically allow virtual copies of the self to be created. These avatars could serve as extensions of a person's will and being, delegating choice and actions to literally like-minded beings.

4.5.5 Omniscience: Distributed Intelligence

Our cognitions are already linked to each other in a kind of hive mind through smart devices and the global Internet. Decentralized and dispersed, the hive mind's perceptions and ideas can spread more widely and rapidly than ever before. A great proportion of humanity is already attached to a pocket or wrist-worn computer system. In the future years, brain chips or other enhancements may connect us even further and more permanently.

Brain–computer interfaces may one day transfer not only thoughts but sensory experiences from one mind to another. These interfaces utilize computers to translate the brain waves comprised of synchronized pulses of electrical activity into discernible mental states. Facebook recently published a paper describing their work on a headset that can transfer a small catalog of a person's thoughts directly to a computer screen (Cuthbertson 2019a, b). Elon Musk's Neuralink also has brain–computer interface projects in development (Royal Society 2019). Brain-to-brain interfaces are also in development that allow for the direct coupling of activity between two brains. In 2002, Kevin Warwick implanted electrodes into his and his wife Irena's nervous systems and conducted the first experiments of direct electric communication between two human beings (Warwick et al. 2003, 2004). Successful brain-to-brain interfaces have allowed for direct brain-to-brain communication between two rats, two humans, and even between a human and a rat (Pais-Vieira 2013; Rao et al. 2014; Yoo et al. 2013). A conscious thought between two human brains was transmitted from India to France through the use of a brain-to-brain interface (Grau et al. 2014). The US military's Defense Advanced Research Projects Agency is working on developing an implantable neural interface in collaboration with a consortium of private companies (Defense Advanced Research Projects Agency 2016).

These and other advances that allow brains to be intimately connected to each other, reservoirs of knowledge, and artificial intelligences may one day be embodied in neural enhancements that make a seemingly endless sea of knowledge possible. Those who posses such enhancements and access to this knowledge may be defined by fundamentalists as something other than *H. sapiens.*

4.5.6 Manufactured Omnipotence: Divine Engineers

Advances in nanotechnology, gene editing, tissue engineering, synthetic biology, and 3D printers are already allowing to manipulate matter and build the creations of our imaginations as never before. Atomically precise manufacturing (APM) may one day occur where atoms and molecules are assembled into minuscule machines capable of crafting slightly larger machines, which would in turn create even larger machines, and so on, resulting in tremendous potential to create (Drexler 1987).

Of course, whether these capacities are permanently embodied, significantly inseparable, and, for some, inheritable may demarcate whether we consider them enhancements or tools. If a subpopulation of us makes significant strides into fantastic in any one of these areas through enhancements, a fundamentalist may deem this subpopulation to have crossed into the realm of a posthuman future.

4.6 Posthumanity

To understand what we might become and the implications of this transformation, we first must, as the Oracle at Delphi advised, know ourselves. On a planetary level, this means focusing our definition of what it means to be human, both as a species and as entities who exist in the environment of Earth. Our definitions of humanity must start with the fundamental question of what it means to be alive and the graduations separating living matter from organic matter. From our starting point as living entities, we must then examine various multidisciplinary definitions of what it means to be human, which all present valuable but incomplete pictures of the whole, like the parable of the blind men gaining valuable but incomplete information about the elephant from feeling its various parts. This definitional examination includes issues such as whether human consciousness is an epiphenomenon that separates us in kind and not merely degree from other living entities. Once this definitional framework is established, a review of recent and predicted advances in genetic and biomechanical enhancement to the human body will clarify issues regarding when the Rubicon is crossed from human to posthuman.

At our traditional starting point are the reproductivists, who believe that a subpopulation of us becomes something else only when it is unable to, or unlikely to, reproduce and produce fertile offspring with the founder population. Another conception of the end of humanity predicts that it will come with a significant change in our biological composition. The fundamentalists believe that the posthuman Rubicon is crossed when our defining functions are modified past a certain threshold by enhancement technology or other processes. Beyond these compositionalists and functionalists are the fundamentalists, those who believe that humanity only reaches the posthuman when it deviates profoundly on a fundamental level from our current conception of human life. Even if our essential biological compositions and functions remained the same, these fundamentalists believe that only a radical change in our experience of life would usher in the death of humanity and the birth of the posthuman.

Why does all this matter to us? Perhaps in our search for meaning through our transient existences, we believe that defining what we are will somehow tell us who we are. We live at a time where unprecedented technical expansion has led to a psychological alienation from the way we lived for eons. Turning this technology inward to change ourselves, when we scarcely understand ourselves, may represent the preeminent psychological dislocation. The questions of the old world and questions of rights and obligations, individualism and collectivism, and origins and destinations will become magnified in their import and joined with new questions. Will enhancements help us pull the curtain from our waking dream? Will they help us become artificial gods and not men? Who will decide what we will become? All these questions rely on understanding the foundation of what we are as a species, so that we may then not only design, but consider the implications of what we propose to come. Without such an understanding, we will be merely carried upon the rivers of technological progress to unknown and potentially dangerous lands. For there is no preexisting Rubicon that will tell us when we've crossed the bank of human to the land of the posthuman, we must create and define that Rubicon for ourselves.

References

Abood, S. (2019). Martian environmental psychology: The choice architecture of a Mars mission and colony. In K. Szocik (Ed.), *The human factor in a mission to Mars: An interdisciplinary approach* (pp. 11–14). Cham, Switzerland: Springer.

Allday, J. (2019). *Space-time: An introduction to Einstein's theory of gravity*. Boca Raton, FL: Taylor & Francis.

Aristotle. (330 B.C.). *Metaphysics*.

Barber, P. (1988). *Vampires, burial, and death*. New Haven, CT: Yale University Press.

Barker, R., & Gilroy, S. (2017). Life in space isn't easy, even if you are green. *Biochemist, 39*(6), 10–13.

Boehm, A. M., Khalturin, K., Anton-Erxleben, F., Hemmrich, G., Klostermeier, U. C., Lopez-Quintero, J. A., et al. (2012). FoxO is a critical regulator of stem cell maintenance in immortal Hydra. *Proceedings of the National Academy of Sciences of the United States of America, 109*(48), 19697–19702.

Boethius. (524). *The Consolation of Philosophy.*

Bremermann, H. J. (1982). Reliability of proliferation controls. The Hayflick limit and its breakdown in cancer. *Journal of Theoretical Biology, 97*(4), 641–662.

Cohan, P. (2013, June 20). Google's engineering director: 32 years to digital immortality. *Forbes*. https://www.forbes.com/sites/petercohan/2013/06/20/googles-engineering-director-32-years-to-digital-immortality/#55bc1a8721e5.

Costanzo, J. P., Reynolds, A. M., do Amaral, M. C., Rosendale, A. J., & Lee, R. E. (2015). Cryoprotectants and extreme freeze tolerance in a subarctic population of the wood frog. *PLoS One, 10*(2), e0117234.

Cuthbertson, A. (2019a, September 10). People will communicate 'not only without speaking but without words—Through access to each other's thoughts at a conceptual level'. *Independent*. https://www.independent.co.uk/life-style/gadgets-and-tech/news/brain-computer-interface-neuralink-elon-musk-telepathy-a9097821.html.

Cuthbertson, A. (2019b, September 10). Brain-computer interface will make people telepathic, scientists say. *Independent*. https://www.independent.co.uk/life-style/gadgets-and-tech/news/brain-computer-interface-neuralink-elon-musk-telepathy-a9097821.html.

Damasio, H., Grabowski, T., Frank, R., Galaburda, A., & Damasio, A. (1994). The return of Phineas Gage: Clues about the brain from the skull of a famous patient. *Science, 264*(5162), 1102–1105.

Darwin, C. (1871). *The descent of man, and selection in relation to sex*. London: John Murray.

de Queiroz, K. (2005). Ernst Mayr and the modern concept of species. *PNAS, 102*(1), 6600–6607.

Defense Advanced Research Projects Agency (2016, January 9). Bridging the bio-electronic divide: New effort aims for fully implantable devices able to connect with up to one million neurons. https://www.darpa.mil/news-events/2015-01-19.

Descartes, R. (1637). *Discourse on the method*. Leiden, Netherlands.

Dodd, D. M. B. (1989). Reproductive isolation as a consequence of adaptive divergence in *Drosophila pseudoobscura*. *Evolution, 43*(6), 1308–1311.

Douglas, D. (2007, June 25). Middle Eastern women may have vitamin D deficiency. *Reuters*. https://www.reuters.com/article/idUSHAR56610220070625.

Drexler, E. (1987). *Engines of creation: The coming era of nanotechnology*. New York: Anchor Books.

Emspak, J. (2017, July 15). Chinese scientists just set the record for the farthest quantum teleportation. *Space*. https://www.space.com/37506-quantum-teleportation-record-shattered.html.

Estep, P. (2010). Declining asexual reproduction is suggestive of senescence in hydra: Comment on Martinez, D., "Mortality patterns suggest lack of senescence in hydra." Exp Gerontol 33, 217–25. *Experimental Gerontology, 45*(9), 645–646.

Fildes, J. (2009). Artificial brain '10 years away'. *BBC News*. https://news.bbc.co.uk/2/hi/8164060.stm.

Freeman, D. (2017, February 28). Will Mars colonists evolve into this new kind of human? *NBC News*. https://www.nbcnews.com/storyline/the-big-questions/mars-colonists-might-evolve-entirely-new-type-human-n708636.

Freeman, P. (2009). *Julius Caesar*. New York, NY: Simon & Schuster.

Fujisawa, T. (2003). Hydra regeneration and epitheliopeptides. *Developmental Dynamics, 226*(2), 182–189.

Futuyma, D. J. (1986). *Evolutionary biology* (p. 450). Sunderland, MA: Sinauer.

Gascoigne, B. (2003). *A brief history of the dynasties of China*. London: Robinson Publishing.

Gibran, K. (1923). *The prophet*. New York: Knopf.

Glenn, A., Kurzban, R., & Raine, A. (2011). Evolutionary theory and psychopathy. *Aggression and Violent Behavior, 16,* 371–380.

Goldsworthy, A. (2006). *Caesar: Life of a colossus.* New Haven, CT: Yale University Press.

Grau, C., Ginhoux, R., Riera, A., Nguyen, T. L., Chauvat, H., Berg, M., et al. (2014). Conscious brain-to-brain communication in humans using non-invasive technologies. *PLoS ONE, 9*(8), e105225.

Hamilton, E. (1942). *Mythology: Timeless tales of gods and heroes.* New York: Little, Brown and Company.

Harlow, J. (1868). Recovery from the passage of an iron bar through the head. *Publications of the Massachusetts Medical Society, 2,* 327–347.

Hawkey, A. (2005). Physiological and biomechanical considerations for a human Mars mission. *Journal of the British Interplanetary Society, 58*(3–4), 117–130.

Hayflick, L. (2007). Biological aging is no longer an unsolved problem. *Annals of the New York Academy of Sciences, 1100*(1), 1–13.

[illegible] M. ([illegible]). [illegible] human beings unique. *BBC*. https://www.bbc.com/future/article/2015070[illegible]-the-small-list-of-things-that-make-humans unique.

Jerison, H. (1973). *Evolution of the brain and intelligence.* New York: Academic Press.

Keim, B. (2007). Could dressing conservatively make Muslim women sick? *Wired.* https://www.wired.com/wiredscience/2007/06could-burkas-ma/.

Ker, R., Bennett, M., Bibby, S., Kester, R., & Alexander, R. (1987). The spring in the arch of the human foot. *Nature, 325*(7000), 147–149.

Kirkwood, T. B. L. (1977). Evolution of aging. *Nature, 270,* 301–304.

Klass, P. (1996, September 29). The artificial womb is born. *The New York Times.*

Kurzweil, R. (2005). *The singularity is near: When humans transcend biology.* New York: Penguin Books.

Kuwabara, Y., Okai, T., Imanishi, Y., Muronosono, E., Kozuma, S., Takeda, S., et al. (1987). Development of extrauterine fetal incubation system using extracorporeal membrane oxygenator. *Artificial Organs, 11*(3), 224–227.

Laertius, D. (circa 200–250 A.D.). *Lives and Opinions of Eminent Philosophers.*

Latimer, B., & Lovejoy, C. (1989). The calcaneus of *Australopithecus afarensis* and its implications for the evolution of bipedality. *American Journal of Physical Anthropology, 78*(3), 369–386.

Lieberman, D., Raichlen, D., Pontzer, H., Bramble, D., & Cutright-Smith, E. (2006). The human gluteus Maximus and its role in running. *Journal of Experimental Biology, 209,* 2143–2155.

Locke, J. (1689). *An essay concerning human understanding.* London: The Baffett.

Martin, G. M. (1971). Brief proposal on immortality: An interim solution. *Perspectives in Biology and Medicine, 14*(2), 339–340.

Martínez, D. E. (1998). Mortality patterns suggest lack of senescence in hydra. *Experimental Gerontology, 33*(3), 217–225.

Mayr, E. (1942). *Systematics and the origin of species from the viewpoint of a zoologist.* New York: Columbia University Press.

Mishal, A. (2001). Effects of different dress styles on vitamin d levels in healthy young Jordanian women. *Osteoporosis International, 12*(11).

Mitteroecker, P., Huttegger, S. M., Fischer, B., & Pavlicev, M. (2016). Cliff-edge model of obstetric selection in humans. *PNAS, 113*(51), 14680–14685.

Nielsen, J., Hedeholm, R. B., Heinemeier, J., Bushnell, P. G., Christiansen, J. S., Olsen, J., et al. (2016). Eye lens radiocarbon reveals centuries of longevity in the Greenland shark (*Somniosus microcephalus*). *Science, 353*(6300), 702–704.

Pais-Vieira, M., Lebedev, M., Kunicki, C., Wang, J., & Nicholelis, M. A. L. (2013). A brain-to-brain interface for real-time sharing of sensorimotor information. *Scientific Reports, 3*(1319).

Parnia, S. (2014). *Erasing death: The science that is rewriting the boundaries between life and death.* New York: HarperCollins.

Partridge, E. A., Davey, M. G., Hornick, M. A., McGovern, P. E., Mejaddam, A. Y., Vrecenak, J. D., et al. (2017). An extra-uterine system to physiologically support the extreme premature lamb. *Nature Communications, 8,* 15112.

Plutarch. (1470). *Parallel Lives.*

Pontzer, H. (2007). Predicting the energy cost of terrestrial locomotion: A test of the LiMb model in humans and quadrupeds. *Journal of Experimental Biology, 210*(3), 484–494.

Prisco, G. (2012, December 12). Uploaded e-crews for interstellar missions. *Kurzweil Accelerating Intelligence.* https://www.kurzweilai.net/uploaded-e-crews-for-interstellar-missions.

Rao, R. P. N., Stocco, A., Bryan, M., Sarma, D., Youngquist, T. M., Wu, J., & Prat, C. S. (2014). A direct brain-to-brain interface in humans. *PLoS ONE, 9*(11), e111332.

Ratiu, P., & Talos, I. (2004). The tale of Phineas Gage, digitally remastered. *New England Journal of Medicine, 351,* e21.

Rhodes, F. H. T. (2016). *Origins: The search for our prehistoric past.* New York: Cornell University Press.

Rodrigo, E. (2011). *The physics of stargates: Parallel universes, time travel, and the enigma of wormhole physics.* New York: Eridanus Press.

Royal Society. (2019, September) iHuman: Blurring lines between mind and machine issued. *DES6094.* https://royalsociety.org/-/media/policy/projects/ihuman/report-neural-interfaces.pdf.

Ruff, C., McHenry, H., & Thackeray, J. (1999). Cross-sectional morphology of the SK 82 and 97 proximal femora. *American Journal of Physical Anthropology, 109*(4), 509–521.

Saadi, H. F., Dawodu, A., Afandi, B. O., Zayed, R., Benedict, S., & Nagelkerke, N. (2007). Efficacy of daily and monthly high-dose calciferol in vitamin D-deficient nulliparous and lactating women. *The American Journal of Clinical Nutrition, 85*(6), 1565–1571.

Schrödinger, E. (1935). Discussion of probability relations between separated systems. *Mathematical Proceedings of the Cambridge Philosophical Society, 31*(4), 555–563. https://doi.org/10.1017/S0305004100013554.

Schrödinger, E. (1936). Probability relations between separated systems. *Mathematical Proceedings of the Cambridge Philosophical Society, 32*(3), 446–452. https://doi.org/10.1017/S0305004100019137.

Shay, J., & Wright, W. (2000). Hayflick, his limit, and cellular ageing. *Nature Reviews. Molecular Cell Biology, 1*(1), 72–76.

Silk, J., Bronsan, S., Henrich, J., Lambeth, S., & Shapiro, S. (2013). Chimpanzees share food for many reasons: The role of kinship, reciprocity, social bonds and harassment on food transfers. *Animal Behavior, 85*(5), 941–947.

Smith, D. L. (2012). *Less than human* (p. 96). New York: St. Martin's Press.

Spoor, F., Wood, B., & Zonneveld, F. (1994). Implications of early hominid labyrinthine Morphology for evolution of human bipedal locomotion. *Nature, 369*(6482), 645–648.

Steudel-Numbers, K. (2006). Energetics in *Homo erectus* and other early hominins: The consequences of increased lower-limb length. *Journal of Human Evolution, 51*(5), 445–453.

Stuijt, A. (2009, May 8). Women could endanger their health by wearing burqas. *Digital Journal.* https://www.digitaljournal.com/article/272307.

Suetonius. (121). *The Twelve Caesars.*

Suzuki, I., Gacquer, D., Van Heurck, R., Kumar, D., Wojno, D., Wojno, M., et al. (2018). Human-specific NOTCH2NL genes expand cortical neurogenesis through delta/notch regulation. *Cell, 173*(6), 1370–1384.

Thorne, K. S. (1994). *Black holes and time warps* (p. 499). London: W. W. Norton & Company.

Vernot, B., Tucci, S., Kelso, J., Schraiber, J. G., Wolf, A. B., Gittelman, R. M., et al. (2016). Excavating Neandertal and Denisovan DNA from the genomes of Melanesian individuals. *Science, 352*(6282), 235–239.

Wacker, M., & Holick, M. (2013). Sunlight and vitamin D: A global perspective for health. *Dermatoendocrinology, 5*(1), 51–108.

Wai, L. K. (2004). Telomeres, telomerase, and tumorigenesis—A review. *MedGenMed, 6*(3), 19.

Warneken, F. (2015). Precocious prosociality: Why do young children help? *Child Development Perspectives, 9*(1), 1–6.

Warwick, K., Gasson, M., Hutt, B., Goodhew, I., Kyberd, P., Andrews, B., et al. (2003). The application of implant technology for cybernetic systems. *Archives of Neurology, 60*(10), 1369–1373.

Warwick, K., Gasson, M., Hutt, B., Goodhew, I., Kyberd, P., Schulzrinne, H., & Wu, X. (2004). Thought communication and control: A first step using radiotelegraphy. *IEE Proceedings on Communications, 151*(3), 185–189.

Weisberger, M. (2018, July 27). Worms frozen for 42,000 years in Siberian permafrost wriggle to life. *Live Science*. https://www.livescience.com/63187-siberian-permafrost-worms-revive.html.

White, T. D., Asfaw, B., Degusta, D., Gilbert, H., Richards, G. D., Suwa, G., & Howell, F. C. (2003). Pleistocene *Homo sapiens* from Middle Awash, Ethiopia. *Nature, 423,* 742–747.

Wilmer, H., Sherman, L., & Chein, J. (2017). Smartphones and cognition: A review of research exploring the links between mobile technology habits and cognitive functioning. *Frontiers in* [illegible]

Yang, [illegible], Colletti, P., & Toga, A. W. (2009). Localization of deformations within the amygdala in individuals with psychopathy. *Archives of General Psychiatry, 66*(9), 986–994.

Yoo, S., Kim, H., Filandrianos, E., Taghados, S. J., & Park, S. (2013). Non-invasive brain-to-brain interface (BBI): Establishing functional links between two brains. *PLoS ONE, 8*(4), e60410.

Chapter 5
Limitations for Extraterrestrial Colonisation and Civilisation Built and the Potential for Human Enhancements

Martin Braddock

Abstract Human beings are poorly adapted to live and work in space for long periods of time. This situation poses as yet insurmountable challenges for deep space travel and the colonisation of other worlds including those in our solar systems such as the Moon or Mars. Research road maps have been built by multi-disciplinary teams working internationally to explore the development of artificial gravity and protection of the body against radiation including the potential for human enhancement by techniques such as gene editing. Terrestrial technological advances in the development of prosthetic limbs and tissue-engineered organs which may be accompanied by the production of exoskeletal structure offer further opportunities for human augmentation. Finally, progress in the generation of brain–computer interface-based communication systems in patients who are unable to communicate may offer more futuristic applications for building a society of hybrid human avatars as future colonists of New World civilisations.

5.1 Introduction

The desire to explore is part of the human survival instinct embedded in our psyche and drove the American and Russian space programmes of the 1960s and 1970s, resulting in the first successful Moon landing on 20 July 1969. Since then, the eyes of astronomers and the space agencies have been on travelling to Mars with the futuristic vision of establishing and maintaining a human colony on the Red Planet. With the potential to send astronauts to Mars by 2030, humankind will need to address the considerable scientific, technical, medical and philosophical challenges of building a New Society in this alien environment.

M. Braddock (✉)
Newton's Astronomical Society at Woolsthorpe, Woolsthorpe Manor, Water Lane, Woolsthorpe By Colsterworth, Grantham, Lincolnshire NG33 5PD, UK
e-mail: martinbraddock@rocketmail.com

Science4U.co.uk, Radcliffe-on-Trent, Nottinghamshire NG12 2LA, UK

K. Szocik (ed.), *Human Enhancements for Space Missions*, Space and Society,
https://doi.org/10.1007/978-3-030-42036-9_5

5.2 Defining the Challenges

This chapter will briefly address some major medical challenges astronauts will face on their journey to Mars as an example of a near-term aim for colonisation. Even when the Earth and Mars are at their closest proximity of around 55–60 million kilometres, it will take approximately nine months to complete the journey. Although this is longer than the six months most astronauts spend on the International Space Station (ISS), over 220 individuals have completed their missions on board the ISS which provides us with a unique data source enabling us to learn much from astronaut experience of coping with and managing the environment to best protect the physical and mental health of the first generation of humans to establish this new frontier.

5.3 Journey Time

The optimal time to travel to Mars is during opposition, which is when Mars is closest to Earth and which occurs approximately every two years. During opposition, Mars can be as close as 55 million kilometres from Earth and many space agencies time their missions to coincide with this orbital alignment. The duration of the journey to Mars is between 150 and 300 days, dependent upon the speed of the craft and alignment of the planets. Mariner 4, launched by the National Aeronautics and Space Administration (NASA) on 28 November 1964, was travelling for 228 days before arriving at the planet on 14 July 1965, whereas Mariner 7, launched on 27 March 1969, reached Mars on 5 August 1969 having required 131 days of space travel. Mariner 9, the first spacecraft to orbit around Mars successfully, was launched on 30 May 1971 and arrived at Mars on 13 November 1971 after a voyage lasting 167 days. This journey time of approximating between 150 and 300 days has been maintained for over 50 years of Mars exploration.

5.4 The Environment in Space and Ergonomic Constraints

Although space is an alien domain, we have gained great experience for survival and for living and working in an extraterrestrial environment since Yuri Gagarin became the first human in space in April 1961, completing an Earth orbit in a flight lasting 108 min in the Vostok 1 spacecraft. By 24 September 2019, according to United States Air Force definition, 572 people have been in space with the longest single mission comprising nearly 484 days (Garcia 2018) on board the ISS. During both of these examples and the several hundred other missions undertaken to date, spacecraft ergonomics is constrained by three factors which impact the daily lives of astronauts and present considerable ergonomic challenges for future travel, which include preparing for planetary colonisation.

5.4.1 Microgravity

The first factor is the environment of weightlessness or microgravity (μg) due to the free fall of spacecraft in orbit around the Earth or away from the Earth's gravitational attraction in deeper space. The major harmful effects of μg on human physiology and psychology and potential mitigation measures have been well documented and are shown in Table 5.1.

On the ISS, all astronauts follow mandatory rigorous exercise regimes (Braddock 2018), which in part are able to counteract the negative effects of μg on, for example, soft and hard tissue wasting (e.g. Braddock 2017). Nevertheless, despite adherence to exercise regimens and compliance with taking pharmaceutical agents known to counteract bone loss on Earth, data from historical ISS missions (Nagaraja and Risin 2013) or ISS and Mir space station missions (Sibonga et al. 2015) show that astronauts lost up to 1.6% and 1% of bone and lean muscle mass per month, respectively (Nagaraja and Risin 2013), or up to 22% of bone mass over the mission (Sibonga et al. 2015). After a typical mission of six months duration on the ISS, astronauts are unable to walk unaided on return to Earth. In addition, the environment of μg has negative effects on the cardiovascular, central nervous and respiratory systems, weakens the immune response and may affect vision in some astronauts (Braddock 2017). Very recently, a study evaluating periodic ultrasound tests of 11 astronauts on board ISS showed that blood flow had either stagnated or reversed in the left internal jugular vein (IJV) in 6 crew members and an inclusive thrombus (blood clot) and a potential partial thrombus were found in 2 astronauts after return to Earth (Marshall-Goebel et al. 2019).

Such is the urgent requirement to find countermeasures to the effects of μg, and NASA with input from space agencies from Europe and Japan and leading worldwide academic institutions have developed a road map for artificial gravity research (Clement 2017).

5.4.2 Radiation

The second factor which imposes a more serious challenge to space travel and for Mars colonisation is astronaut exposure to radiation (Chancellor et al. 2014; Cucinotta 2014; Blue et al. 2019). Unlike on Earth, where the magnetosphere and Van Allen Belts form protective layers from the harmful effects of galactic cosmic radiation (GCR), in space no such protection is available. Astronauts are exposed to GCR which is comprised of protons, helium alphas and heavy ions with very high energies. Current shielding is ineffective as the ions easily penetrate spacecraft and may produce secondary radiation such as X-rays and neutrons. GCR is radiation of high linear energy transfer (LET) which means radiation is deposited in a non-uniform manner along a track. Astronauts also experience solar particle events, which

Table 5.1 Current and potential future countermeasures for physical and psychological effects of space on astronauts and colonists

Effect	Current countermeasures	Future countermeasures
Visual disturbances Motion sickness	Vision—no known countermeasure. Pharmacological intervention	No immediate countermeasure
Sleep quality	Alignment with time exposure to light/dark and adaptation of work schedules	Optimisation and monitoring of cycles, acquisition of data for long (>1 yr) residency in space
Brain function	Adaptation	Study of behaviour for long (>1 yr) residency in space
Hard and soft tissue loss	Resistance training, nutrition, pharmacological intervention	Optimisation of exercise [illegible] predict muscle and bone loss
Healing capacity	Risk-based monitoring to reduce injury, pharmacological intervention	Limited data available on healing rates in humans in space. Translation of animal model to human healing rates in microgravity requires further exploration
Respiratory function	Limit number of EVAs and optimise hyperoxia-inducing regime	Construct 3D human lung models and characterise lung biomarkers under different conditions
Immune function and infection	Vaccination, screening, pre-mission Isolation, diet maintenance	Immunological, microbiome and genomic profiling before mission, monitoring during mission
Cardiovascular function	Resistance training, monitoring, pharmacological intervention, pressure suits	Understanding what level of function Is required and assessment of intervention risk benefit
Psycho-social function	Crew selection, adequate menial stimulation, contact with Earth, pharmacological intervention	Monitoring complex human behaviour in space missions and other isolation environments. Optimisation of exercise regimes
Gastro-intestinal function	Diet maintenance, pharmacological intervention	Microbiome profiling before mission, monitoring during mission

Selected physical and psychological effects are illustrated together with current and proposed future-looking countermeasures comprise electrons, protons and alpha particles of less energy than GCR, which can be shielded and have low LET being deposited in a uniform manner along the track.

Despite many decades of knowledge of the effects of ionising radiation on living systems on Earth from studies carried out in peacetime and as a consequence of warfare, there is a lack of precision in predicting the health risk to astronauts on both short-term (<3 months) and long-term (>3 months) missions, and recent studies have attempted to address the limitations (Chancellor et al. 2018; Blue et al. 2019). It has been recommended that astronauts are limited to career lifetime radiation exposures that would induce no more than a 3% risk of exposure-induced death and this translates into a radiation dose dependent upon astronaut sex and age. Although estimates vary, for a Mars mission of 975 days which includes 536 days in transit, astronauts would receive a radiation dose of approximately 2 SieVert (Sv) (Sion 2011). This compares with background radiation exposure on Earth of between 1 and 2 milliSv per year and a recommended 2.5 and 1.75 Sv lifetime exposures for a 35-year-old male and a 35-year-old female astronaut, respectively. It is clear that the current advisory limit for astronaut exposure to radiation will be challenged, may limit mission duration and would very likely be exceeded on longer missions (Rask et al. 2011; Scoles 2017) and in the establishment of a colony on Mars unless new mitigation measures are developed.

5.4.3 *Isolation*

The third factor which provides both a physical and especially a psychological challenge for astronauts is to live and work for long periods of time in a confined environment (Kanas 2010, 2011). In space, there is no 'going outside' for a break and although the ISS provides a spacious habitat for the crew, there is the inevitable monotony of being in the same location for many months and with the same people. This may lead to mood changes and periods of anxiety or depression which could affect personal performance and that of the crew in executing the mission. This overall risk, as are others that relate to effects of microgravity, is documented in NASA's Human Research Roadmap (Williams 2019) and describes a number of factors, some interrelating on the manifestation of behavioural and psychiatric disorders. Astronauts have also experienced both visual and olfactory hallucinations with two examples being flashes of light so-called dancing fairies reported by numerous astronauts including Don Pettit on the ISS in 2012 and a phantom odour reported by a crew of the Soyuz-21 mission on the Salyut-5 space station in 1976. The concern that the odour may signify a system malfunction led to an early crew return to Earth although on checking all systems in the craft back on Earth, no odour or any technical issues were discovered. The effects of working in a confined environment, in some cases much smaller than the ISS as in the Apollo missions of the 1960s and 1970s, have been reviewed (e.g. Kanas et al. 2011; Pagel and Chouker 2016; Alcibiade et al. 2018) countermeasures (Sandal 2001) and mitigation strategies for enabling function in confinement which include pharmaceutical intervention with antidepressants (Stingl et al. 2015), securing routes to enable expression of state of mind (Sipes and Fielder 2007) and adequate rest (Czeisler and Barger 2017). We are well placed to understand the effects of isolation on human mental health by drawing on data from Mars analogue studies which include Mars500 (Basner et al. 2013), Flashline Mars Arctic Research Station (FMARS; Binsted et al. 2010) and the Mars Desert Research Station (Rai and Kaur 2012). Most evidence for understanding human psychology, how conflict arises and how cooperation either does or does not sustain effective group dynamics comes from studies of scientist postings in Antarctic research stations and has recently been reviewed (Sandal et al. 2018). This subject, better known as the collective of confined environment psychology, may be a major challenge for these expeditions where occupants may split into divisive factions, a finding also reported in the closed ecosphere named Biosphere 2 (Nelson 2010). The selection of colonists should consider selection of whole groups who work well together rather than an ad hoc assembly of people who may be highly qualified but potentially fail to work together.

5.5 Potential for Human Enhancement

Human enhancement (HE) may simply be defined as the natural, artificial or technological alteration of the human body in order to enhance physical or mental capabilities, and the challenges and opportunities have recently been discussed (e.g. Austin and Buchanan 2018).

5.5.1 Currently Used Human Enhancements

Taking the definition of HE provided above, it may be recognised that HE has been with humankind for centuries with the invention of crude devices to correct hearing loss and failing vision through to more recent examples of orthodontics, elective and non-elective surgery including cosmetic surgery and performance-enhancing drugs to enhance both physical and mental performances. Selected commonly enabled terrestrial human enhancements are illustrated in Table 5.2 and include both non-life-threatening and life-threatening enhancements. Non-life-threatening enhancements may be divided into the development of prostheses, sensory aids such as visual (spectacles and contact lenses) and aural (hearing aids) and the growing trend in health monitoring devices (Peake et al. 2018), performance-enhancing drugs (Momaya et al. 2015; White and Nouen 2017) and a number of elective surgical procedures including cosmetic and reconstructive surgery. Human beings have evolved natural enhancements, and recent technological advances have recognised innate human ability to withstand extreme terrestrial environments which may confer a species advantage

Table 5.2 Selected terrestrial human enhancements

Procedure	Agent or target	Modality
Lifestyle enabling (non-life threatening)	Prostheses	Physical
	Sensory aids (glasses, contact lenses, hearing-aids)	Physical
	Health monitoring devices, wearable electronics	Physical
	Performance enhancing drugs: physical and menial	Pharmaceutical Interventional
Non-elective surgery (non-life threatening)	Most major organs, may include cosmetic surgery	Physical Invasive or non-Invasive procedural
Organ replacements (life threatening)	Most major organs	Physical procedural
Elective surgery (life threatening)	Most major organs	Physical procedural

Procedures are illustrated ranging from lifestyle to elective surgical enhancement together with the agent or target and modality

phenotype (e.g. Illardo and Nielsen 2018) and which may help in design considerations for deep space missions (Bartone et al. 2019) and planetary colonisation (Braddock et al. 2019a).

In addition to these natural adaptive pathways, astounding developments in multiple disciplines create the possibility for either or both genetic and physical enhancements and this next section will highlight some of the latest advances, together with suggestions on where caution is advised.

5.5.2 Gene Editing

Central to many potential genetic enhancements is the concept of gene editing. This topic has been reviewed in this book and extensively in both the scientific literature (e.g. De Lecuona et al. 2017; Adli 2018; Cyranoski 2019) and non-scientific literature where the boundaries between science fact and fiction are blurred. Unlike gene therapy, which is transient and for which there are now both products (Approved Cellular and Gene Therapy Products 2019; Gene Therapy Medicinal Products 2019) and multiple human clinical trials (Gene Therapy Clinical Trials Databases 2019), gene editing is ultimately envisaged as a permanent genetic change in a human being which can either be generation-specific or be carried through generations via germline gene editing. Potential options to achieve gene-mediated human enhancement are described in Fig. 1. For the purpose of this chapter, we will make some simple assumptions:

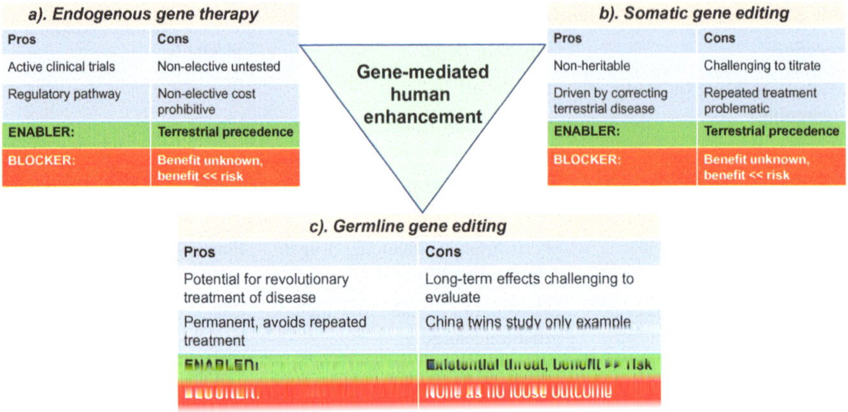

Fig. 1 Potential options for gene-mediated human enhancement. Three potential routes for gene-mediated human enhancement are described: **a** endogenous gene therapy, **b** somatic gene editing and **c** germline gene editing. For each of the three modalities, the pros and cons are presented together with one enabler and one blocker factor

- Gene editing is a scientific possibility and will *eventually* become a potential option as a non-elective modality for corrective treatment which may avoid severe reduction in quality of life or fatal disease to humans on Earth.
- Elective gene editing in humans to confer a phenotypic *advantage* in healthy individuals requires a compelling rationale, a thorough understanding of benefit/risk and the necessary regulatory, legal and ethical guidelines to be proposed, and internationally recognised standardised procedures to be adopted, monitored and enforced. It will be many years before this is an option.
- Germline gene editing is a *future ambition*, will proceed elective gene editing and will be a requirement for long-term human colonisation of space where the environment demands human enhancement. It will be many years before this is an option and will not be available unless humankind is presented with severe existential threat.

Some molecular targets which may be candidates for gene editing are shown in Table 5.3. Based on non-clinical in vitro and in vivo studies, a number of genes have been identified which, when exogenously expressed, may confer some element of radio-resistance. These include damage suppressor (Dsup; Hashimoto et al. 2016; Hashimoto and Kunieda 2017; Gupta et al. 2016), interleukin-17 (IL-17; Li et al. 2015) and p53 (Gudkov and Komarova 2003; Lee et al. 2013).

Of these potential candidates, Dsup which is a tardigrade-unique DNA-associating protein isolated from the bacterium *Deinococcus radiodurans* suppresses the occurrence of DNA breaks in the host organism and has been shown to increase radiotolerance when expressed in cultured human HEK293 cells irradiated with X-rays at 1 (Hashimoto and Kunieda 2017) or 4 Gy (Hashimoto et al. 2016). Interestingly, an Mn^{2+} decapeptide complex has successfully shown protection of mice from a

Table 5.3 Mitigation strategies for risk to astronaut health—candidate molecular enhancement targets

Risk for mitigation	Desired outcome	Molecular larget	Reference
Radiation exposure	Radioprotection	Dsup MDP IL-17 p53	Hashimoto et al. (2016), Hashimoto and Kunieda (2017) Gupta et al. (2016) Li et al. (2015) Gudkov and Komarova (2003), Lee et al. (2013)
Reduced gravity	Prevention of hard tissue loss	LRP5, LRP6 BMPs	Boyden et al. (2002), Xu et al. (2014a, b), Siamwala et al. (2015), Burgers et al. (2016), Norwitz et al. (2019)
Reduced life support	Reduced demand for oxygen	EPAS1 EELN1 PPARA SENP1 ANP32D	Simonson et al. (2010, 2012), Hanaoka et al. (2012), MacInnis and Rupert (2011). Peng et al. (2011). van Patot and Gassmann (2011). Zhou et al. (2013)
Psychological effects of confinement	Reduced body odour production	ABC 11	Martin et al. (2010), Rodriguez et al. (2013)
Human longevity	Life extension	Geroprotectors	Blagosklonny (2006, 2007), Harrison et al. (2009), Marilena (2011), Schlicker et al. (2011), Ergin et al. (2013), Kasiotis et al. (2013), Moskalev et al. (2015), Nouvelle et al. (2016), Blagosklonny (2019), Konopka et al. (2019), Song et al. (2019)

Selected molecular targets which may deliver enhancement are shown alongside the risk to mitigated and the desired outcome

radiation dose of 9.5 Gy, where it was shown to protect against the effects of ionising radiation-induced damage to bone marrow and hematopoietic stem cells (Gupta et al. 2016). IL-17 has been shown to induce radio-resistance in B lymphoma cells where the mechanism of action appears to involve suppressed expression of the transcription factor p53 and the subsequent inhibition of radiation-triggered apoptosis (Li et al. 2015). Lastly, the third example is p53 itself where it has been known for many years from the oncology literature that it plays a role in radiotherapy resistance in human cancer (Gudkov and Komarova 2003) as a tumour suppressor protein. The premise for conferring radio-tolerance if not radio-resistance in humans may underestimate the complex interactions these examples provide. In all three cases, it is highly unlikely that any one target alone will confer radio-protection which would be the ultimate goal for human enhancement.

The search for other molecular targets may capitalise on previous studies employing genome-wide association analyses of radiation resistance in the fruit fly *Drosophila melanogaster* (Vaisnav et al. 2014) or transposon-directed insertion sequencing of bacterial strains (Byrne et al. 2014). In parallel, advances in drug discovery will isolate molecular targets which may confer drug resistance and which may be amenable to both astronaut radio protections. The relentless drive on Earth to subvert radio-protective mechanisms for multiple tumour types for the treatment of patients with cancer may provide further insight. Recently, a multidisciplinary team led by the NASA Ames Research Centre has published a road map for an integrated approach to understanding how radiation resistance could be conferred upon astronauts (Cortese et al. 2018). The team outlines future research directions which includes upregulation of endogenous DNA repair and outside of the theme of gene-mediated enhancement discusses the concept of isotopic substitution of organic molecules as a further potential radio-protective mechanism.

Other molecular targets which may provide candidates for enhancement include those which are known to play major roles in the development and maintenance of hard tissue, such as the skeleton, which is as described earlier in Sect. 5.4.1, and shown in Table 5.3 are negatively affected by the environment of μG. Proteins such as bone morphogenetic protein (BMP; Xu et al. 2014a, b; Siamwala et al. 2015) and low-density lipoprotein receptor-related proteins 5 and 6 (LRP5, LRP6; Boyden et al. 2002; Xu et al. 2014a, b; Burgers et al. 2016; Norwitz et al. 2019) are key molecules in bone anabolism and maintenance. Mice harbouring a heterozygous LRP6 deletion have an impaired ability to repair fractures (Burgers et al. 2016), and more interestingly a mutation in LRP5 was shown to be associated with high bone density (Boyden et al. 2002). As was described for targets that may confer radio-resistance, the biology of BMP and LRP signalling through the Wnt pathway is highly complex and it is unlikely that a single entity will be sufficient to maintain Earth like values of bone mass, even if proven safe to engineer. Other proteins which have been shown to positively influence μG-induced hard and soft tissue atrophy which may become targets for gene editing or gene therapy include osteoprotegerin, myostatin, activin and sclerostin and have been reviewed elsewhere (Braddock 2019; Ryder and Braddock 2020).

Additional instances where human enhancement may be desirable could be in atmospheric adaptation. There are a number of genes associated with human habitation at high altitude, generally taken to be >2500 m which in part came to prominence at the beginning of the Himalayan climbing era in the early twentieth century. Tibetans, Andeans and Ethiopians all possess genetic determinants that have evolved at different times which confer advantages for gaseous exchange in the respiratory system. Genes harbouring mutations in EPAS1, EGLN1, SENP1, PPARA and

ANP32D (Simonson et al. 2010, 2012; Hanaoka et al. 2012; MacInnis and Rupert 2011; Peng et al. 2011; van Patot and Gassmann 2011; Zhou et al. 2013) have all been shown to play roles in adaptation to hypoxia.

The final example, which may appear cosmetic in the first instance, is the discovery that a human population exists which naturally does not produce body odour (Prokop-Prigge et al. 2015). Human axillary (armpit) odour is formed by bacterial action on odour precursors that originate from apocrine sweat glands. Caucasians and Africans possess a strong axillary odour, whereas many Asians have only a faint acidic odour. Although this may appear trivial, within the confines of a tightly enclosed environment either in space or in a colony, disagreeable body may contribute to a stressful environment with the potential to even act as a trigger for disharmony, conflict and a loss of team performance. The gene ABCC11 encodes an apical efflux pump and is critical for the formation of axillary odour. A single-nucleotide polymorphism, prominent among Asian people, leads to a nearly complete loss of the typical odour components in axillary sweat (Martin et al. 2010; Rodriguez et al. 2013).

Taken together, there are a number of plausible targets which in theory have the potential to mitigate the effects of numerous parameters in the space environment and may constitute mechanisms for human enhancement. However, all of these targets demand transient if not permanent expression, in some cases in multiple human tissues and the reality of achieving successful genetic manipulation of human beings, as outlined in Fig. 1 is likely many decades away.

5.5.3 Tissue Engineering and Regenerative Medicine

The shortage of healthy donor organs on Earth has stimulated the field of tissue engineering and regenerative medicine (Lieben 2016). Of particular relevance to the topic of prolonged habitation in space or in the establishment and maintenance of a colony is the potential for three-dimensional bioprinting (3D bioprinting). 3D bioprinting is an additive manufacturing methodology which employs simultaneous layer-by-layer deposition of cell types and cyto-compatible biomaterials such as hydrogels which together provide a supporting structure capable of generating organoids or potentially whole functional organs. It has been successfully used on Earth to produce ear-shaped cartilage for auricular reconstruction (Zhou et al. 2018). The area has been recently reviewed, and the state of the bioprinting art on Earth and potential applications for regenerative medicine in space is discussed (Ghidini 2018). In February 2019, the world's first bio-fabrication experiment was reported (three-dimensional bioprinting in space, 2019) having been performed during Expedition 57 on board the ISS. 3D Bioprinting Solutions, in partnership with the Russian Federal Space Agency (Roscosmos), has successfully tested formative bioprinting of three-dimensional tissue spheroids comprising mouse thymocytes using low concentrations of paramagnetic nanoparticles which are more amenable to the protection of cell culture viability. The thyroid gland is particularly sensitive to radiation exposure and is an ideal first target organ for study, which together with human cartilage provided reference points for Earth-based laboratory work. In addition to 3D Bioprinting Solutions, other companies will be trialling 3D bioprinting technology in space throughout 2019 and 2020 who include system manufacturer nScrypt and spaceflight equipment developer Techshot (Jackson 2018). The long-term premise is to envisage human enhancement as a consequence of not solely organ replacement, but as incremental improvements on organic tissue by the potential incorporation of inorganic molecules such as scaffolds which may confer extra strength and resilience to forces or, more likely the lack of force such as gravity and radio-resistance.

5.5.4 *Life Extension Strategies*

I will now turn our attention to the possibility of mechanisms to induce life extension and name this ‘pseudo-enhancement’. This falls under the very broad category of geroprotectors for which there are several hundred that have shown to extend lifespan in non-clinical model organisms. They include compounds which attenuate redox pathways as antioxidants, vitamins, phytochemicals and hormones (Blagosklonny 2006, 2007; Harrison et al. 2009; Marilena 2011; Schlicker et al. 2011; Ergin et al. 2013; Kasiotis et al. 2013; Moskalev et al. 2015; Nouvelle et al. 2016; Blagosklonny 2019; Konopka et al. 2019; Song et al. 2019), and such is the need to coordinate efforts a database has been compiled to curate potential therapeutic interventions in ageing and age-related disease, which by extension may slow down the ageing process and extend functional lifespan (Moskalev et al. 2015). As of 8 December 2019, there are 259 agents listed in the database. Of note are clinical studies in progress with metformin (Metformin in Longevity Study (MILES)) and rapamycin (Blagosklonny 2006, 2007, 2019), drugs which have been implicated in attenuating the ageing process in model systems (Ehninger et al. 2014; Konopka et al. 2019; Song et al. 2019). Full clinical trial results are yet to be published.

5.5.5 *Exoskeleton*

The concept of an exoskeleton was first proposed in the late nineteenth century (Yagin 1890) and was designed to assist human movement, being dependent upon locomotory ability. Since then, the concept of powered or semi-powered exoskeletons has received great attention (e.g. Ferris 2009; Cortes et al. 2016; Awad et al. 2017; Hill et al. 2017; Zhang et al. 2017; Agrawal et al. 2018) in medical applications for patients who have lost use of limbs through acute trauma or paralysis and may undergo robot-assisted rehabilitation (Low et al. 2016; Lajeunesse et al. 2016), military applications such as the development of exosuits to make traversing tough terrain more feasible and industrial applications to reduce the risk of upper body injury sustained through extension movements or lifting heavy objects (Brown et al. 2003). This field is a further illustration of where terrestrial-driven progression of technology, primarily to assist patients with disabilities, may have applications for space exploration and colonisation and NASA has described the application of a system called X1 (Rea et al. 2013), originally developed for patients with paraplegia, to be added to the inventory of countermeasures for the effects of μG described earlier in this chapter.

5.5.6 *Brain–Computer Interface*

The next part of this chapter very briefly focusses on the potential for brain–computer interface (BCI; Jebari 2013; Gonfalonieri 2018). In the context of human enhancement and in one of its simplest and most pragmatic applications, we will pay attention to examples where the thought process may control movement and where communication may be restored. In patients who have suffered paralysis or are tetraplegic, multi-joint motor control has been restored to some extent using a neurally controlled arm (Hochberg et al. 2012; Collinger et al. 2013; Aflalo et al. 2015; Micera 2017; Salas et al. 2018). Other studies have shown the potential to induce artificial proprioceptive and cutaneous sensations by intracortical micro-stimulation which can replicate some aspects of the body's natural physiological abilities. Helping patients with tetraplegia or completely locked-in syndrome who are unable to communicate via cortical stimulation and decoding of question and answer (yes or no) speech activity has made significant progress over the last 3–5 years (Chaudhary et al. 2017; Moses et al. 2019). Taken together, this short summary highlights some terrestrial examples of where micro-stimulation of brain compartments is able to illicit a response in patients whose motor function is impaired or missing. In the context of positive human enhancement in astronauts who have normal cognitive function, what enhancements may be possible?

One scenario could be for BCIs to create one or more connections between multiple brains via a bespoke personal BCI or a BCI hub, which may be a centralised interface accessed by all astronauts or colonists. A specific example is the brain–brain interface DARPA-funded Silent Talk project (DARPA 2010), which seeks to effect communication of individuals in a military setting without vocal communication by analysis neural signals (Denby et al. 2009a) and the technology of 'silent speech' has been successfully demonstrated (Denby et al. 2009b). Should the development of such technology be possible, for example, in the setting of a lunar or Martian colony, it may make real-time communication faster, and if successfully coupled with thought control operative processes such as habitat construction or experiments on the surface of the Moon or Mars, it may reduce the risk to astronauts of radiation exposure by allowing them to stay within a safe enclosure or exclusion limit.

5.5.7 *Human Avatars—The Ultimate Upgrade?*

Could we envisage a human avatar that is a conglomerate of technology? Perhaps today it is hard to imagine though scientific and medical progress in multiple disciplines is set to converge even more rapidly than today, in part because of increasing existential threat to humankind. There is growing evidence that activities of humankind as of approximately 12,000 years ago have increased and continue to increase extinction rates of many species and that we are experiencing the 6th extinction level event (ELE) or the Holocene extinction (Pimm et al. 2014; Ceballos et al. 2015]. Recent reports in 2019 (Svoboda 2019) of stark warnings of climate change to come and the urgent needs for remedial action are a reminder that life on Earth, human and non human, is not guaranteed to persist. The consequences of this latest ELE add urgency to considering other planets to which mankind can migrate and settle, in part as a potential staging post for further exploration and in part as a fail-safe mechanism should Earth become inhospitable to support life as we know it today.

Despite the potential, yet theoretical concerns on the development of artificial general intelligence (AGI), entrepreneur Elon Musk and the late Sir Stephen Hawking are joined by Astronomer Royal Sir Martin Rees, physicists Max Tegmark and Michio Kaku among many others who believe that a future for deep space exploration and colonisation to exoplanets may involve the creation of a human hybrid avatar and that this may be a logical extension of the human species (Tegmark 2017; Rees 2018; Kaku 2018).

Although AGI is not in our society today, what is in our society and is a subject of considerable debate is defined by the digital afterlife industry (DAI). DAI relates to the vast quantity of digital data that are and will continue to be generated at an exponential rate and could be collected and 'reconstituted' after a person dies. This is already an issue for the social platform Facebook, where it is estimated that 1.7 million people per year in the USA alone will have passed away and yet their profiles will still be active.

The subject of data ownership is an area of intense legal debate and out of scope for this paper; however, what appears to be the case is that the technology platform rather than human person who generated the digital footprint is the owner of the data. This may have some relevance to distant space exploration as a reconstituted digital astronaut or other subject matter expert may constitute a potential future e-crew. Moreover, there is a concurrent philosophical debate on whether a digital reconstitution would have human rights and to demonstrate support for AI in the country, in 2017 a robot named Sophia was given Saudi Arabian citizenship (Weisberger 2019). Another more recent example relates to that of James Dunn, a 24-year-old man who died from the skin condition epidermolysis bullosa and skin cancer (De Quetteville 2019). With the help of Pete Trainor from the company Us ai, they were able to develop a chatbot named Bo who was able to replicate aspects of James' thoughts from many conversation James had in 'training' the bot. This is one of several landmark cases which show the potential of the DAI but also seek to call for regulation so that the wishes of both those physically dead and alive are respected and that those who wish to 'live' in a virtual world are accorded the appropriate level of protection and privacy (Ohman and Floridi 2017, 2018).

More immediately, if permanent outposts and/or colonies are created on solar system bodies, the development of AI systems and assistants is likely to extend to these locations. There are several examples of where AI is being used to make discoveries on space missions (Chien and Wagstaff 2017) where it is proposed that autonomy will be a key technology for the future exploration of our solar system and conditions are not permissible for human habitation and where robotic spacecraft may be even out of communication with their human mission controllers (Chien and Wagstaff 2017; Campa et al. 2019; Braddock et al. 2019b).

5.6 Summary

In this chapter, I aimed to briefly summarise the considerable medical challenges in supporting prolonged space exploration ultimately with the vision of colonisation of a body in the solar system, taking Mars as a candidate planet. Physical challenges of transporting human beings to Mars, let alone build a colony fit to support habitation and procreation to enable the establishment and maintenance of a self-sustaining colony, are out of the scope of this chapter though are equally, if not more challenging to achieve. Let as assume technology will support both travel to and existence on Mars. Humankind is presented with several options for settlement of our species distant from Earth. The near-term option accepts that physical human beings will settle on Mars and simply adapt to a very different lifestyle from that enjoyed on Earth, where severe physical and mental constraints on terrestrial quality of life will be the price paid for the privilege to set the New Frontier. This is likely to have minimal chance of success it will be decades away given that lack of our knowledge of the effects of reduced gravity on human existence, reproduction and growth and the effects of incident radiation in addition to psychological effects of isolation we can model with limited success on Earth. Our second option is to provide future colonists with enhancements which I will call a SpaceApp at either or both genetic and physical levels which will make life physically, though not necessarily psychologically bearable. Great progress has been made in understanding and developing the potential of gene therapy with products available for patients with a range of disease (Approved Cellular and Gene Therapy Products 2019; Gene Therapy Medicinal products 2019). Similarly, rapid scientific progress is being made with understanding and developing techniques for gene editing in model systems and in the first instance somatic cell gene editing in humans. In February 2019, it was announced that the first patient was treated with CTX001, an autologous CRISP/Cas9 gene-edited hematopoietic stem cell therapy for severe haemoglobinopathies. Two clinical trials with CTX001 are currently recruiting patients with either transfusion-dependent β-thalassemia or severe sickle cell disease (ClinicalTrials.gov 2019). As for any new modality to be tested, to be developed and ultimately to be commercialised in patients, it will be essential to establish both the efficacy and safety of this new treatment, likely before embarking on a broader range of indications with additional candidate gene editing agents. It should be remembered that gene editing today is to be trialled in patients for which there are few if any other options and that using gene editing to enhance human characteristics in otherwise healthy individuals is and will be a subject of both clinical and ethical debates (Baylis 2017; Conboy et al. 2018; Howard et al. 2018; Rath 2018). In Fig. 1, I summarise three potential routes to achieving gene-mediated human enhancement with some pros and cons of the modalities. In exploring advances for human enhancement, one should ask the question 'what would instigate elective enhancement in healthy individuals?'. For example, consider the molecular targets shown in Table 3.3 and let us select for illustrative purposes either DSUP or MTP agents shown to confer radio protection in vitro or in a murine model exposed to ionising radiation. What justification would there be to attempt to engineer humans to become radio-tolerant if not radio-resistant? Ethically and legally how could this be tested in a clinical environment where the data should be used to drive greater understanding of both the technology and target efficacy and safety? The question is even more germane in the context of germline gene editing given the widespread critique of the gene-edited Chinese twins reported earlier in 2019 (Cyranowski 2019) where the long-term effects, particularly on the immune system, are completely unknown and irreversible. Indeed, perhaps the only justification, at least in the foreseeable future, is that the existential threat to the survival of the human species is so great, and it is an essentially a no-lose outcome. Thankfully, humankind is not in that position in 2019.

Gene-mediated enhancement is one potential SpaceApp, and there are many others. These include the potential to replace organs which may deteriorate in space by, for example, 3D bioprinting and the administration of medicines which may have some ability to prolong human lifespan although mitigation against the effects of living and working in space or in a lunar or Martian colony will still be required. Recent advances in the development of exoskeletal technology and exosuits, initially for patients with paraplegia, may have great utility in space and enable, for example, construction of habitats where assisted mechanical strength is required. Finally, our last SpaceApp refers to the equally advancing field of brain–computer interfaces, again being developed on Earth for patients unable to communicate as a consequence of a medical condition or traumatic event. Lastly, I pose the question do we really need fully physical human beings for either space exploration, or even space colonisation and, given enormous strides in the development of digital technologies, is the future of the human race a hybrid human avatar, a question posed in some depth over the last two years (Tegmark 2017; Rees 2018; Kaku 2018)? To complete with one final option, we should consider whether space exploration, if not colonisation, should be fully automated (Campa et al. 2019) given the immediacy of the potential prizes of space mining and the scientific and medical challenges that have been presented in this chapter.

The next stages of the technological revolution are happening and unstoppable in many disciplines, as was the Industrial Revolution and other paradigm shifts in the advancement of our species throughout the ages. Innovation and adoption of new technology, especially that likely to cause a step change for humanity, require concerted collaboration between optimists and pessimists with critical reality checking, objective communication and an understanding of risk and acceptance that the fate of humankind is not yet predetermined. However, humankind may be at a crossroads where we have ethical challenges to face, difficult questions to ask and decisions to make on our future that have never been made before. Human beings should be bold and work together across our disciplines with dialogue and open collaboration, ensure regulation is in place to understand and plan mitigation strategies and make the very best of a truly life-changing opportunity for our species on Earth and perhaps, in the not too distant future, elsewhere in the solar system.

References

Adli, M. (2018). The CRISPR tool kit for genome editing and beyond. *Nature Communications, 9*, Article number: 1911.

Aflalo, T., Kellis, S., Klaes, C., et al. (2015). Decoding motor imagery from a posterior parietal cortex of a tertrplegic human. *Science, 348*, 906–910.

Agrawal, A., Dube, A. N., Kansara, D., Shah, S., & Sheth, S. (2018). Exoskeleton: The friend of mankind in context of rehabilitation and enhancement. *Indian Journal of Science and Technology.*, *9*(S1), 1–8.

Alcibiade, A., Del Mastro, A., Schlacht, I. L., et al. (2018). Stress and human factors from Antarctica to Mars. In *International Conference on Applied Human Factors and Ergonomics* (pp. 183–194). Springer.

Approved Cellular and Gene Therapy Products. (2019). United States Food and Drug Administration (2019). Retrieved on December 12, 2019 from https://www.fda.gov/vaccines-blood-biologics/cellular-gene-therapy-products/approved-cellular-and-gene-therapy-products.

Austin, A., & Buchanan, A. (2018). Better than human: The promise and perils of biomedical enhancement. *Medical Law Review, 26*, 357–361.

Awad, L. N., Bae, J., O'Donnell, K., et al. (2017). A soft robotic exosuit improves walking in patients after stroke. *Science Translational Medicine, 9*, eaai9084.

Bartone, P. T., Roland, R. R., Bartone, J. V., et al. (2019). Human adaptability for deep space missions: An exploratory study. *Journal of Human Performance in Extreme Environments, 15*, Article 5.

Basner, M., Dinges, D. F., Mollicone, D., et al. (2013). Mars 520-d mission simulation reveals protracted crew hypokinesis and alterations of sleep duration and timing. *Proceedings of the National Academy of Sciences USA, 110*, 2365–2640.

Baylis, F. (2017). The potential harms of human gene editing using CRISPR-Cas9. *Clinical Chemistry, 64*, 489–491.

Binsted, K., Kobrick, R. L., Griofa, M. O., Bishop, S., & Lapierre, J. (2010). Human factors research as part of a Mars exploration analogue mission of Devon island. *Planetary and Space Science, 58*, 994–1006.

Blagosklonny, M. V. (2006). Aging and immortality: Quasi-programmed senescence and its pharmacologic inhibition. *Cell Cycle, 5*, 2087–2102.

Blagosklonny, M. V. (2007). An anti-aging drug today: From senescence-promoting genes to anti-aging pill. *Drug Discovery Today, 12*, 218–224.

Blagosklonny, M. H. (2019). Rapamycin for longevity: Opinion article. *Aging, 11*, 8048–8067.

Blue, R. S., Chancellor, J. C., Antonsen, E. L., et al. (2019). Limitations in predicting radiation-induced pharmaceutical instability during long-duration spaceflight. *Npj Microgravity, 5,* 15.

Boyden, L. M., Mao, J., Belsky, J., et al. (2002). High bone density due to a mutation in the LDL-receptor-related protein 5. *New England Journal of Medicine, 346,* 1513–1521.

Braddock, M. (2017). Ergonomic challenges for astronauts during space travel and the need for space medicine. *Journal of Ergonomics, 7,* 1–10.

Braddock, M. (2018). Exercise and ergonomics on the International Space Station and Orion spacecraft. *Journal of Ergonomics Research, 1,* 2. https://doi.org/10.4172/JEOR.1000104.

Braddock, M. (2019). From target identification to drug development in space: Using the microgravity assist. *Current Drug Discovery Technologies, 16,* 1–11.

Braddock, M., Wilhelm, C.P., Romain, A., Bale, L., & Szocik, K. (2019a). Application of socio-technical systems models to Martian colonisation and society build. *Theoretical Issues in Ergonomics Science.* https://doi.org/10.1080/1463922X.2019.1658242.

Braddock, M., Campa, R., & Szocik, K. (2019b). Ergonomic constraints for astronauts. Challenges and opportunities today and for the future. In *Proceedings of the International Conference on Ergonomics and Human Factors 2019* (1st ed.), Stratford Upon Avon, 29 April–1 May 2019.

Brown, M., Tsagarakis, N., & Caldwell, D. G. (2003). Exoskeletons for human force augmentation. *Industrial Robot: An International Journal, 30,* 592–602.

Burgers, T. A., Vivanco, J. F., Zahatnansky, J., et al. (2016). Mice with a heterozygous *Lrp6* deletion have impaired fracture healing. *Bone Research, 4,* 16025.

Byrne, R. T., Chen, S. H., Wood, E. A., Cabot, E. L., & Cox, M. M. (2014). *Escherichia coli* genes and pathways involved in surviving extreme exposure to ionizing radiation. *Journal of Bacteriology, 196,* 3534–3545.

Campa, R., Szocik, K., & Braddock, M. (2019). Why space colonisation will be fully automated. *Technological Forecasting and Social Change.* https://doi.org/10.1016/j.techfore.2019.03.021.

Ceballos, G., Ehrlich, P. R., Barnosky, A. D., et al. (2015). Accelerated modern human-induced species losses: Entering the sixth mass extinction. *Science Advances, 1*(5), e1400253.

Chaudhary, U., Xia, B., Silvoni, S., Cohen, L. G., & Birmbauer, N. (2017). Brain-computer interface-based communication in the completelet locked-in state. *PLOS Biology, 15,* e1002593.

Chancellor, J. C., Scott, G. B. J., & Sutton, J. P. (2014). Space radiation: The number one risk to astronaut health beyond low Earth orbit. *Life, 4,* 491–510.

Chancellor, J. C., Blue, R. S., Cangel, K. A., Aunon-Chancellor, S. M., Rubins, K. H., Katzgraber, H. G., & Kennedy, A. R. (2018). Limitation in predicting the space radiation health risk for exploration astronauts. *npj Microgravity, 4,* 8. https://doi.org/10.1038/s41526-018-0043-2.

Chien, S., & Wagstaff, K. L. (2017). Robotic space exploration agents. *Science Robotics, 2,* eaan4831.

Clement, G. (2017). International roadmap for artificial gravity research. *npj Microgravity 3,* 29. https://doi.org/10.1038/s41526-017-0034-8.

ClinicalTrials.gov. Retrieved on December 13, 2019 from https://clinicaltrials.gov/ct2/results?cond=&term=CTX001&cntry=&state=&city=&dist=.

Collinger, J. L., Wodlinger, B., Downey, J. E., et al. (2013). High-performance neuroprosthetic control by an individual with tetraplegia. *Lancet, 381,* 557–564.

Conboy, I., Murthy, N., Etienne, J., & Robinson Z. (2018). Making gene editing a therapeutic reality. *F1000 Research, 21*(7), pii: F1000 Faculty Rev-1970. https://doi.org/10.12688/f1000research.16106.1. eCollection 2018.

Cortes, C., Unzueta, L., de Los Reyes-Guzman, A., Ruiz, O. E., & Florez, J. (2016). Optical enhancement of exoskeleton-based estimation of glenohumeral angles. *Applied Bionics and Biomechanics 2016,* Article ID 5058171.

Cortese, F., Klokov, D., Osipov, A., et al. (2018). Vive la radioresistance!: Converging research in radiobiology and biogerontology to enhance human radioresistance for deep space exploration and colonisation. *Oncotarget, 9.* https://doi.org/10.18632/oncotarget.24461.

Cucinotta, F. A. (2014). Space radiation risks for astronauts on multiple international space station missions. *PLoS ONE, 9,* 396099.

Cyranoski, D. (2019). The CRISPR-baby scandal: What's next for human gene editing. *Nature, 566,* 440–442.
Czeisler, C. A., & Barger, L.K. (2017). Sleep-awake actigraphy and light exposure during spaceflight-long. Retrieved on October 16, 2019 from https://www.nasa.gov/mission_pages/station/research/experiments/294.html.
DARPA. Department of Defense Fiscal Year (FY). (2010). Budget Estimates May 2009 Research, Development, Test and Evaluation, Defence-Wide Volume 1—Defense Advanced Research Projects. Agency Exhibit R-2a, PB 2010 Defense Advanced Research Projects Agency RDT&E Project Justification. https://www.darpa.mil/Docs/2010PBDARPAMay2009.pdf.
De Lecuona, I., Casado, M., Marfany, G., Baroni, M. L., & Escarrabill, M. (2017). Gene editing in humans: Towards a global and inclusive debate for responsible research. *Yale J Biol Med., 90,* 673–681.
Denby, B., Schultz, T., Honda, K., Hueber, T., Gilbert, J. M., & Brumberg, J.S. (2009a). Silent Speech Interfaces. *Speech Communication, 52,* 2010.
Denby, B., Schultz, T., Honda, K., et al. (2009b). Silent speech interfaces. *Speech Communication, 52,* 270–287.
De Quetteville, H. (2019). This young man died in April. So how did our writer have a conversation with him last month? *The Telegraph.* Retrieved on February 17, 2019, https://www.telegraph.co.uk/technology/2019/01/18/will-digital-soul/.
Ehninger, D., Neff, F., & Xie, K. (2014). Longevity, aging and rapamycin. *Cellular and Molecular Life Sciences, 2014,* 4325–4346.
Ergin, V., Hariry, R. E., & Karasu, C. (2013). Carbonyl stress in aging process: Role of vitamins and phytochemicals as redox regulators. *Aging and Disease, 4,* 276–294.
Ferris, D. P. (2009). The exoskeletons are here. *Journal of Neuro Engineering and Rehabilitation, 6,* 17.
Garcia, M. (2018). *International space station facts and figures.* Retrieved on October 16, 2019 from https://nasa.gov/feature/facts-and-figures.
Gene Therapy Clinical Trials Databases. (2019). Retrieved on December 12, 2019 from https://www.genetherapynet.com/clinical-trials.html.
Gene Therapy Medicinal Products. (2019). Federal Institute for Vaccines and Biomedicines, Paul Ehrlich Institut. Retrieved on December 12, 2019 from https://www.pei.de/EN/medicinal-products/atmp/gene-therapy-medicinal-products/gene-therapy-node.html.
Ghidini, T. (2018). Regenerative medicine and 3D bioprinting for human space exploration and colonisation. *Journal of Thoracic Disease, 10*(Suppl 20), S2363–S2375.
Gonfalonieri, A. (2018). *A beginner's guide to brain-computer interface and convolutional neural networks.* Retrieved on December 9, 2019 from https://towardsdatascience.com/a-beginners-guideto-brain-computer-interface-and-convolutional-neural-networks-9f35bd4af948.
Gudkov, A. V., & Komarova, E. A. (2003). The role of p53 in determining sensitivity to radiotherapy. *Nature Reviews, 3,* 117–129.
Gupta, P., Gayen, M., Smith, J. T., et al. (2016). MDP: A *Deinococcus* Mn^{2+}—Decapeptide complex protects mice from ionizing radiation. *PLoS ONE, 11,* e0160575.
Hanaoka, M., Droma, Y., Basnyat, B., et al. (2012). Genetic variants in EPAS1 contribute to adaptation to high-altitude hypoxia in Sherpas. *PLoS ONE, 7*(12), e50566.
Harrison, D., Strong, R., Sharp, Z., et al. (2009). Rapamycin fed late in life extends lifespan in genetically heterogeneous mice. *Nature, 460,* 392–395.
Hashimoto, T., Horikawa, D. D., Saito, Y., et al. (2016). Extremotolerant tardigrade genome and improved radiotolerance of human cultured cells by tardigrade-unique protein. *Nature Communications, 7,* 12808.
Hashimoto T., & Kunieda T. (2017). DNA protection protein, a novel mechanism of radiation tolerance: Lessons from Tardigrades. *Life, 7,* 26. https://doi.org/10.3390/life7020026.
Hill, D., Holloway, C. S., Ramirez, D. Z. M., et al. (2017). What are user perspectives of exoskeleton technology? A literature review. *International Journal of Technology Assessment in Health Care, 33,* 160–167.

Hochberg, L. R., Bacher, D., Jarosiewicz, B., et al. (2012). Reach and grasp control by people with tetraplegia using a neutrally controlled robotic arm. *Nature, 485,* 372–376.

Howard, H. C., van El, C. G., Forzano, F., et al. (2018). One small edit for humans, one giant edit for humankind? Points and questions to consider for a responsible way forward for gene editing in humans. *European Journal of Human Genetics, 26,* 1–11.

Jackson, B. (2018). *ESA lays groundwork for 3D bioprinting bone in space.* https://3dprintingindustry.com/news/esa-lays-groundwork-for-3d-bioprinting-bone-in-space-158410/. Retrieved on December 1, 2019.

Illardo, M., & Nielsen, R. (2018). Human adaptation to extreme environmental conditions. *Current Opinion in Genetics and Development, 53,* 77–82.

Jebari, K. (2013). Brain machine interface and human enhancement—An ethical review. *Neuroethics, 6,* 617–625.

Kaku, M. (2018). *The future of humanity. Terraforming Mars, interstellar travel, immortality and our destiny beyond Earth.* Penguin books.

Kanas, N. (2010). Expedition to Mars: Psychological, interpersonal and psychiatric issues. *Journal of Cosmology, 12,* 3741–3747.

Kanas, N. (2011). From Earth's orbit to the outer planets and beyond. Psychological issues in space. *Acta Astronautica, 68,* 576–581.

Kasiotis, K. M., Pratsinis, H., Kletsas, D., & Haroutounian, S. A. (2013). Resveratrol and related stilbenes: Their anti-aging and anti-angiogenic properties. *Food and Chemical Toxicology, 61,* 112–120.

Konopka, A. R., Laurin, J. L., Schoenberg, H. M., et al. (2019). Metformin inhibits mitochondrial adaptations to aerobic exercise training in older adults. *Aging Cell, 18,* e12880.

Lajeunesse, V., Vincent, C., Routhier, F., Careau, E., & Michaud, F. (2016). Exoskeletons' design and usefulness evidence according to a systematic review of lower limb exoskeletons used for functional mobility by people with spinal cord injury. *Disability and Rehabilitation: Assistive Technology, 11,* 535–547.

Lee, C.-L., Blum, J. M., & Kirsch, D. G. (2013). Prole of p53 in regulating tissue response to radiation by mechanisms independent of apoptosis. *Trends in Cancer Research, 2,* 412–421.

Li, Q., Xo, X., Zhong, W., et al. (2015). IL-17 induces radiation resistance of B lymphoma cells by suppressing p53 expression and thereby inhibiting irradiation-triggered apoptosis. *Cellular and Molecular Immunology, 12,* 366–372.

Lieben, L. (2016). Regenerative medicine: The future of 3D printing of human tissues is taking shape. *Nature Reviews Rheumatology, 12.* https://doi.org/10.1038/nrrheum.2016.29.

Low, K. H., Liu, X., Goh, C. H., & Yu, H. (2016). Locomotive control of a wearable lower exoskeleton for walking enhancement. *Journal of Vibration and Control, 12,* 1311–1336.

MacInnis, M. J., & Rupert, J. L. (2011). ome on the Range: Altitude adaptation, positive selection, and Himalayan genomics. *High Altitude Medical Biology,* 133–139.

Marilena, G. (2011). Can ageing be slowed? Hormetic and redox perspectives. *Journal of Medicine and Life, 4,* 346–351.

Marshall-Goebel, K., Laurie, S. S., Alferova, I. V., et al. (2019). Assessment of jugular venous flow stasis and thrombosis during spaceflight. *Journal of the American Medical Association Open, 2,* e1915011.

Martin, A., Saathoff, M., Kuhn, F., et al. (2010). A functional ABCC11 allele is essential in the biochemical formation of human axillary odour. *Journal of Investigative Dermatology, 130,* 529–540.

Metformin in Longevity Study (MILES). Retrieved on December 9, 2019 from https://clinicaltrials.gov/ct2/show/NCT02432287.

Micera, S. (2017). Restoring multi-joint motor control. *Nature Biomedical Engineering, 1,* 0073.

Momaya, A., Fawal, M., & Estes, R. (2015). Performance-enhancing substances in sports: A review of the literature. *Sports Medicine (Auckland, N. Z.), 45,* 517–531.

Moses, D. A., Leonard, M. K., Makin, J. G., & Chang, E. F. (2019). Real-time decoding of question-and-answer speech dialogue using human cortical activity. *Nature Communications, 10,* 3096.

Moskalev, A., Chernyagina, E., de Magalhaes, J. P., et al. (2015). Geroprotectors.org: A new, structured and curated database of current therapeutic interventions in aging and age-related disease. *Aging, 7,* 616–628.

Nagaraja, M. P., & Risin, D. (2013). The current state of bone loss research: Data from spaceflight and microgravity simulators. *Journal of Cellular Biochemistry, 114,* 1001–1008.

Nelson, M. (2018). *Pushing our limits: Insights from Biosphere 2.* University of Arizona Press https://doi.org/10.2307/j.ctt1zxsmg9

Norwitz, N. G., Moto, A. S., Misra, M., & Ackerman, K. E. (2019). LRP5, bone density, and mechanical stress: A case report and literature review. *Frontiers in Endocrinology, 19,* Article 184.

Novelle, M. G., Ali, A., Diéguez, C., Bernier, B., & de Cabo, R. (2016). Metformin: A hopeful promise in aging research. *Cold Spring Harbour Perspectives on Medicine, 6,* a025932.

Ohman, C., & Floridi, L. (2017). The potential economy of death in the age of information: A critical approach to the digital afterlife industry. *Minds and Machines, 27,* 639–662.

Ohman, C., & Floridi, L. (2018). An ethical framework for the digital afterlife industry. *Nature Human Behaviour, 2,* 318–320.

Pagel, J. I., & Chouker, A. (2016). Effects of isolation and confinement on humans implications for manned space explorations. *Journal of Applied Physiology, 120,* 1449–1457.

Peake, J. M., Kerr, G., & Sullivan, J. P. (2018). A critical review of consumer wearables, mobile applications and equipment for providing biofeedback, monitoring stress and sleep in physically active populations. *Frontiers in Physiology, 9,* Article 743.

Peng, Y., Yang, Z., Zhang, H., et al. (2011). Genetic variations in Tibetan populations and high-altitude adaptation at the Himalayas. *Molecular Biology and Evolution, 28,* 1075–1081.

Pimm S. L., Jenkins C. N., Abell, R., et al. (2014). The biodiversity of species and their rates of extinction, distribution and protection. *Science, 344.* https://doi.org/10.1126/science.1246752.

Prokop-Prigge, K. A., Mansfield, C. J., Parker, M. R., et al. (2015). Ethnic/racial and genetic influences on cerumen odor profiles. *Journal of Chemical Ecology, 41,* 67–74.

Rai, B., & Kaur, J. (2012). Human factor studies on a Mars analogue during crew 100b international lunar exploration working group EuroMoonMars crew: Proposed new approaches for future human space and interplanetary missions. *North American Journal of Medical Sciences, 4,* 548–557.

Rask, J., Vercoutere, W., Navarro, B. J., & Krause, A. (2011). *Space fairing: The radiation challenge.* Retrieved on October 16, 2019 from https://www.nasa.gov/pdf/284273main_Radiation_HS_Mod1.pdf.

Rath, J. (2018). Safety and security risks of CRISPR/Cas9. In D. Schroeder, J. Cook, F. Hirsch, S. Fenet, & V. Muthuswamy (Eds.), *Ethics dumping.* Springer Briefs in Research and Innovation Governance. Cham: Springer.

Rea, R., Beck, C., Rovekamp, R., Diftler, M., & Neuhaus, P. (2013). X1: A robotic exoskeleton for in-space countermeasures and dynamometry. *American Institute of Aeronautics and Astronautics.* https://doi.org/10.2514/6.2013-5510.

Rees, M. (2018). *On the future prospects for humanity.* Princeton University press.

Rodriguez, S., Steer, C. D., Farrow, A., Goldring, J., & Day, I. N. M. (2013). Dependence of deodorant usage on ACCC11 genotype: Scope for personalised genetics in personal hygiene. *Journal of Investigative Dermatology, 7,* 1760–1767.

Ryder, P., & Braddock, M. (2020). Harnessing the space environment for the discovery and development of new medicines. In *Handbook of space pharmaceuticals.* Wiley (in press).

Salas, M. A., Bashford, L., Kellis, S., et al. (2018). Proprioceptive and cutaneous sensations in human elicited by intracortical micro-stimulation. *eLife, 7,* e32904.

Sandal, G. M. (2001). Psychosocial issues in space: Future challenges. *Gravitational and Space Research, 14,* 47–54.

Sandal, G. M., van deVijver, F. J. R., & Smith, N. (2018). Psychological hibernation in Antarctica. *Frontiers in Psychology, 9,* 2235. https://doi.org/10.3389/fpsyg.2018.02235.

Schlicker, C., Boanca, G., Lakshminarasimhan, M., & Steegborn, C. (2011). Structure-based development of novel sirtuin inhibitors. *Aging (Albany NY), 3,* 852–872.

Siamwala, J. H., Rajendran, S., & Chatterjee, S. (2015). Strategies of manipulating BMP signalling in microgravity to prevent bone loss. *Vitamins & Hormones, 99,* 249–272.

Sipes, W., & Fielder, E. (2007). *Current psychological report for US astronauts on the International Space Station.* Retrieved on October 16, 2019 from https://ntrs.nasa.gov/archive/nasa/casi.ntrs.nasa.gov/20070035862.pdf.

Scoles, S. (2017). NASA likely to break radiation rules to go to Mars. Retrieved on October 16, 2019 from https://www.pbs.org/wgbh/nona/next/space/nasa-mars-radiation-rule/.

Sibonga, J. D., Spector, E. R., Johnston, S. L., & Tarver, W. J. (2015). Evaluating bone loss on ISS astronauts. *Aerospace Medicine & Human Performance, 86*, A38–A44.

Simonson, T. S., Yang, Y., Huff, C. D., et al. (2010). Genetic evidence for high-altitude adaptation in Tibet. *Science, 329*(5987), *72–75*.

Simonson, T. S., McClain, D. A., Jorde, L. B., & Prchal, J. T. (2012). Genetic determinants of Tibetan high-altitude adaptation. *Human Genetics, 131*(4), 527–533.

[illegible] (2011). Can astronauts survive radiation on prolonged space missions? *Bulletin of the Canadian Radiation Protection Association, 31*, 20–26.

Song, J., Jiang, G., Zhang, J., et al. (2019). Metformin prolongs lifespan through remodelling the energy distribution strategy in silkworm, Bombyx mori. *Aging, 11,* 240–248.

Stingl, J. C., Welker, S., Hartmann, G., Damann, V., & Gerzer, R. (2015). Where failure is not an option—personalised medicine in astronauts. *PLoS ONE, 10,* e0140764.

Svoboda, M. (2019). *12 major climate change reports for 2019.* Retrieved on December 13, 2019 from https://www.yaleclimateconnections.org/2019/10/12-major-climate-change-reports-from2019/.

Tegmark, M. (2017). *Life 3.0. Being human in the age of artificial intelligence.* Penguin Press. ISBN 9780141981802.

Three dimensional bioprinting in space. https://bioprinting.ru/en/press-center/publications/nasa-organaut/. Retrieved on August 2, 2019.

Vaisnav, M., Xing, C., Ku, H.-C., et al. (2014). Genome-wide association analysis of radiation resistance in *Drosophila melanogaster*. *PLoS ONE, 9,* e104858.

van Patot, M. C., & Gassmann, M. (2011). Hypoxia: Adapting to high altitude by mutating EPAS-1, the gene encoding HIF-2α. *High Altitude Medical Biology, 12,* 157–167.

Weisberger, M. (2019) Lifelike 'Sophia' robot granted citizenship to Saudi Arabia. *Live Science.* Retrieved on February 15, 2019 from https://www.livescience.com/60815-saudi-arabia-citizen-robot.html.

White, N. D., & Noeun, J. (2017). Performance-enhancing drug use in adolescence. *American Journal of Lifestyle Medicine, 11,* 122–124.

Williams, T. (2019). *Risk of adverse cognitive or behavioral conditions and psychiatric disorders.* NASA Human Research Roadmap. Retrieved on January 1, 2020 from https://humanresearchroadmap.nasa.gov/Risks/risk.aspx?i=99.

Xu, G.-Y., Qiu, Y., Mao, H.-J. (2014a). Common polymorphism in the *LRP5* gene may increase the risk of bone fracture and osteoporosis. BioMed Research International Article ID 290531.

Xu, H., Duan, J., Ning, D., et al. (2014b). Role of Wnt signalling in fracture healing. *BMB Reports, 47,* 666–672.

Yagin, N. (1890). *Apparatus for Facilitating Walking.* U.S. Patent 440,684 filed February 11, 1890 and issued November 18, 1890.

Zhang, J., Fiers, P., Witte, K. A., et al. (2017). Human-in-the-loop optimization of exoskeleton assistance during walking. *Science, 356,* 1280–1284.

Zhou, D., Udpa, N., Ronen, R., et al. (2013). Whole-genome sequencing uncovers the genetic basis of chronic mountain sickness in Andean highlanders. *American Journal of Human Genetics, 93,* 452–462.

Zhou, G., Jiang, H., Yin, Z., Liu, Y., Zhang, Q., et al. (2018). In vitro regeneration of patient-specific ear shaped cartilage and its first clinical application for auricular reconstruction. *EBioMedicine, 28,* 287–302.

Chapter 6
Human Enhancements: New Eyes and Ears for Mars

Mark Shelhamer

Abstract What are the realistic, feasible, and ethical body "modifications" that could be made to astronauts in order to aid their performance and survival on other planets? Here we consider some possible human enhancements that might help early Martian settlers, considering aspects that are within the realm of near-term feasibility, especially if planetary colonization is in play and the first settlers must perform extraordinarily demanding work in an extraordinarily demanding environment.

Artificial enhancements to human capabilities might help initial settlers on Mars as they adjust to the new and unprecedented living conditions, while undertaking risky and demanding work. It is worthwhile to consider first the major environmental factors inherent to spaceflight that have an adverse effect on human health or performance. This greatly depends on what we mean by *spaceflight*, since the criticalities and priorities of these risks are different for time spent in deep space or planetary orbit as opposed to on a planetary or lunar surface, for example. Nevertheless, we can consider the following to be significant risks across a wide range of spaceflights for the foreseeable future (Mindock et al. 2017):

- Altered gravity level. This can lead directly to a range of physiological problems from the equivalent of terrestrial disuse: cardiovascular deconditioning, muscle atrophy, bone demineralization. Current evidence suggests that these can be mitigated to a great extent by exercise (Hackney et al. 2015). Less well-understood are those problems related to the balance (vestibular) system of the inner ear, and those resulting from chronic head-ward fluid shifts.
- Radiation. The dominant forms of radiation in deep space are galactic cosmic rays (GCR: high-energy protons and atomic nuclei moving a relativistic speeds) and solar particle events (mostly fast protons from the sun). Radiation can have long-term effects such as an increase in the lifetime risk of cancer, as well as short-term effects such as vascular degeneration and even immediate effects on cognitive function (Chancellor et al. 2014).

M. Shelhamer (✉)
Johns Hopkins University School of Medicine, Baltimore, USA
e-mail: mshelhamer@jhu.edu

K. Szocik (ed.), *Human Enhancements for Space Missions*, Space and Society,
https://doi.org/10.1007/978-3-030-42036-9_6

- Isolation and confinement. Small groups of people living together in small spaces for long periods of time, apart from the normal comforts and familiarity of home (Earth), experience a variety of stress effects (Palinkas 2007). Settlers might resolve some of the problems, at least over time, through the addition of people through reproduction and new arrivals, and the creation of larger habitats. Nevertheless, the first voyagers will be faced with these problems and their peak performance will depend on dealing with the psychological consequences, especially as they apply to the teamwork and inter-personal relationships needed to establish viable colonies.
- Remoteness. The major impact of remoteness from the safe haven of Earth (or even Earth orbit) is the overwhelming need for autonomy on the part of the crew. Communication delays alone, due to the finite speed of light (up to 20 min one way at Mars), will make it mandatory that settlers and explorers have the tools to deal with emergency situations on their own. Again, capabilities in this area will improve over time as the local infrastructure and skill sets expand.
- Toxic environment, dust. Environmental toxicity can arise in a number of ways, including the normal consequences of groups of people living in close quarters. Depending on the capabilities of the life-support system, various other toxic situations might arise, such as mutations of microorganisms and elevated CO_2 levels. The surface soil of a planetary body may present additional hazards, as it may be abrasive or cause allergic and inflammatory reactions. (Khan-Mayberry et al. 2011; Liu and Taylor 2011)

Clearly, given this range of hazards and risks, it would be helpful to provide settlers with all the advantages possible to survive and thrive in a new environment. Here, we consider those body modifications and enhancements that are within the realm of current or foreseeable technology: practicality and feasibility limit the scope of this discussion. A gray area exists between things that might be considered true modifications (such as corneal implants) and things that might be considered artificial augmentations (such as implantable pumps). Here we emphasize the former, with minor consideration of the latter.

6.1 Vestibular System and Balance

The vestibular system consists of sensors that transduce angular and linear motions of the head. The latter (the otolith organs) also detect the magnitude and direction of gravity, which is equivalent to a linear acceleration force. Through this sensing process, many common body functions are properly coordinated for the 1g environment of Earth. These functions include maintenance of upright posture (standing) and walking (Goldberg et al. 2012).

In the usual gravity level of Earth, head tilt is sensed by the otolith organs of the vestibular system. The otolith organs are linear accelerometers, which detect linear motions of the head, and also the magnitude and direction of gravity (a linear acceleration force). In the weightlessness of space, there is no net gravitoinertial force vector ("zero gravity"), and the otolith organs can therefore not transduce head tilts. In other gravity levels, the amount of head tilt transduced by the otolith organs may not match the amount of actual tilt. In either case, spatial disorientation and motion sickness may result, with attendant deficits in the ability to perform tasks such as piloting of a spacecraft during landing. Even walking might be impaired, as it involves small periodic tilts of the head with each step. It is this lack of veridical otolith information about head motions that is thought to be a major contributor to motion sickness upon initial entry into space, and again on initial return to the gravity of Earth. The consequences can be significant (Reschke et al. 1994). What might be done to improve this situation?

Fortunately, the development of vestibular prostheses is an active area of current research (Golub et al. 2014). Analogous to a cochlear implant to recover some hearing in the profoundly deaf, a vestibular prosthesis is implanted in the head, senses head motion, and relays that information to the appropriate nerves. Most work to date has been directed at replacing missing sensory information on head *rotations* from the semicircular canals of the vestibular system. However, the same technology could also replace missing or altered otolith function. In this manner, a head tilt in a planetary gravity setting—which would normally be transduced and interpreted as being of more or less than the actual amount—could be made to appear (as far as nerve firing is concerned) as being of the correct value. That is, the prosthesis could alter the information sent to the nerves to indicate the actual level of head tilt that the brain is accustomed to, as if it had occurred in a 1g setting as on Earth. This otolith-derived information would then coincide with simultaneous information from the other senses as to the degree of head tilt, and all would be fine (no inter-sensory conflict). The prosthesis could be programmed to gradually change its activity and eventually taper off its contribution to nerve firing, allowing a slow (and less traumatic) adaptation of the nervous system to a new gravity environment.

The situation in weightlessness is more complicated, since there is no gravitational vector to begin with, and therefore no otolith information on head tilt to be modified. In fact, accelerometers in the prosthesis will be as useless for transducing tilt as will be the otolith organs themselves, and for the same reason. Therefore, the prosthesis would have to derive tilt information from a different set of sensors—those which transduce rotations—and track these rotational motions very accurately in three dimensions to maintain a sense of orientation (tilt). Furthermore, "tilt" must be defined as relative to some local up/down vector, since gravity usually provides this information ("tilt" on Earth typically means a change in orientation *with respect to* the gravity vector). A useful up/down surrogate would be that provided by immediate visual cues: the usual ceiling is "up," for example. Needless to say, there is no need for such a prosthetic capability in Earth-bound patients, so this would have to be developed specifically for spaceflight use.

6.2 Visual System

Vision is the sense that dominates in normal life, and methods being developed for its maintenance and recovery in pathology could potentially be altered for augmentation and enhancement. The case is less clear here than it is for the vestibular system, where a clear deficit can be identified as a result of the physics of altered gravity levels. The prevailing evidence indicates that visual function itself is not altered in any dramatic way when in space. (This does not include the important case of changes in visual acuity that occur presumably as a result of an increase in intracranial pressure due to fluid shifts in weightlessness (Lee et al. 2018). This is considered to be a consequence of fluid shifting, not a direct change in visual function due to space itself.) Nevertheless, given the importance of the visual sense, especially in demanding situations such as will be faced by space settlers, it is worth considering how this sense might be protected or augmented.

A rather mundane but nevertheless significant problem in weightless spaceflight is the presence of debris that does not settle to the floor. This can be a problem also on planets with low gravity levels, where debris does not settle as quickly as on Earth. The problem is that this debris can be easily inhaled and get into the eyes (not to mention equipment and sensitive interconnections of various kinds). This was a problem on the Apollo lunar missions, where lunar regolith was prevalent inside the lunar module after EVAs (Scheuring et al. 2008). In this case, a simple solution would be to implant a lens or corneal replacement membrane to protect the eye itself from debris exposure and abrasion.

A more ambitious enhancement to the visual system would involve a retinal implant. Like cochlear implants and vestibular prostheses, a retinal implant is embedded in the body itself, transduces the relevant sensory stimuli, and encodes it for transmission to the appropriate set of nerves for further processing by the central nervous system. Retinal implants are challenging due to the many individual photoreceptors in the retina, the large amount of information that is transmitted to the many fibers in the optic nerve, and the visual processing that takes place in the retina itself. Nevertheless, progress has been made and retinal implants are able to convey some minimal information on light and shape (Luo and Da Cruz 2014).

One wonders, then, if these implants can be used to augment and enhance vision, not just recover lost visual function. Any such enhancement would be many years away, and an implementation would have to be developed in which the implant enhanced existing function while not interfering with it: the sensors on an implant should not block the normal path of light to the retina, and the neural interface should not interfere with the neural firing that results from normal visual function. This is a considerable challenge, yet we can conjecture as to the types of visual enhancements that might be made available. The capability that comes most readily to mind is expansion of the range of visual sensitivity to a wider range of wavelengths. Astronomy has benefited greatly from the expansion from visual wavelengths (those visible to unaided humans) to infrared, ultraviolet, radio, microwave, and X-ray portions of the electromagnetic spectrum. Likewise, although sensors already exist to provide imaging in those portions of the spectrum not directly perceptible by the human visual system, the ability to actually *see* in the normal sense in these wavelength ranges could provide great benefits for safety and efficiency. Infrared vision, for example, would permit seeing objects by their heat signature, thereby aiding vision in the midst of dust storms (on planets like Mars) and darkness. Sensitivity in other wavelengths could provide indications of high radiation, including that from unpredicted solar flares or leaks around radiation equipment. Of course these new capabilities would have to be weighed against the possibility of overwhelming the nervous system's capability to sort through sensory streams and focus attention where needed in the face of an abundance of distractions.

6.3 Biome

A rising volume of research is confirming the recent realization of the importance of the gut microbiome (the fungi, bacteria, and other microorganisms in the intestines). It turns out that the microbiome is important not only for digestion and proper nutrition, as might be expected, but that is has very broad interactions with other body systems, including cognition (Gareau 2014; Shreiner et al. 2015). Disruptions to the microbiome can occur when diet is altered, medications are ingested, or in the presence of a variety of environmental stressors. All of these are present in extended spaceflight, and changes to the microbiome have been found in spaceflights with durations on the order of months (Voorhies and Lorenzi 2016). The implications of these changes are not fully understood, but they do raise concerns. This is exacerbated by the fact that the normal turnover of the biome as a result of environmental exposure and contact with different people, in normal life, will also not be present in space. The resulting lack of diversity and regular turnover could lead to alterations in many body functions. We might propose, as a speculative notion, that an implantable pump (akin to implantable insulin pumps) might provide a steady infusion of new microbiome components (like probiotics) directly into the gut in order to maintain a vigorous non-stagnant population. This would be a form of continuous automated fecal transplant; this is a clinically accepted method for improving microbiome function in some diseases (Bojanova and Bordenstein 2016).

6.4 CSF Shunt

As noted, one of the dominant physiological features of spaceflight is a shift of body fluids toward the head (Lee et al. 2018). This is thought to lead to observed changes in visual acuity, through cerebrospinal fluid (CSF) pressure on the back of the eye changing its shape and therefore optical properties. What is of greater concern is that this might be just the initial manifestation of accumulating neural damage due to slightly but chronically increased CSF pressure on the brain (Lawley et al. 2017). If this worst-case scenario is confirmed, and simpler mitigations cannot be found, then the situation may call for a more heroic measure: a cerebral shunt. Such a shunt would drain excess CSF from vesicles within the brain or from the space surrounding the brain, to another location where it can be more readily absorbed by the body (Drake et al. 2000). This is commonly used in terrestrial cases of increased intracranial pressure (e.g., hydrocephalus). Of course, this would be most needed in weightlessness where fluid shift is a major problem; to the extent that the shunt relies on gravity to provide the drainage, the shunt will be ineffective in the very environment in which it would be most needed. Some form of implantable pump would thus be needed in association with any shunt of this type.

6.5 BCI for HSI

Perhaps most speculative of all of the approaches considered here is one to address the need for close interaction and teaming between humans and machines: human–system interaction (HSI). A venture as complex and demanding as human spaceflight requires cooperative teaming between crew members and robotic assistants of various kinds. This will be even more the case for long-term missions for settlement and colonization. These assistants could be EVA partners, maintenance assistants, medical/surgical robots, or the general array of sensors and monitors that watch over the spacecraft and habitat. Especially in the case of humans working closely with robot partners in close quarters, trust and situational awareness will be needed on both sides (Booher 2003; Mishkin et al. 2007).

One approach to this general issue is the use of brain–computer interfaces (BCI). This is a very active line of research, and great progress has been made in many areas of BCI (Wolpaw and Wolpaw 2012). One might imagine an interface of noninvasive surface electrodes to detect brain-wave patterns (electroencephalogram: EEG), coupled with an embedded processor, which would permit the human user to interact with semi-automated partners through thought processes alone. This is a long way off and would require a great deal of technology development in order to have the desired level of reliability. It would, however, greatly aid in the ability of humans and robots to work together with a shared mental model of what overall task needs to be accomplished and how the individual component tasks must be dynamically parsed and shared among the members of a human–machine team.

6.6 What Is Not Covered Here

When one considers the possibility of space colonization and settlement in the far future, the prospects are daunting. The first settlers, unlike current space explorers, would expect to remain and live in the new environment (planetary or orbital) for the rest of their lives. Future generations could gradually adapt to the new environment, creating a new space-faring species of human. In the meantime, however, the initial groups will need to expend substantial effort in creating habitable spaces, with no assistance from preceding groups. A number of more extreme enhancements might be considered for these first settlers and their offspring, given the particularly demanding and unique nature of their predicament. These would not be undertaken lightly, with side effects that might be unforeseen and detrimental. Genetic modification (or use of genetic information for crew selection) is one such area (Powell and Buchanan 2011). Natural tolerance to deep-space radiation, for example, might be enhanced with gene editing or other methods. Radiation protection might be obtained more easily, if transiently, with ingestion of melanin (Cordero 2017). It might also be argued that—at least in weightlessness—the lack of lower limbs is a decided advantage, since they are not needed for locomotion, become excess mass and volume, and reduce g tolerance (Gibson 2006). More than this, however, is that the lower body is a large reservoir for body fluids, which then shift toward the head in the absence of gravity, causing a number of problems that are only now beginning to be understood. Genetic modification or breeding for short or absent lower limbs would be one apparent remedy.

On the other hand, there are apparatus and techniques that might be needed for only a short time during an initial adaptation phase, and which might be so benign and temporary that they would not qualify as enhancements in the normal sense. One example is an exoskeleton (Bogue 2009). These are skeleton-like structures that attach to the outside of the body to aid and augment natural motions which might be impaired through disease or challenged by substantial external loads (as in military special forces personnel who often must carry heavy loads of equipment into the field). Obviously such a device would be of great value to compensate for bone and muscle decrements after extended flight in weightlessness, and perhaps even more so if tasks are to be carried out on a planetary surface which exhibits more than the normal 1g of Earth. Another device that could be useful in the weightless phases of flight is lower-body negative pressure (LBNP). This entails drawing a slight vacuum about the lower body, from the waist down, in order to draw fluids that have shifted to the upper body back to the lower body (Goswami et al. 2008). This is a direct countermeasure to the undesired fluid shift that occurs in weightlessness, but at the moment the required apparatus is cumbersome and heavy. A portable version that allows normal movement could be helpful during long transits in deep space or in orbit.

Finally, there are augmentations to normal human function that would be especially useful but for which it is difficult to foresee any feasible implementation. One would be the enhancement of empathy—as a means of improving inter-personal relations in confined and isolated settings—perhaps through automated detection of the emotional states of others. For example, facial-recognition and voice stress-analysis software might function continuously and provide information on others' emotional state to a person through a small visual display or other modality. This could alert the wearer/user to subtle indications in others that care is needed to maintain a relationship on good terms.

6.7 Ethical Considerations

One must wonder: If it is necessary to go to these extremes to settle space, is it worth it (Gibson 2006)? Are we creating, or leading to the creation of, a space-based subspecies? If so, is this an ethical way to mitigate the risks to these pioneers? Or does it mean that we are overstepping a bound—not that humans should not undertake the venture, but perhaps that we should wait until advances in conventional technology such as propulsion more readily enable our exploratory impulses. We must also consider the extent to which these modifications would make it prohibitive for the pioneers to return to Earth, if they should so desire.

References

Bogue, R. (2009). Exoskeletons and robotic prosthetics: A review of recent developments. *Industrial Robot, 36,* 421–427.

Bojanova, D. P., & Bordenstein, S. R. (2016). Fecal transplants: What is being transferred? *PLoS Biology, 14,* e1002503.

Booher, H. R. (2003). *Handbook of human systems integration.* Hoboken: Wiley.

Chancellor, J. C., Scott, G. B., & Sutton, J. P. (2014). Space radiation: The number one risk to astronaut health beyond low earth orbit. *Life, 4,* 491–510.

Cordero, R. J. (2017). Melanin for space travel radioprotection. *Environmental Microbiology, 19,* 2529–2532.

Drake, J. M., Kestle, J. R., & Tuli, S. (2000). CSF shunts 50 years on—Past, present and future. *Child's Nervous System, 16,* 800–804.

Gareau, M. G. (2014). Microbiota-gut-brain axis and cognitive function. In *Microbial endocrinology: The microbiota-gut-brain axis in health and disease* (pp. 357–371). Berlin: Springer.

Gibson, T. M. (2006). The bioethics of enhancing human performance for spaceflight. *Journal of Medical Ethics, 32,* 129–132.

Goldberg, J. M., Wilson, V. J., Angelaki, D. E., Cullen, K. E., Buttner-Ennever, J., & Fukushima, K. (2012). *The vestibular system: A sixth sense.* Oxford: Oxford University Press.

Golub, J. S., Ling, L., Nie, K., Nowack, A., Shepherd, S. J., Bierer, S. M., et al. (2014). Prosthetic implantation of the human vestibular system. *Otology & Neurotology, 35,* 136–147.

Goswami, N., Loeppky, J. A., & Hinghofer-Szalkay, H. (2008). LBNP: Past protocols and technical considerations for experimental design. *Aviation, Space, and Environmental Medicine, 79,* 459–471.

Hackney, K. J., Scott, J. M., Hanson, A. M., English, K. L., Downs, M. E., & Ploutz-Snyder, L. L. (2015). The astronaut-athlete: Optimizing human performance in space. *The Journal of Strength & Conditioning Research, 29,* 3531–3545.

Khan-Mayberry, N., James, J. T., Tyl, R., & Lam, C. W. (2011). Space toxicology: Protecting human health during space operations. *International Journal of Toxicology, 30,* 3–18.

Lawley, J. S., Petersen, L. G., Howden, E. J., Sarma, S., Cornwell, W. K., Zhang, R., et al. (2017). Effect of gravity and microgravity on intracranial pressure. *The Journal of Physiology, 595,* 2115–2127.

Lee, A. G., Mader, T. H., Gibson, C. R., Brunstetter, T. J., & Tarver, W. J. (2018). Space flight-associated neuro-ocular syndrome (SANS). *Eye, 32,* 1164–1167.

Liu, Y., & Taylor, L. A. (2011). Characterization of lunar dust and a synopsis of available lunar simulants. *Planetary and Space Science, 59,* 1769–1783.

Luo, Y. H., & Da Cruz, L. (2014). A review and update on the current status of retinal prostheses (bionic eye). *British Medical Bulletin, 109,* 31–44.

Mindock, J., Lumpkins, S., Anton, W., Havenhill, M., Shelhamer, M., & Canga, M. (2017). Integrating spaceflight human system risk research. *Acta Astronautica, 139,* 306–312.

Mishkin, A., Lee, Y., Korth, D., & LeBlanc, T. (2007). Human-robotic missions to the Moon and Mars: Operations design implications. In *2007 IEEE Aerospace Conference* (pp. 1–10).

Palinkas, L. A. (2007). Psychosocial issues in long-term space flight: Overview. *Gravitational and Space Research, 14,* 25–33.

Powell, R., & Buchanan, A. (2011). Breaking evolution's chains: The prospect of deliberate genetic modification in humans. *The Journal of Medicine and Philosophy: A Forum for Bioethics and Philosophy of Medicine, 36,* 6–27.

Reschke, M. F., Bloomberg, J. J., Harm, D. L., & Paloski, W. H. (1994). Space flight and neurovestibular adaptation. *The Journal of Clinical Pharmacology, 34,* 609–617.

Scheuring, R. A., Jones, J. A., Novak, J. D., Polk, J. D., Gillis, D. B., Schmid, J., et al. (2008). The Apollo Medical Operations Project: Recommendations to improve crew health and performance for future exploration missions and lunar surface operations. *Acta Astronautica, 63,* 980–987.

Shreiner, A. B., Kao, J. Y., & Young, V. B. (2015). The gut microbiome in health and in disease. *Current Opinion in Gastroenterology, 31,* 69–75.

Voorhies, A. A., & Lorenzi, H. A. (2016). The challenge of maintaining a healthy microbiome during long-duration space missions. *Frontiers in Astronomy and Space Sciences, 3,* 23.

Wolpaw, J., & Wolpaw, E. W. (Eds.). (2012). *Brain-computer interfaces: Principles and practice.* Oxford, USA: Oxford University Press.

Chapter 7
Human Enhancement from the Overview Effect in Long-Duration Space Flights

Andrew B. Newberg and David B. Yaden

> *From this distant vantage point, the Earth might not seem of any particular interest. But for us, it's different. Consider again that dot. That's here. That's home. That's us. On it everyone you love, everyone you know, everyone you ever heard of, every human being who ever was, lived out their lives...There is perhaps no better demonstration of the folly of human conceits than this distant image of our tiny world. To me, it underscores our responsibility to deal more kindly with one another, and to preserve and cherish the pale blue dot, the only home we've ever known.*
> —*Carl Sagan*

Abstract The purpose of this chapter is to review the potential metabolic, neurotransmitter, and morphological changes that may occur in the human central nervous system (CNS) during long-duration habitation of Earth's Moon. When there are either permanent space habitats or long-duration interplanetary missions, we must determine if there will be any detrimental reversible or irreversible effects on the brain from this prolonged exposure to, for example, the lunar environment. Like space, it has many troublesome characteristics, including electromagnetic fields, radiation, and one-sixth gravity, which may have effects on the function and morphology of the CNS. The potential for changes in a lunar environment can, to an extent, already be anticipated from research on microgravity, including alterations in the neurovestibular system, cephalic fluid shifts, loss of total body fluid, changes in electrolyte concentrations, decreases in muscular and skeletal mass, alterations in sensory perception, changes in proprioception, and changes in human behavior. Important issues are at stake, including human health and adaptation to a lunar environment for work, recreation, and eventually, the construction of permanent human communities.

A. B. Newberg (✉)
Department of Integrative Medicine and Nutritional Sciences, Thomas Jefferson University, Philadelphia, USA
e-mail: Andrew.Newberg@jefferson.edu

D. B. Yaden
Department of Psychology, University of Pennsylvania, Philadelphia, USA
e-mail: dbyaden@gmail.com

K. Szocik (ed.), *Human Enhancements for Space Missions*, Space and Society,
https://doi.org/10.1007/978-3-030-42036-9_7

7.1 Introduction

One of the most important ways of enhancing human beings is to provide new perspectives and imbue such perspectives with intense emotions and a sense of knowing the world in a more fundamental way. Such experiences have been at the heart of most religious and spiritual traditions. These experiences are frequently termed, "spiritual" or "mystical." In fact, these types of experiences are the goal of most traditions as a way of knowing the world or knowing God in a fundamental or absolute manner. Common elements include a loss of the sense of self, a sense of oneness or connectedness, a powerful feeling of joy, and the sense that the experience carries with it a more intense feeling of epistemological reality than any other perspective. While such experiences occur through years of intense spiritual practice such as prayer or meditation, they can occur spontaneously, via the use of psychedelic substances, and, as we shall describe in this chapter, from the overview effect associated with space flight.

The overview effect was first identified by White (1987) based on his interviews of numerous astronauts who frequently described some type of "truly transformative experience involving senses of wonder and awe, unity with nature, transcendence, and universal brotherhood" (Vakoch 2012). Because of the work of White, the space community now recognizes that many people who observe the Earth from space report feeling overcome with emotion, seeing themselves and their world differently, and returning to Earth with a renewed sense of purpose and meaning in life. It is also important to note that it is not just being in space, but viewing and contemplating the Earth from a distance that results in such an overview effect.

In reviewing a number of the descriptions of the overview effect, there are four main themes that emerge: (1) an appreciation and perception of beauty, (2) an unexpected and overwhelming emotional reaction, (3) a decreased self-salience and increased sense of connection, and (4) a short- and long-term change in attitude, self-identification, and worldview (Yaden et al. 2016). And from a human enhancement perspective, experiencing the overview effect can result in a heightened sense of well-being, peace, and both emotional and cognitive enrichment.

Astronauts have attributed both short- and long-term cognitive and emotional benefits to these overview effect experiences (Mitchell and Williams 1996). However, such benefits have not been measured directly since this has not been the goal of most research on the effects of spaceflight. In fact, most literature on the effects of spaceflight address potential dangers such as psychological problems and conflicts within crews that have affected mission performance (Holland 2000). Thus, there are few mentions of positive psychological aspects of spaceflight in the current literature.

But the enhancement derived from the overview effect can go far beyond the mere individual looking out on the Earth or at the stars from a deep-space vantage point. The rest of humanity also stands to experience an enhanced sense of knowledge, wisdom, and purpose in the face of these overview effects. For example, one of the most famous photographs ever taken was "Earthrise" by the Apollo astronauts from the moon. For the first time, the idea of how humanity was positioned in the universe appeared in plain sight not just for astronauts, but for everyone. Humans have long understood cognitively that the Earth is not the center of the solar system, let alone the universe, but images like Earthrise take the understanding from something cognitive to something visceral, elevating it from something understood to something deeply felt. The simultaneous complexity and fragility of our lives on Earth was no longer a mere concept, but a perspective that could be apprehended personally. Decades before the rise of space travel, British astronomer Fred Hoyle predicted that "once a photograph of the Earth, taken from the outside, is available…a new idea as powerful as any in history will be let loose" (Kelley 1988, p. 430).

If a single photograph can produce such a compelling response in people who have not actually been in space, it seems reasonable to conclude that the actual experience is many times more powerful for the individual actually in space. A number of astronauts have reported a self-transcendent sense of awe from this experience (Linenger 2000; Mitchell and Williams 1996). Astronaut Edgar Mitchell describes the overview effect as "a spontaneous epiphany experience," writing that "instead of being an intellectual experience, it was a personal feeling… that was accompanied by a sense of joy and ecstasy."

Examining the subjective qualities of space exploration could strengthen our understanding of human functioning in extreme environments and how such environments might lead to profound insights associated with the overview effect. Further, the overview effect represents a unique class of experiences valuable to modern psychologists that pertain directly to human enhancement. From an astronaut's perspective, the overview effect might be one of the most positive and important aspects of spaceflight. Exploring the overview effect also raises meaningful questions about the relationship between these types of intense subjective experiences and well-being. Astronauts' descriptions of this experience suggest something deeper and more intense than an acknowledgement of beauty; their language reflects feelings of wonder, reverence, and humility. Among other things, viewing the Earth from space is typically identified as a powerful feeling of awe.

Social psychologists characterize awe as an intense emotion resulting from the perception of something vast, as well as the subsequent need to adjust to the experience (Keltner and Haidt 2003). D'Aquili and Newberg (1999) discussed awe in the context of Rudolf Otto's book, *The Idea of the Holy*, such that awe is a combination of fear and fascination. A feeling of awe can arise in situations both natural (witnessing a fierce storm) and social (being in the presence of a powerful figure), and overlaps with a range of states that include wonder, fear, and curiosity. Awe also has physiological correlates that likely include activation of brain structures such as the amygdala and insula that regulate intense emotional responses. In addition, the experience of vastness can be *perceptual*, as in literally seeing something large such as the Grand Canyon, or *conceptual*, as in contemplating eternity, infinity, the divine, or, in the case of the overview effect, the fragility and complexity of life on a small planet in the immensity of space. The overview effect also leads to ways of trying to make sense of this awe inspiring perception. The human brain thus tries to find a sense of meaning or context for the experience. From the perspective of enhancement, such intense experiences can lead people to describe improvements in their sense of meaning and purpose in life, improvements in their interpersonal relationships, their psychological health, and their spiritual life (Newberg and Waldman 2016).

Awe is directly associated with a number of psychological and cognitive oriented enhancements. Experiences of awe are associated with improvements in the sense of well-being (Rudd et al. 2012) and appear to increase altruistic, pro-social concerns and behaviors (Piff et al. 2015). Awe is considered part of a larger class of positive emotions that can result in a *broadening* of attention and a *building* of psychological and social resources (Fredrickson 2001). Positive emotions, and even dispositional optimism, have been suggested to enhance physical health (Fredrickson et al. 2000), facilitate better collaboration in groups (Fredrickson 2001), and even enhance creativity (Isen et al. 1987).

The emotion of awe by itself might not be sufficient to explain the long lasting changes many astronauts report after experiencing the overview effect (Cohen et al. 2010). The experience has "psychedelic qualities" that seem to typify a pattern of subjective experience in which individuals transcend their ordinary individual boundaries and becoming one with something greater than the self. Such self-transcendent experiences (STEs) are characterized by (1) powerful feelings of empathy or unity with an "other" and (2) reduced awareness of the self as a discrete entity. People frequently describe feeling deeply connected with other individuals, mankind, God, or the entire universe. STEs occur along a spectrum of intensity, ranging from states of mild states of mindfulness and flow to awe and mystical experiences (Yaden et al. 2017a). STEs are generally perceived to be positive or spiritual or both, and are often transformative, with subjects reporting them to be among the most important experiences in their lives (Griffiths et al. 2006, 2008; Hood et al. 2009; Miller and C'de Baca 2001).

The overview effect might best be understood as an example of a visually induced STE coupled with intense feelings of awe (Yaden et al. 2016). Activation of the visual system ultimately is evaluated by areas such as parietal lobe which helps to establish a spatial representation of the self. We have previously proposed, and found support for based on neuroimaging studies, that a reduction of activity in the parietal lobes is associated with an altered sense of self and a sense of connectedness or oneness of the self with others, with God, or with the universe (d'Aquili and Newberg 1999). The vastness of space likely results in a reduction of visual input into the parietal lobes and leads to a perceptually mediated decrease in parietal lobe activity, thus resulting in a feeling of self-transcendence.

Although these experiences are usually temporary and also occur only during the time of spaceflight itself, they are profound and memorable, and produce visceral feelings of compassion and personal connection with all of the Earth and its inhabitants. Astronauts such as Edgar Mitchell and Yuri Artyukhin state that this sense of unity is "felt" rather than merely imagined. Artyukhin's described it as "not simply an observation. With it comes a strong sense of compassion and concern for the state of our planet and the effect humans are having on it.".

Such experiences have traditionally been associated with religious and spiritual environments, such as religious rituals, prayer, meditation, or the use of psychedelic drugs (Yaden et al. 2017b). Religious and spiritual experiences are particularly associated with feelings of self-transcendence in which subjects perceive to contact the divine or spirit world as part of an altered state of consciousness (James 1902). The work by Maslow (1964) on "peak experiences" in addition to the descriptions of mystical experiences and experiences such as the overview effect makes a strong case that the experience of perceiving something larger and more powerful than oneself is connected to these alterations in self-awareness, and that there might be an intimate connection between experiences of awe and self-transcendence.

The overview effect associated with spaceflight has certain unique elements that might be particularly valuable as a "tool" toward personal enhancement and enhancement of the human species more broadly. Importantly, the overview effect appears to be quite potent as a reliable, secular, non-pharmacological stimulus for self-transcendence. Further research on the overview effect might shed light on how intense states of awe, and stimuli capable of evoking extreme conceptual vastness, can induce visceral and engrossing changes in human consciousness that lead to profound enhancements in emotions, cognitions, and performance.

The overview effect reportedly results in deep conceptual changes, even after the experience ends, and people who have this experience attempt to integrate it into their broader worldviews. From a neurological perspective, intense emotional experiences activate the limbic areas of the amygdala and hippocampus that ultimately transform memories that lead to new ways of thinking about one's self and one's place in the universe. This can have profound implications for human enhancement since astronauts describe broader shifts in their conception of Earth and of humanity's place in the greater scheme of things. Cognitive shifts such as these indicate that extreme subjective experiences—and the altered states of awareness that they evoke—can enhance the ways in which individuals understand and approach concepts, and even affect how strongly concepts that are already familiar register for them.

These changes go beyond immediate self-perception; according to Rusty Schweikart, "When you go around the Earth in an hour and a half, you begin to recognize that your identity is with that whole thing" (White 1987). Schweikart's "recognition" of his unity with the Earth as a whole was not just an immediate subjective experience, but also a realization.

Besides the pro-social and altruistic effects of awe and self-transcendent experience, there seems to be an added altruistic element that results from changes associated with the overview effect. Astronaut Edgar Mitchell speaks to this altruistic urge, "In outer space, you develop an instant global consciousness, a people orientation, an intense dissatisfaction with the state of the world, and a compulsion to do something about it." These changes are conscious and reflect not just a heightened sensitivity to empathy and positive emotions, but also a "newfound *commitment to concrete and long-term values*, such as respect for life or betterment of the human condition" (Yaden et al. 2016). The emotions experienced when looking at the Earth from space, and particular far space, seem to affect the concepts that observers adopt and appreciate, including those that are ideological in a broader context about the nature of life, humanity, and the universe.

Studying the psychological enhancements associated with spaceflight and the overview effect are an important consideration for future manned space flight, especially at a time when more people than ever will be entering space. The life-enhancing and salutogenic qualities of one's experience in space could be a primary motivating force for individuals to volunteer for manned missions to space, and could potentially form an important buffer against some of the psychological risks of astronaut missions which include loneliness, isolation, and intense stressors (Newberg and Alavi 1998).

It will also be important to evaluate whether there are cross-cultural differences regarding the ways in which the overview effect is experienced and interpreted, including differences in religious and philosophical beliefs. That said, spaceflight is one of the few types of human endeavors that can be a true source of collective inspiration and ultimately lead to collective enhancement of human beings. There might even be substantial predispositions beneath the human urge to explore that might be enhanced by the pursuit of the overview effect and the life-enhancing qualities it possesses.

Studying the overview effect should be of interest to anyone interested in psychology, human enhancement, or space exploration. If the drive to understand our world is indeed a drive to understand ourselves, then part of that understanding derives from our encounter with the vast and unknown. From the perspective of deep spaceflight, this experience includes the powerful overview effect that results when we look back and see ourselves and our home differently.

We conclude with the words of astronaut Ron Garan and T. S. Elliot:

> As I looked back at our Earth from the orbital perspective, I saw a world where natural and man-made boundaries disappeared, I saw a world becoming more and more interconnected and collaborative, a world where the exponential increase in technology was making the impossible possible on a daily basis. Thinking about the next 50 years, I imagined a world where people and organizations set aside their differences and work together toward their common goals. They set aside their differences and realize that each and every one of us is riding through the Universe together on this Spaceship we call Earth. They realize that because we are all interconnected, we are all in this together and because we are all family, the only way to solve the problems we all face is together. (Ron Garan)

> We shall not cease from exploration
>
> And the end of all our exploring
>
> Will be to arrive where we started
>
> And know the place for the first time. (T. S. Eliot)

References

Cohen, A. B., Gruber, J., & Keltner, D. (2010). Comparing spiritual transformations and experiences of profound beauty. *Psychology of Religion and Spirituality, 2*(3), 127.

d'Aquili, E. G., & Newberg, A. B. (1999). *The mystical mind: Probing the biology of religious experience*. Minneapolis, MN: Fortress Press.
Fredrickson, B. L. (2001). The role of positive emotions in positive psychology: The broaden-and-build theory of positive emotions. *American Psychologist, 56*(3), 218–226.
Fredrickson, B. L., Mancuso, R. A., Branigan, C., & Tugade, M. M. (2000). The undoing effect of positive emotions. *Motivation and Emotion, 24*(4), 237–258.
Griffiths, R. R., Richards, W. A., Johnson, M. W., McCann, U. D., & Jesse, R. (2008). Mystical-type experiences occasioned by psilocybin mediate the attribution of personal meaning and spiritual significance 14 months later. *Psychopharmacology (Berl), 22*(6), 621–632.
Griffiths, R. R., Richards, W. A., McCann, U., & Jesse, R. (2006). Psilocybin can occasion mystical-type experiences having substantial and sustained personal meaning and spiritual significance. *Psychopharmacology (Berl), 187*(3), 268–283.
Holland, A. W. (2000). Psychology of spaceflight. *Journal of Human Performance in Extreme Environments, 5*(1), 1.
Hood, R. W., Jr., Hill, P. C., & Spilka, B. (2009). *Psychology of religion: An empirical approach*. New York: Guilford Press.
Isen, A. M., Daubman, K. A., & Nowicki, G. P. (1987). Positive affect facilitates creative problem solving. *Journal of Personality and Social Psychology, 52*(6), 1122.
James, W. (1902). *The varieties of religious experience*. Harlow: Longmans, Green & Co.
Kelley, K. W. (Ed.). (1988). *The home planet*. Boston: Addison Wesley Publishing Company.
Keltner., & Haidt, J. (2003). Approaching awe, a moral, spiritual, and aesthetic emotion. *Cognition and Emotion, 17*(2), 297–314.
Linenger, J. M. (2000). *Off the planet*. New York: McGraw-Hill.
Maslow, A. H. (1964). *Religions, values, and peak-experiences*. London: Penguin Books Limited.
Miller, W. R., & C'de Baca, J. (2001). *Quantum change: When epiphanies and sudden insights transform ordinary lives*. New York: Guilford Press.
Mitchell, E., & Williams, D. (1996). *The way of the explorer* (V. I.). New York: Putnam.
Newberg, A. B., & Alavi, A. (1998). Changes in the central nervous system during long-duration space flight: Implications for neuroimaging. *Advances in Space Research, 22*(2), 185–196.
Newberg, A. B., & Waldman, M. R. (2016). *How enlightenment changes your brain: The new science of transformation*. New York, NY: Penguin Random House.
Piff, P. K., Dietze, P., Feinberg, M., Stancato, D. M., & Keltner, D. (2015). Awe, the small self, and prosocial behavior. *Journal of Personality and Social Psychology, 108*(6), 883–899.
Rudd, M., Vohs, K. D., & Aaker, J. (2012). Awe expands people's perception of time, alters decision making, and enhances well-being. *Psychological Science, 23*(10), 1130–1136.
Vakoch, D. A. (Ed.). (2012). *Psychology of space exploration: Contemporary research in historical perspective*. Government Printing Office.
White, F. (1987). *The overview effect: Space exploration and human evolution*. Boston: Houghton-Mifflin.
Yaden, D. B., Haidt, J., Hood, R. W., Jr., Vago, D. R., & Newberg, A. B. (2017a). The varieties of self-transcendent experience. *Review of General Psychology, 21*(2), 143–160.
Yaden, D. B., Iwry, J., Slack, K. J., Eichstaedt, J. C., Zhao, Y., Vaillant, G. E., & Newberg, A. B. (2016). The overview effect: Awe and self-transcendent experience in space flight. *Psychology of Consciousness: Theory, Research, and Practice, 3*, 1–11.
Yaden, D. B., Le Nguyen, K. D., Kern, M. L., Belser, A. B., Eichstaedt, J. C., Iwry, J., … Newberg, A. B. (2017b). Of roots and fruits: A comparison of psychedelic and nonpsychedelic mystical experiences. *Journal of Humanistic Psychology, 57*(4), 338–353.

Chapter 8
Science and Ethics in the Human-Enhanced Exploration of Mars

Gonzalo Munévar

Abstract This chapter will discuss several possibilities for human enhancement in the exploration of Mars in light of scientific and ethical considerations. The troublesome enhancements are basically of two kinds: implants and genetic interventions. A rather uncontroversial proposal is that we equip space explorers with the right kinds of suits, goggles and other equipment that will allow them to operate safely in harsh space environments. Implants, however, require surgical risks that may be unwarranted. Genetic interventions are themselves of two kinds: somatic genetic changes and germline changes. Somatic genetic changes may be riddled with unknown and potentially devastating side effects to the explorers who undergo them. Germline genetic changes may be passed on to future generations of human beings, a move that would be very difficult to justify. Of course, sending space explorers into harsh environments does require us to provide for their safety. But the most likely exploration of Mars in the next several decades will have as its greatest safety challenge the protection from radiation. No foreseeable enhancement can do that. We will mainly need safe suits and buildings. If extant life were to be found on Mars, then it would be crucial to determine whether it could dangerously infect human beings. In that case, we would need to "enhance" the explorers by giving them immunity. Other enhancements often mentioned would at best make some difficult tasks easier, but generally that would not be enough to overcome the burden of danger they create. This chapter will also examine the scientific plausibility of many of the proposed biotechnological advances. There is a reason to believe that they may be unwarranted extrapolations from present genetic research, for example, which brings their plausibility into question.

G. Munévar (✉)
Lawrence Technological University, Southfield, MI, USA
e-mail: gmunevar@ltu.edu

K. Szocik (ed.), *Human Enhancements for Space Missions*, Space and Society,
https://doi.org/10.1007/978-3-030-42036-9_8

8.1 Introduction

Science and ethics have much to tell us about the suggestion that human beings should undergo enhancements in order to explore space. In this chapter, I will limit myself to a discussion of such enhancements in the exploration of Mars, but its conclusions could be extended to other explorations of space that we may wish to carry out in the upcoming decades. In short, I will argue that science in the context of ethics does not, in general, support the use of such enhancements.

In carrying out the exploration of Mars, human beings may face two main threats to their physiology: the deleterious effects of living long periods in microgravity, mainly traveling to and returning from Mars, and the dangerous levels of radiation they will encounter in transit and on Mars itself. In the [illegible], I will [illegible] the two main kinds of [illegible] that have been proposed to help humans explore space: implants and genetic interventions. I will then consider how such proposals might be tailored to apply to the exploration of Mars specifically.

Subsequently, I will develop an important objection to both types of proposal: Insofar as Mars is concerned, they both commit the fallacy of false dilemma. More precisely, they both attempt to get around possible ethical objections by suggesting that, however troublesome their potential secondary effects, proceeding along such lines can prevent far more serious health or other problems that threaten the human explorers. Nevertheless, I will argue the presumably dire situations have alternative solutions that do not expose humans to such risky violations of their biological integrity.

8.2 Implants

It is not uncommon to read in the space literature proclamations to the effect that future space explorers will have to become cyborgs, as we can see in such titles as "A Cyborg Space Race" (*Astrobiology Magazine*, 2010) and "Cyborg Astronauts Needed to Colonize Space" (*Science and Astronomy*, 2010). This way of thinking began with an essay by Clynes and Kline (1960), followed by a NASA study called "The Cyborg Study: Engineering Man for Space" (1963), which considered the possibility not only of drugs and hibernation but also of organ replacements.

As of now, 56 years later, medical technology has developed deep brain stimulation (DBS) to treat Parkinson's disease, and implantable insulin pumps and heart pacemakers, as well as metal replacements for hips and other bones. In this respect, some human beings might be said, quite fancifully, that they are cyborgs to some degree. But it is one thing to implant into people devices that will help them manage a very serious medical condition—a morally justified action—and quite another to do so in people who are not ill, just so they can perform a job better—a morally questionable action.

To get around this ethical problem, some might wish to argue that exploring in space will make people very ill, and thus if they are going to be in such extraordinary circumstances, of their own accord, it would be morally reasonable to offer them the means to manage the medical conditions that space exploration subjects them to.

Let us consider the first cause of such adverse medical conditions: long exposure to microgravity. One of the serious problems that astronauts would encounter in a voyage lasting around seven months, such as the trips to Mars and back, is the loss of muscle mass. NASA's Rodent Research-6 (**RR-6**), at the International Space Station, has examined a drug and a delivery system that may prevent muscle loss in mice (*Johnson Space Center*, 2018). Presumably, astronauts are potential beneficiaries of the drug compound and the nano-channel delivery implant. Perhaps the research might someday also help patients with musculoskeletal diseases here on Earth. The implant, about the size of a grain of rice, would go beneath the skin and allow for a steady, controlled delivery of the drug compound using diffusion, for months, even years. When the drug is depleted, the device can be refilled using two needles through the skin.

Once we consider implants of this sort, which space agencies have in mind today, as opposed to flirting with science fiction, the case for enhancements may appear more reasonable. But a detailed look makes matters more complicated again. Microgravity induces not only osteoporosis and muscle atrophy, but also cardiac problems and disruption of the transport of fluids throughout the body. In addition, it affects many types of cells, including those of the immune system, which can have serious consequences (Pietsch et al. 2011). To prevent all these and other problems, we would then have to develop drugs and implanted devices for their delivery. This seems far too much to impose on any human being, even dedicated astronauts. The possible interactions between all these drugs could themselves pose new problems, to say nothing of the load on the organism from all these foreign objects implanted in it.

Moreover, the operations to implant the devices are quite risky. The most ordinary, down to Earth, of the implants mentioned above are perhaps those used in hip replacements. They surely do not seem to compare to anything like DBS. But, if I may be forgiven an anecdote, a relative's recent hip replacement led to two embolisms in her lungs, a rather common risk from the operation. Such embolisms are sometimes fatal or may cause a great and permanent deterioration of the patient's quality of life. And let us remember that human beings will perform such operations and that we are not perfect, not even if we work for NASA. In the case of that relative, the surgeon implanted the wrong metal femur head. She now has a leg almost two cm longer than the other.

Things may go wrong to a far greater extent in proposals such as Elon Musk's to implant electrodes in the brain that would allow astronauts to easily control computers with their brains. This is based on research that led to electrodes inserted in the motor areas of the brains of paralyzed people so as to enable them to control robotic arms (Hochberg et al. 2006). Those arms allowed them to bring a cup of water to their lips and do other simple things that most of us take for granted. Such risky but laudable medical endeavors are beset by a variety of problems. Making them small enough so they will not be too invasive makes them fragile. If they break inside the brain, there could be serious problems. Even if not damage is done by them, they will have to be replaced after some time, for, as neurosurgeon Chengyuan Wu points out, brain cells will respond to the implanted wires as they would to tissue damage, "causing scar tissue" (interviewed in *Popular Science*, 2019). And if hip replacement surgery presents problems, operations in the brain can be very dangerous. There is always the risk of bleeding and infection, for example. As Wu says, "What if you get a hemorrhage in the motor cortex? You could paralyze patients otherwise moving fine." It is difficult to see how it could be ethical to implant such devices in the brains of healthy people—brains not suffering from any ill effects of space travel, incidentally. Besides, thanks to advances in neuroimaging and artificial intelligence, systems that place the electrodes on the skull, not inside the brain, are becoming more and more sophisticated and coming closer to the level of performance of the invasive systems (Edelman et al. 2019). Even if the noninvasive systems fall somewhat short, the avoidance of risk, let alone fatal risk, would make them preferable.

Since Musk's device would be implanted in healthy astronauts, not to manage adverse conditions created by space travel, it cannot appeal to such a justification. Even less likely to so appeal are implants that will afford us superhuman abilities and intelligence. I do not believe that neuroscience bears out such speculations, but a discussion of them would take us too far afield. At any rate, lacking such implants does not present any impediments to the exploration of Mars.

8.3 Genetic Interventions

It would probably be inevitable that someone would argue that the best way to handle the inhospitable environment of outer space is to change human explorers so they adapt to it. By this, it is usually meant that we should change the genetic endowment of astronauts. Ever since the 1970s, there has been much talk and hope of genetic interventions in medicine. The success has been rather modest, but recently a new technique called "clustered interspaced short palindromic repeats," CRISPR-Cas9, seems to be causing a revolution in the field. Now genetic surgeons practically expect genetic cures of lung problems, blood disorders, cancer and many other diseases, thanks to the combination of genetic insertion, dilution and disruption.

Of great relevance to the exploration of Mars is the possibility of introducing resistance to radiation in humans by genetic intervention. As we saw earlier, radiation is considered probably the biggest problem astronauts will face traveling to and from Mars and on the surface of Mars. A study published in *Nature Communications* described the identification of a tardigrade-unique DNA-associated protein. A protein called Dsup from the tardigrade species R. varieornatus, which exhibits extraordinary tolerance against high-dose radiation, was inserted into human cells in vitro (Hashimoto et al. 2016). Tardigrades, also called "water bears," are very small, pudgy aquatic animals with four pairs of legs. Some species are able to tolerate dehydration, extreme low and high temperatures, high pressure and the space environment, as well as exposure to radiation. According to the investigators, human cells with associated Dsup in their DNA were able to reduce radiation damage from X-rays by 40%. This surprising result has led to speculations that after exhaustive experimentation in animals, in a few decades, perhaps only two or so, we might have astronauts enhanced with tardigrade DNA exploring Mars!

Some do not wish to wait so long. The University of California campuses at San Francisco and Berkeley have received a grant from the Department of Defense for the sum of USD 20 millions to work together using CRISPR-Cas9 to find a genetic alteration that would protect radiation patients, soldiers and eventually astronauts from exposure to radiation. The ideal is to change gene expression in a manner that is temporary and reversible and cannot be passed on to future generations. Achieving this goal would be important for our purposes in this paper, since it would comply with an ethical requirement that astronauts should be able to return and lead a normal life (which otherwise could be affected by permanent changes to their DNA). It would also restrict itself to somatic changes and not to germline changes that would affect future generations, an issue also beset with ethical challenges.

As explained by Henderson (2019), when using CRISPR-Cas9, programmable RNAs guide the Cas9 protein to specific locations in the genome. There they change the proteins and molecules that surround DNA, thus changing gene expression. Scientists at UC Berkeley and UC San Francisco will look throughout the entire human genome for genes that protect against radiation when turned on or off by CRISPR-Cas9, and then for the key proteins involved. Presumably this long-term, large-scale experiment may lead to the desired genetic protection against radiation.

CRISPR-Cas9 is the main but not the only potential tool for genetic intervention. For example, type VI CRISPR-Cas systems contain Cas13, which has been demonstrated for knockdown and editing of RNA (Cox et al. 2017). This particular system, RNA Editing for Programmable A to I Replacement (REPAIR), has the generic name of "base editing." In another important study (Gaudelli et al. 2017), it provides the ability of fixing point mutations (about 32,000 known point mutations cause illness in humans). As an illustration, in this second study the investigators were able to change the base pair A-T into a G-C base pair (remember that bases pair off, A with T and G with C, across the two strands of the double helix). This sort of genetic intervention may also have broad application in scientific research and medicine, although so far it has not been proposed as a way of preventing radiation damage in space explorers.

It does seem, nevertheless, that CRISPR-Cas9 is poised to become not only the gene editing tool of choice in clinical contexts but also in those specifically searching for protection from radiation in the exploration of Mars. Until recently, the examination of the genetic alterations induced by Cas9 has been restricted to the vicinity of the target site, with some consideration also of distal off-target sequences. Such an examination concluded that CRISPR-Cas9 was largely specific.

Unfortunately, a new and important study (Kosicki et al. 2018) reports "significant on-target mutagenesis, such as large deletions and more complex genomic rearrangements at the targeted sites in mouse embryonic stem cells… and a human differentiated cell line." The investigators found that the DNA breaks introduced by "single-guide RNA/Cas9" often resolve into deletions found over many thousands of bases. Moreover, they also identified crossover events and lesions distal to the cut site. Most distressingly, the genomic damage they observed may have pathogenic consequences.

The CRISPR genome editing technology may have ushered a revolutionary era in biology, finally enabling scientists to cash in the promises made by recombinant DNA in the 1970s, forging new varieties of plants and animals and making possible new therapies for a large number of diseases. Nevertheless, the results of this study force us to proceed with extreme caution when it comes to human beings, space explorers included, as implied by the comments made in the previous paragraph. In plain numbers, it is very concerning that in 20% of cells CRISPR-Cas9 causes deletions and rearrangements that can vary from hundreds to thousands of DNA letters long, according to the lead investigator in the study, Allan Bradley (interviewed in *New Scientist Health*, 2018). "There's a risk of causing cancer sometime in a patient's lifetime," says Bradley. "We need to understand more before rushing into human clinical trials."

It is important to point out that this study is considered highly reliable by the CRISPR community. For example, according to Gaetan Burgio (interviewed by *New Scientist Health*, 2018), "I do believe the findings are robust…. This is a well-performed study and fairly significant." Burgio, an expert on the technique, had debunked previous challenges to the safety of CRISPR.

The study by Bradley and his team examined genetic changes induced by CRISPR-Cas9, and it did not address much newer proposed techniques such as base editing. Its conclusions are so powerful that one suspects that its findings are likely to apply to the newer techniques as well, e.g., its findings on deletions on thousands of bases associated with RNA/Cas9 DNA breaks. We might agree that the deletion or modification of defective genes might be ethically acceptable when the person's life is at stake. But this sort of justification cannot be extended to CRISPR-Cas9 or base editing of *non-defective genes*. Apart from all the risks outlined above, consider that we have only 26,000 genes. Of necessity, then, genes must enter into complex relations with many other genes in order to influence our phenotype. If a gene is defective and we modify it so it will work properly, those genetic relationships may actually be safeguarded. But if the gene is not defective and we change it, who knows how much we will disrupt the normal web of genetic connections.

As we saw in the ***Introduction***, genetic interventions are of two kinds: somatic and germline. Somatic genetic changes may be riddled with unknown and potentially devastating side effects to the explorers who undergo them. Germline genetic changes may be passed on to future generations of human beings, even though we cannot assume that they will choose to become space explorers (except in the unusual case discussed below). Thus, the attempts to justify somatic genetic enhancements, which fall short, are even less likely to justify germline enhancements.

Of course, sending space explorers into harsh environments does require us to provide for their safety. But the most likely exploration of Mars in the next several decades will have as its greatest safety challenge the protection from radiation. No foreseeable enhancement can do that without creating serious problems. If extant life were to be found on Mars, then it would be crucial to determine whether it could dangerously infect human beings. In that case, we would need to "enhance" the explorers by giving them immunity. If we ever colonize Mars, a proven, safe way to pass that immunity on to the first colonists' descendants, who will be born and live their lives on Mars, might be warranted. Other enhancements often mentioned would at best make some difficult tasks easier, but generally that would not be enough to overcome the burden of danger they create. In the next section, I will discuss how to safeguard the safety of human explorers of Mars without invasive enhancements.

8.4 False Dilemmas

How can human explorers be protected from the effects of radiation and of microgravity? Let me take up radiation first.

8.4.1 Radiation

As NASA explains, once space explorers bound for Mars leave the protection of the Earth's magnetosphere, they will be exposed to two kinds of radiation (NASA's Goddard Space Flight Center, 2017). The first will be a stream of solar particles, with occasional larger bursts in the wake of giant explosions, such as solar flares and coronal mass ejections, consisting mainly of protons. These can be shielded by the walls of the spacecraft and on Mars by the walls of dwellings and vehicles.

Cosmic rays, which come from other stars, sometimes from other galaxies, are a much bigger threat and not as easy to stop. In addition to protons, they occasionally include helium and some heavier elements, all accelerated near the speed of light. When they strike the walls of a spacecraft or the skin of an astronaut, they cause a shower of subatomic particles into the structure. This secondary radiation can be very dangerous. It is thought that there are two ways to stop this radiation from affecting the space travelers. One is to increase the mass of the spacecraft so as to present a thicker shield. This approach is generally considered expensive and requiring a great increase in propellant as well. A second way is to use new, special materials that will stop cosmic rays. This approach is quite promising, as we will see below. But it seems to me there is a third way: to redesign the spacecraft so that many of your supplies are part of the exterior walls of the ship.

This last suggestion bears some further explanation. Hydrogen, the most abundant element in the universe, blocks both protons and neutrons extremely well. It is found in water and in plastics like polyethylene. Thus, to the standard, sturdy spacecraft walls we could add contiguous compartments with polyethylene containers full of water. We would thus store much of our water supplies. Later in the trip, the explorers may need to replace those containers, as they use them, with others filled with recycled water from the advanced life support systems. And as NASA suggests, the trash could also end up as part of the walls, in convenient plastic containers.

As hinted at earlier, NASA is developing an ideal shielding material: hydrogenated boron nitride nanotubes (hydrogenated BNNTs), small "nanotubes made of carbon, boron and nitrogen, with hydrogen interspersed throughout the empty spaces left in between the tubes." This material is excellent at absorbing radiation and extremely strong, even at very high temperatures. In addition to hydrogen, boron contributes by stopping neutrons. As if this technological research were not remarkable enough, yarn has been successfully made out of BNNTs, flexible enough to be integrated into the fabric of space suits that protect space explorers from radiation! It thus seems that the present development and testing of hydrogenated BNNTs promise to offer shielding from radiation, as well as structural strength, in spacecraft, habitats, vehicles and space suits used in the exploration of Mars.

Science is also investigating the possibility of endowing spacecraft with force fields that would protect them from radiation, as Earth's magnetosphere protects us. The power required to create force fields strong enough is too large for our present space technology. At any rate, as we have seen above, space technology is developing, or has already, means by which to protect space explorers from radiation.

Now, the idea of inflicting all the dangers of genetic interventions on space explorers of Mars had as justification the threat posed by radiation. But as we saw in the previous section, those dangers may be actually far too large. Nevertheless, those are not the only two options. This was a false dilemma. Space technology is offering us much more sensible and noninvasive solutions to the problem of radiation. In view of this alternative, genetic enhancements cannot be morally justified.

8.4.2 *Microgravity*

As we have seen, microgravity does affect the health of space travelers, but the proposed implants may pose even greater threats. These are not the only options open to humans as we travel to and from Mars, however. There is a third option: artificial gravity.

Artificial gravity comes mainly in two varieties: spacecraft rotation and constant acceleration. I will take up spacecraft rotation first, an idea published in 1911 by the great rocketry pioneer Konstantin Tsiolkovsky and later developed by O'Neill (1977) and other scientists.

There are many versions of this idea for the exploration of Mars. My own is simple. A powerful rocket engine is connected to its right to the large human compartment of a spacecraft and to its left to an equally large supply compartment. The two compartments, in the same plane, are far apart as a result, say 200 m or so, and pointing toward Mars, as does the rocket engine. The connection between the compartments and the engine could be a rigid, narrow tunnel, or even a cable. We then make the two compartments rotate by means of small rockets on their sides. The rocket section then takes off from Earth orbit moving toward Mars, as the two compartments rotate around it. It is that rotation that creates the artificial gravity (see Fig. 8.1). As an astronaut stands on the floor, i.e., the hull, as the compartment rotates the centrifugal

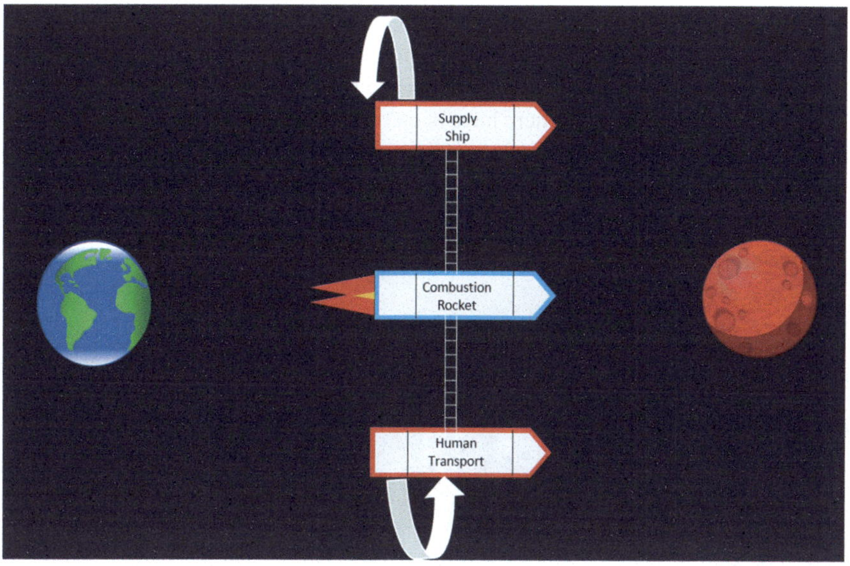

Fig. 8.1 Artificial gravity produced by rotating spacecraft. The interior surface (the "floor") of the outside walls of the human transport and the supply ship would impart an acceleration on astronauts equivalent to gravity. Art courtesy of Phillip McMurray

force acting on him or her would feel as if pointing toward the hull. In other words, the astronaut would feel as if gravity were pulling him or her toward the floor.

The long radius of rotation, around 100 m, is needed so as to avoid Coriolis forces that may cause nausea. With that radius, we might be able to keep the rotation to two per minute, which humans can tolerate. There may be a slight but not very noticeable difference in the artificial gravity felt at the floor and at the top of the human compartment (presumably not a very tall one). Incidentally, we may choose an amount of artificial gravity less than 1 g, perhaps around 38% of 1 g. That way the space explorers will become used to the amount of gravity they will experience on Mars.

Yes, this approach is more complicated than what we now have. It is quite possible, however, that the spacecraft that will fly team after team of human explorers to Mars will be assembled in orbit, since given the large size it will be easier to carry the separate components into orbit, and then the passengers and all the supplies. If so, all we have to do is assemble the very same components in a different arrangement and then add a cable, or some other way of connecting the engine to the two compartments. The small rockets that will provide for the initial rotation, which will then continue in the vacuum of space all the way to Mars, can be adapted from small maneuvering rockets that will likely be standard equipment.

Or we can try spacecraft under constant acceleration, taking advantage of Einstein's principle of equivalence (between acceleration and gravity). This is not feasible with the chemical rockets that dominate space technology today, for they burn a lot of their fuel to provide the initial acceleration and then coast at the initially achieved velocity the rest of the way, with propulsion corrections now and then. A chemical rocket would have to be truly gigantic to keep accelerating all the way to Mars. But for several decades there have been proposals, and some degree of experimentation, with other forms of propulsion: fusion rockets (see Fig. 8.2), ion rockets, and even solar sails (see Figs. 8.3 and 8.4) or electric sails. Indeed, recently a small

Fig. 8.2 Fusion reactors are now contemplated as the heat source that could bring rocket propellant to extremely high temperature (and hence high-velocity exhaust) or expel ultra-hot plasma to provide thrust. Courtesy of © ITER Organization, https://www.iter.org/

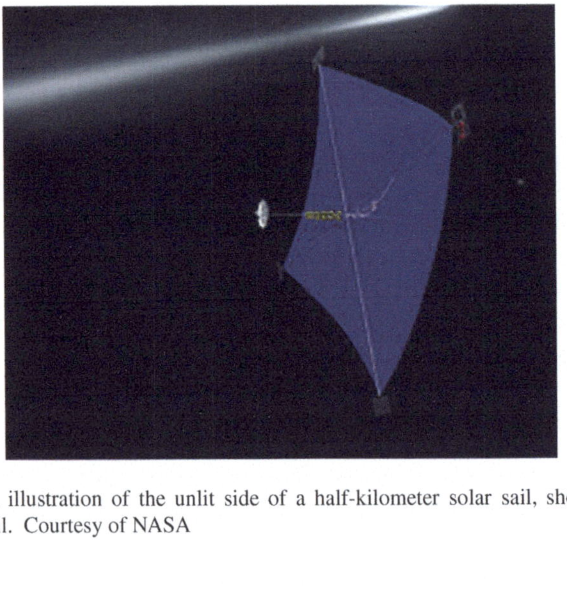

Fig. 8.3 NASA illustration of the unlit side of a half-kilometer solar sail, showing the struts stretching the sail. Courtesy of NASA

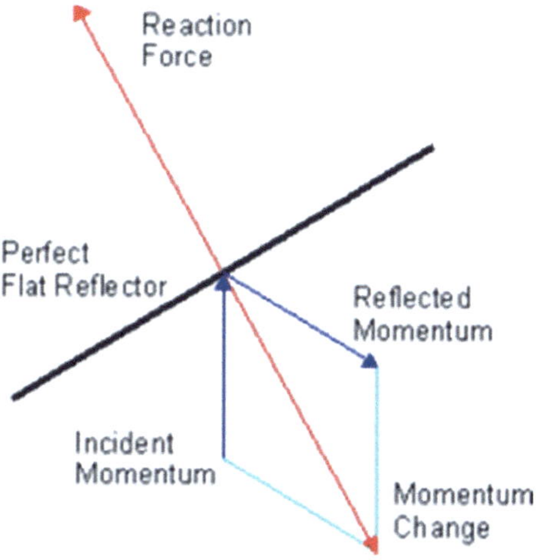

Fig. 8.4 Force on a sail results from reflecting the photon flux. July 24, 2013, 22:01:02. Courtesy of Jerry Wright. Own work

solar sail was successfully deployed in space by The Planetary Society (Cassella 2019).

These technologies are not almost ready-at-hand, unlike those that will enable us to deal with radiation, but they do give us a third alternative on the matter of microgravity. And pursuing them will have far great beneficial consequences for humanity in the not-so-long run. Consider, for example, that constant acceleration is likely to make trips to Mars far shorter than the ones we face in the next few years. That alone would be of great advantage.

8.5 Conclusion

This chapter has examined the scientific plausibility of many of the proposed biotechnological advances that have been suggested either as implants or as genetic modifications of human explorers of Mars. I have argued that there is a reason to believe that such plausibility is in question, particularly when it comes to the damage they can cause to the human body. I have explained in great detail, for example, how the most celebrated genetic technique, CRISPR-Cas9, is besieged by unexpected and unwelcome consequences. Much refinement and experimentation will be required before we can seriously consider using it on healthy human space explorers, for it may cause more problems than it could prevent. In this case, and in general, I have shown that these scientific drawbacks lead to ethical concerns.

Those concerns loom even larger, decisively I think, once we realize that the very reason for even considering those biologically invasive technologies was that we had no alternative. This, I pointed out, presented us with a false dilemma. Science and technology are indeed developing noninvasive approaches to solve the threats that microgravity and radiation pose to the men and women that will sail the heavens to enrich us with the fruits of their exploration of Mars.

References

Cassella, C. (2019). Mission success declared as Solar Sail propels itself from Earth Using only sunbeams. *Science Alert*, August 1.

Clynes, M., & Kline, N. (1960) Cyborgs and Space. *Astronautics*.

Cox, D. B. T., Gootenberg, J. S., Abudayyeh, O. O., Franklin, B., Kellner, M. J., Joung, J., & Zhang, F. (2017). RNA editing with CRISPR-Cas13. *Science, 358*, 1019–1027.

Edelman, B. J., Meng, J., Suma, D., Zurn, C., Nagarajan, E., Baxter, B. S., Cline, C. C., & He, B. (2019). Noninvasive neuroimaging enhances continuous neural tracking for robotic device control. *Science Robotics, 4*(31), eaaw6844. https://doi.org/10.1126/scirobotics.aaw6844.

Frazier, S. (2017). Real Martians: How to protect astronauts from space radiation on Mars. *NASA's Goddard Space Flight Center* Publication. September 30, 2015. Updated: August 7, 2017.

Gaudelli, N. M., Komor, A. C., Rees, H. A., Packer, M. S., Badran, A. H., Bryson, D. I., & Liu, D. R. (2017). Programmable base editing of A•T to G•C in genomic DNA without DNA cleavage. *Nature, 551*, 464–471.

Hashimoto, T., Horikawa, D. D., Saito, Y., Kuwahara, H., Kozuka-Hata, H., Shin-I, T., et al. (2016). Extremotolerant tardigrade genome and improved radiotolerance of human cultured cells by tardigrade-unique protein. *Nature Communications, 7*, 12808. https://doi.org/10.1038/ncomms12808. www.nature.com/naturecommunications

Henderson, H. (2019). From battlefields to cancer wards: CRISPR to combat radiation sickness. *Innovative Genomics Institute News*, June 27.

Herath, A. K. (2010). Cyborg astronauts needed to colonize space. *Science & Astronomy*, September 16.

Hochberg, I. R., Serruya, M. D., Friehs, G. M., Mukand, J. A., Saleh, M., Caplan, A. H., et al. (2006). Neuronal ensemble control of prosthetic devices by a human with tetraplegia. *Nature, 442*, 164–171.

Howard, J. (2018). Investigation to combat muscular atrophy with implantable device. In *International space station program science office*. Johnson Space Center Publication.

Kosicki, M., Tomberg, K., & Bradley, A. (2018). Repair of double-strand breaks induced by CRISPR–Cas9 leads to large deletions and complex rearrangements. *Nature Biotechnology, 36*, 765–771.

Le Page, M. (2018). CRISPR gene editing is not quite as precise and as safe as thought. *New Scientist Health*, 16 July.

Mullen, L. (2010). A cyborg space race. *Astrobiology Magazine*, April 5.

O'Neill, G. K. (1977). *The high frontier: Human colonies in space*. New York, NY: William Morrow & Company.

Pietsch, J., Bauer, J., Egli, M., Infanger, M., Wise, P., Ulbrich, C., & Grimm, D. (2011). The effects of weightlessness on the human organism and mammalian cells. *Current Molecular Medicine, 11*, 350–364.

Tsiolkovsky, K. (1911). Investigation of outer space rocket devices. *The Science Review*.

Wetsman, N. (2019). Brain interfaces aren't nearly as easy as Elon Musk makes them seem. *Popular Science*, July.

Chapter 9
Interstellar Missions and Human Enhancement

Simon P. Worden

Abstract Humanity's first interstellar probes will be launched later this century. Perhaps near the end of the century we will have the ability to place a very small payload on the surface of a planet orbiting a nearby star. Although highly speculative, some experts today think this could enable us to "boot-up" earth life, or modify earth life in these distant environments. This possibility raises some interesting ethical and philosophical issues. If we can begin spreading our life across the galaxy—should we?

One of the humanity's longest held dreams is to expand our civilization into the galaxy. This topic has been the topic of speculative fiction and visionary thought for almost a century. Noted Cosmologist the late Stephen Hawking believed interstellar expansion is the only way to avoid our ultimate extinction as discussed in his last book (Hawking 2018). Such an extinction might be the result of cosmic catastrophes such as a large asteroid strike, a major solar outburst or nearby supernovae. We may also perish through our own invention and folly such as engineered bio-plagues, technological warfare and catastrophic climate change. Many, notably Elon Musk (SpaceX), seek to back up humanity with Martian settlements. But even Martian settlements are vulnerable to many extinction-level events. For these reasons, Hawking and others believe we must expand into the galaxy.

The possibility that alien civilizations exist and might also have sought galactic expansion is the subject of one of the great cosmic mysteries. This paradox was formulated by noted physicist Enrico Fermi. Its premise in that life should be endemic to our galaxy, if not the universe. The theory assumes some life would ultimately evolve into an intelligent, technological species, which in turn would figure out how to expand into interstellar distances. Even at sub-light speeds, it would take only a few tens of millions of years to expand across our galaxy. The Fermi Paradox is that we see no evidence of such expansion despite the possible presence of tens of billions of earth-sized planets with earth-like conditions in the galaxy. There

S. P. Worden (✉)
Breakthrough Prize Foundation, Menlo Park, CA, USA
e-mail: pete@breakthroughprize.org

K. Szocik (ed.), *Human Enhancements for Space Missions*, Space and Society,
https://doi.org/10.1007/978-3-030-42036-9_9

are numerous explanations for this lack of signals or evidence of extraterrestrial intelligence. However, one explanation is that we are the first intelligent, space-faring species—or at least the first in our part of the galaxy.

We need to examine whether, how and when we might consider interstellar expansion—if for no other reason to back up our species. There are, of course, ethical and philosophical issues with such expansion—particularly if there is already life elsewhere, even primitive life. More will be said on this topic later. But now we must consider how and when we might perform such expansion.

Most speculative fiction postulates some means to significantly exceed light speed—often in a human-occupied "starship." But modern physics seems to rule out starships. Others suggest massive, city-sized sub-light "generation ships bridging interstellar space over the course of many centuries. These might be feasible but not any time soon. There may be another way that will emerge in the next few decades.

On April 12, 2016, Investor and Physicist Yuri Milner along with Professor Hawking announced Breakthrough StarShot (Starshot). This program is a multi-decade, privately-funded effort to build and launch scientific probes to the nearest stars at a reasonable fraction of light speed. We are targeting 0.2 c–60,000 km/s which would enable us to visit these nearby systems on human time-scales of a few decades. Conventional or even unconventional propulsion is not practicable. The amount of energy needed to send even a gram at 0.2 c is equivalent to a several kiloton nuclear weapon. To send a typical interplanetary probe weighing 1000 s of kg would require energy equivalent to all the nuclear weapons of a major superpower. To send an interstellar manned mission including the means to stop at the other end would be many orders of magnitude harder. For this reason, we chose a method based on a significant proposal by UC Santa Barbara physicist Philip Lubin (2016). He suggested that gram-class probes, within the reach of today's technology, could be accelerated to 0.2 c using light pressure on a modest-sized light sail from a laser array. The laser array would be quite large—kilometer scale and have an output of 100 GW. This is orders of magnitude larger than possible today—but within our technological capabilities and cost reasonableness within a few decades. But these are gram-class spacecraft and they do not stop at the nearby star systems. They will fly by the target star system at 0.2 c and return a few images and other scientific data of planets orbiting those stars. To see the current system model for this approach, see (Parkin 2018). Figure 9.1 shows a mock-up of the StarChip. Not very useful for interstellar settlement—or is it?

In 2007, several German experts suggested that the StarShot technology might be further evolved to milligram class spacecraft (Heller and Hippke 2017). Such systems, attached to somewhat larger light sails and traveling a factor of ten slower (0.02 c), could potentially use the starlight of the target star system to slow down the probe enough to enter the star system and eventually orbit target planets—and even enter the atmospheres of these worlds. But what use if a few milligrams on an alien planetary surface?

At a workshop sponsored by the Harvard Center for Astrophysics and the Breakthrough Initiatives on December 9, 2018, ("Seeding Life" Workshop Center for Astrophysics, Harvard University December 9, 2018) noted Harvard geneticist

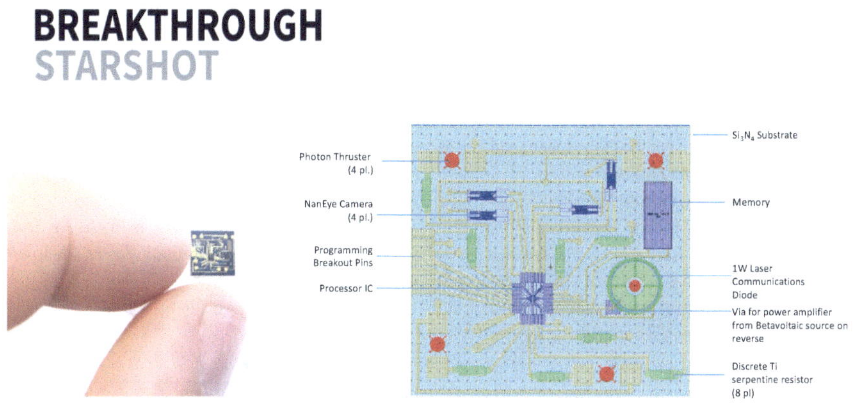

Fig. 9.1 Mock-up of sub-gram interstellar spacecraft— "StarChip." Courtesy Breakthrough Initiatives

George Church proposed an approach using StarShot technologies that could enable our civilizations interstellar expansion. While highly speculative, and what many might regard as "science fiction," he suggests that even nanoscale probes could contain bio-mechanical systems that could, if planted on a suitably resource-rich exoplanet surface with suitable temperature and pressure conditions, "boot-up" a life system. With genetic codes for terrestrial life it could in time incubate higher life forms, including humans. It could also construct electro-optical antennas to receive high data rate transmissions from earth. The high data rates would enable us to transmit, effectively, our intelligences and civilization to these new homes. It is likely, moreover, that the life incubated on these new worlds would be genetically optimized for the environments there—the ultimate human enhancements.

These interstellar human expansions could begin this century. Some such intelligent "seeding" might even be how life emerged on earth billions of years ago as first suggested by noted biologist and DNA co-discover Crick and his colleague Orgel (1973).

But what of the philosophical and ethical implications of these expansion efforts? In particular, we may have little way of knowing what other life including life we might not even recognize exists already on our target worlds. Is it even appropriate to impose our life and system elsewhere? And finally, what are the ethics of encapsulating a human intelligence in a bit stream? These challenges may be more formidable than the technical ones and we should begin to address them now.

References

Crick, F. H., & Orgel, L. E. (1973). Directed panspermia. *Icarus, 19*(3), 341–346.
Hawking, S. (2018). *Brief answers to the big questions*, Blackwell.
Heller, R., & Hippke, M. (2017). Deceleration of high-velocity interstellar photon sails into bound orbits at α centauri. *The Astrophysical Journal Letters, 835*(2).
Lubin, P. (2016). A roadmap to interstellar flight. *JBIS, 69,* 40–72.
Parkin, K. L. G. (2018). The breakthrough starshot system model. *Acta Astronautica, 152,* 370–384.
Reworking. (2014). *Reworking the human genome so people can colonize other planets.* https://phys.org/news/2014-03-reworking-human-genome-people-colonize.html.
SpaceX. (2014). https://www.spacex.com/mars.
Starshot. (2014). https://breakthroughinitiatives.org/initiative/3

Chapter 10
Anti-Aging Medicine as a Game Changer for Long-Lasting Space Missions

Riccardo Campa

Abstract Many studies have highlighted that astronauts involved in long-lasting space missions are at risk of developing pathologies similar to those caused by aging. This finding has led several scholars to raise ethical concerns about plans for exploration and colonization of other celestial bodies. This research, while recognizing the hitherto expressed medical and ethical concerns as founded, highlights the fact that human enhancement, and in particular anti-aging therapies, could be the game changer. As it will be showed, anti-aging medicine is progressing at an increasingly accelerated pace. Quantitative and qualitative methods are implemented to provide evidence of this fact. First, the author presents the results of a scientometric analysis of trends in the field of anti-aging medicine. The available data clearly show that studies in this field grow at an almost exponential rate. Then, the author resorts to qualitative meta-analysis to review several medical innovations such as stem cell transplants, organ regeneration, genetic editing, heterochronic parabiosis, telomere lengthening, growth hormone treatments, and some drug therapies, starting from those based on lithium. Given the amount and quality of studies in the field of regenerative medicine, it seems justified to cultivate a moderate optimism regarding future long-lasting space missions. If safe remedies to rejuvenate the human organism in its entirety are discovered, ethical doubts about space exploration and colonization should *ipso facto* dissolve.

10.1 Background

A review of the most recent literature on the medical aspects of space exploration is puzzling. On the one hand, studies are pointing out that long-term space missions may have anti-aging effects. For instance, a team of researchers observed that "in *Caenorhabditis elegans,* spaceflight suppressed the formation of transgenically expressed polyglutamine aggregates, which normally accumulate with increasing

R. Campa (✉)
Institute of Sociology, Jagiellonian University, Cracow, Poland
e-mail: riccardo.campa@uj.edu.pl

K. Szocik (ed.), *Human Enhancements for Space Missions*, Space and Society,
https://doi.org/10.1007/978-3-030-42036-9_10

age." The researchers also observed that "the inactivation of each of seven genes that were down-regulated in space extended lifespan on the ground" (Honda et al. 2012).

Experiments on nematodes showing that aging slows down in space prompted investigations of spaceflight's anti-aging effects in humans. As is well known, the atmosphere and the geomagnetic field protect life from solar wind and cosmic rays. By analyzing astronauts' heart rate variability (HRV), Otsuka et al. (2019) concluded that "the magnetic field can affect and enhance HRV indices involved in longevity, notably during the daytime and evening, probably in association with an elevated activity of brain's DMN [Default Mode Network, A/N] with help from the circadian clock." The researchers observed that anti-aging HRV indices "were significantly increased on days of higher magnetic activity".

On the other hand, there is copious scientific literature pointing out the deleterious effects of spaceflights on human health. Strollo et al. (2018) state that "astronauts coming back from long-term space missions present with different health problems potentially affecting mission performance, involving all functional systems and organs and closely resembling those found in the elderly". In other words, it seems that space missions have aging effects, rather than anti-aging ones. These apparently aging effects concern several organs, tissues, and systems, such as bones, immune system, blood, cartilage, and eyes. More in detail, Cappellesso et al. (2015) underline that "osteoporosis is one of the established major consequences of long-duration spaceflights in astronauts seriously undermining their health after their returning on Earth" and specify that "astronauts typically lose more bone mass during one month than postmenopausal women on Earth lose in one year".

That same fact was noticed by Cavanagh et al. back in 2005. These researchers showed that "bone loss in the lower extremities and lumbar spine is an established consequence of long-duration human space flight" and suggested possible countermeasures for the problem. It has been suggested that the class of drugs most commonly used to treat osteoporosis in postmenopausal women, that is antiresorptives such as risedronate sodium, alendronate sodium, raloxifene, and zoledronic acid, could be also used to treat astronauts. Besides, the authors mentioned the possible use of supplements such as vitamin D and calcium, and of "anabolic agents such as parathyroid hormone (PTH) and teriparatide (rhPTH [1–34])". The idea of dealing with the problem of bone loss induced by long-term spaceflights by way of genetic enhancement has also been explored, for instance by Szocik et al. (2019a).

Sonnenfeld et al. (2003) offer evidence that immune response is another of the many physiologic functions of mammals affected by exposure to space flight conditions and concluded that "such alterations could lead to compromised defenses against infections and tumors". More recently, Crucian et al. (2018) have confirmed that the immune system is subject to dysregulation during spaceflight and suggested potential countermeasures for deep space exploration missions. The reactivation of latent herpesviruses in astronauts for the duration of a 6-month orbital spaceflight is a well-known phenomenon. It has been observed that spaceflight dysregulates adaptive immunity, heightens innate immunity, and alters the interaction between adaptive and innate immunity. As a consequence, it may happen that astronauts experience persistent hypersensitivity reactions during flight.

Red blood cells are also negatively affected by the condition of space flights. De Santo et al. (2005) point out that "a state of anemia associated with reduced erythropoietin levels and reduced plasma volume" was already observed after the very early crewed missions in space, and that "the reduction in red blood cell mass is driven by a process of selective hemolysis, which has been named neocytolysis".

In this case also, a parallel could be traced between the clinical condition of the astronauts and that of the elderly. It is well known that anemia is very common in seniors. Ferrucci and Balducci (2008) state that "aging is associated with increased incidence and prevalence of anemia" and, "in approximately 30% of cases, anemia in older individuals is due to either relative or absolute erythropoietin deficiency".

Both the astronauts and the elderly tend to experience problems related to articular cartilage. The deleterious effects of microgravity on the human skeleton and more specifically on cartilage are known. "Data from terrestrial hind-limb unloading (HLU) animal experiments and human bed-rest studies demonstrate that reduced mechanical forces associated with joint unloading and immobilization leads to the proteoglycan loss in articular cartilage" (Fitzgerald 2017). Lotz and Loeser (2012) emphasize the effects of aging on articular cartilage homeostasis, mentioning extracellular matrix changes such as "reduced thickness of cartilage, proteolysis, advanced glycation and calcification" and cellular changes such as "reduced cell density, cellular senescence with abnormal secretory profiles, and impaired cellular defense mechanisms and anabolic responses".

Finally, there are many studies on the weakening of sight, in both astronauts after long-term missions and the elderly. There is no need to provide literature confirming sight problems in seniors, as we all observe this phenomenon in everyday life. Several studies provide evidence that analogous problems may affect astronauts. By analyzing data concerning 300 individuals, Mader et al. (2011) show that "approximately 29% and 60% of astronauts on short- and long-duration missions, respectively, experienced a degradation in distant and near visual acuity" and add that "some of these vision changes remain unresolved years after flight". The postflight inquiry documented that "after 6 months of space flight, 7 astronauts had ophthalmic findings, consisting of disc edema in 5, globe flattening in 5, choroidal folds in 5, cotton wool spots (CWS) in 3, nerve fiber layer thickening by OCT in 6, and decreased near vision in 6 astronauts." The effect of microgravity on ocular structures and visual function has been reviewed in detail also by Taibbi et al. (2013).

After reading this literature, one wonders if the prevailing effect of spaceflight is aging or anti-aging. To put it in simple words, by flying for some time in space, does one become older or younger than one would become while remaining firmly on planet Earth? In terms of effects on the human body, microgravity is an environment that is still not completely understood. As we could see, the effects are mixed. Moving in a microgravity environment requires less effort, and this condition may benefit some organs. Besides, human cells and organs tend to adapt to the new condition, starting from gene expression (Wnorowski et al. 2019). Still, it seems that aging effects are largely prevailing. Even those articles that stress the anti-aging effects of spaceflight recognize that all that glitters are not gold. For instance, if it is true that aging in *C. elegans* seems to slow down in space, it is also true that “male Drosophila lived shorter on the ground after spaceflight compared to controls that maintained on the ground through their lifetimes” (Honda et al. 2012).

This does not mean, however, that humans should abandon their dreams of space exploration and colonization. There are at least three good reasons that justify a moderate optimism. The first reason is that we already possess quite effective sets of countermeasures to combat the negative effects of spaceflight (Iwamoto et al. 2005; Payne et al. 2007; Hargens et al. 2013). The second reason is that outer space is a “gold mine” and, consequently, private companies and governmental agencies will keep pursuing space exploration and colonization by using all the possible means offered by emerging technologies, including artificial intelligence and advanced robotics ([illegible] et al. 2019, Campa et al. 2019). The third reason is that health problems related to spaceflight can also be countered by human enhancement and in particular artificial rejuvenation. Anti-aging medicine keeps progressing. And this is the main theme of this chapter. Our purpose is to show how breakthroughs in anti-aging medicine may shortly become a game changer for space exploration and colonization, helping to prevent or promptly recover from damages caused by hostile space conditions.

10.2 A Scientometric Glance on Anti-Aging Research

If one looks for the term “rejuvenation” in Google Scholar, the search engine finds around 300,000 results. It is clear however that this word can be used in very different contexts. A handmade search reveals that not a few of the detected scientific works focus on the rejuvenation of various activities and things, such as business management or libraries, rather than that of the human organism.

If the search is restricted to the combination of the terms “rejuvenation” and “medicine,” it still provides a large number of items—96,000—to be exact. Nonetheless, it seems safer to look for a more specialist term, such as “anti-aging”. In this case, Google Scholar detects 173,000 scientific works which include the word. If the search is limited to the twenty-first century, 76,500 books and articles on the subject are detected. Another 28,600 scientific works emerge, if we set the search on the British spelling of the word that is “anti-ageing”.

The most interesting aspect is the trend. As one can see from Graph 1, in the twenty-first century, the publications including the term “anti-aging” or “anti-ageing” experience a steady growth.

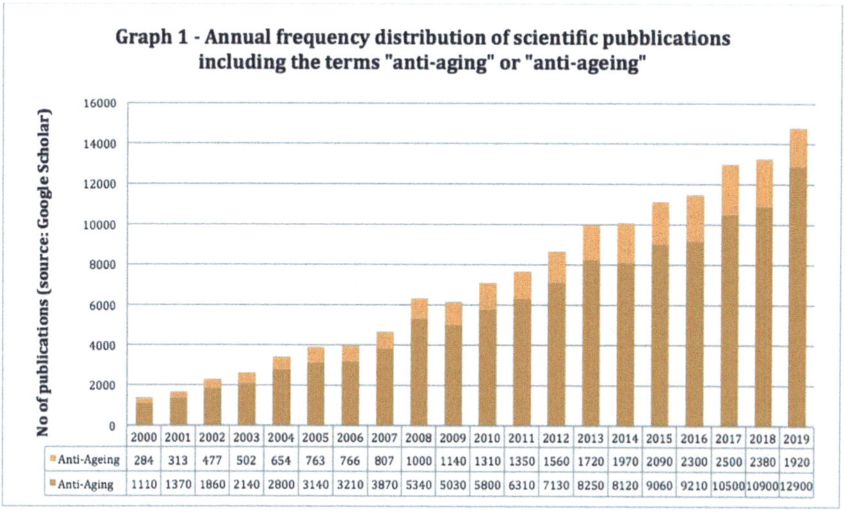

	2000	2001	2002	2003	2004	2005	2006	2007	2008	2009	2010	2011	2012	2013	2014	2015	2016	2017	2018	2019
Anti-Ageing	284	313	477	502	654	763	766	807	1000	1140	1310	1350	1560	1720	1970	2090	2300	2500	2380	1920
Anti-Aging	1110	1370	1860	2140	2800	3140	3210	3870	5340	5030	5800	6310	7130	8250	8120	9060	9210	10500	10900	12900

The figure represents the distribution of the absolute frequencies. Relative frequencies would tell us the proportion of the intellectual effort devoted to this research field. However, in this context, the global effort is more significant than the relative one. After all, it is the global effort that increases the likelihood of finding an effective solution for the aging problem, which is what matters. We can anyway exclude that the growth of publications in anti-aging medicine is a side effect of the general growth of publications. As Graph 2 clearly shows, the distributions of scientific publications globally follow a different pattern. Even when the absolute frequency of publications decreases, those including the word "anti-aging" keep growing.

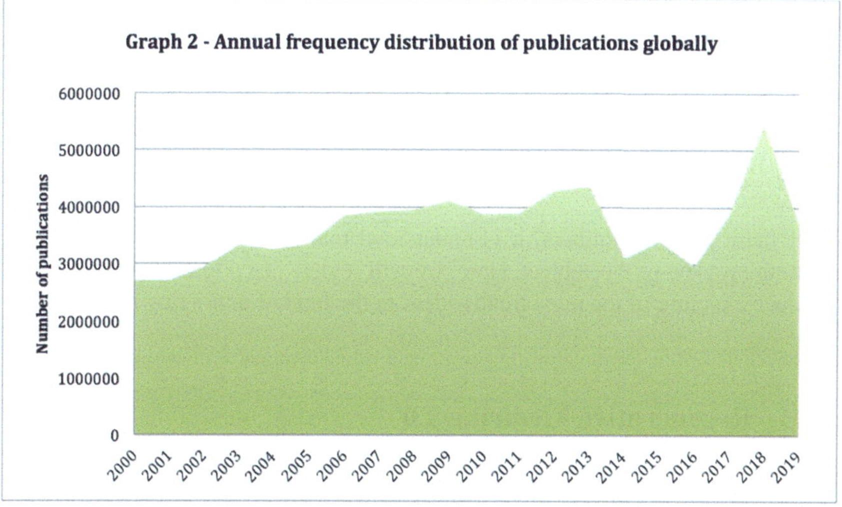

This search could be implemented in different ways. For instance, we could include in the search the scientific literature in languages other than English, and the number of items would massively grow. Or we could restrict the search to the journals with higher impact factors, and the numbers would fall. Google Scholar detects publications in any discipline. Indeed, anti-aging therapies can be discussed by physicians as well as philosophers or theologians. If we restrict the search to medical literature, the global number decreases. However, even if we rely on a more specialized search engine such as PubMed, which primarily accesses the MEDLINE database of references and abstracts on biomedical topics and life sciences, the number of articles remains remarkable. The medical publications including the term "anti-aging" are 3726 globally. Those published in the twenty-first century are 3617, while only 109 articles were published in the twentieth century. The search for the British word "anti-ageing" gives a similar result, even if on a smaller scale. The total amount of articles is 569, while those published in the twenty-first century are 548. The struggle against senescence has reportedly gained momentum in the new millennium. Graph 3 shows that the growth of medical publications is even more accentuated than the growth of all publications including the terms "anti-aging" or "anti-ageing." If Graph 1 represents a steady growth, the curve in Graph 3 takes the shape of a seemingly exponential function.

Graph 3 - Annual frequency distribution of medical publications including the terms 'anti-aging' or 'anti-ageing'

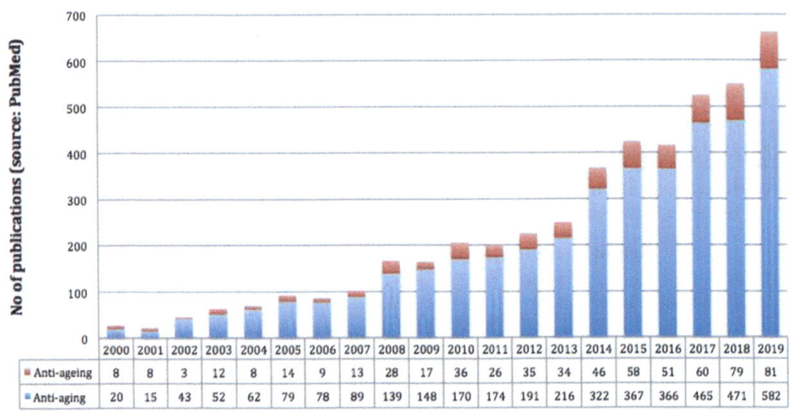

	2000	2001	2002	2003	2004	2005	2006	2007	2008	2009	2010	2011	2012	2013	2014	2015	2016	2017	2018	2019
Anti-ageing	8	8	3	12	8	14	9	13	28	17	36	26	35	34	46	58	51	60	79	81
Anti-aging	20	15	43	52	62	79	78	89	139	148	170	174	191	216	322	367	366	465	471	582

If these are the numbers, it is understood that we cannot provide a complete literature review of the subject. Here, we will review just a few exemplary cases, to produce a picture of the most trendy ideas in the field of anti-aging medicine.

10.3 Regenerative Medicine 2.0

With regard to stopping or even reversing the senescence process, a first research line considered very promising is that of regenerative medicine. The basic idea is to regenerate organs and tissues by infusing reprogrammed stem cells in the human body. The fundamental breakthrough in this field dates back to the beginning of the eighties when pluripotent stem cells are derived from the embryos of mice in two independent studies (Martin 1981; Evans and Kaufman 1981). After the discovery of the CRISPR/Cas-9 system, which allows us to reprogram the DNA as if it were software, we have entered a new phase of regenerative medicine (Li et al. 2018).

To take an example of regenerative medicine 2.0, in 2019, a group of researchers led by Ergin Beyret showed that CRISPR/Cas9 therapy can suppress aging, improve health, and prolong lifespan in mice. More precisely, researchers have developed a new gene therapy capable of suppressing the accelerated aging observed in mice with Hutchinson-Gilford progeria syndrome, a rare genetic disease that also affects humans (Beyret et al. 2019). The ultimate goal is to use this knowledge to halt the senescence process of the human organism.

Many independent research projects focus on the regeneration of single organs and tissues. Also, in 2019, the combined use of new genetic editing technologies and reprogrammed stem cell infusion has been successfully applied to the regeneration of skull bones (Truong et al. 2019).

As for the prospects in the near future, one thinks about the use of mesenchymal stem cells for tooth regeneration. A team of Chinese biologists remarked that "a number of basic researches, preclinical studies and clinical trials have investigated that dental stem cells efficiently improve formation of dental specialized structure and healing of periodontal diseases, suggesting a great feasibility and prospect of these approaches in translational medicine of dental regeneration" (Shuai et al. 2018).

Regenerating the entire organism, piece by piece, is the goal on which the efforts of thousands of researchers around the world converge. Advances in this field will become fully apparent when regenerative medicine is successfully applied to repair the visible parts of the body, especially teeth, hair, skin, and muscles. An octogenarian with the body's internal and external organs completely regenerated, on a functional and aesthetic level, would de facto be a young person.

New skin regeneration techniques have been studied for years, to replace traditional plastic surgery. The book *Skin Tissue Engineering and Regenerative Medicine* by Albanna and Holmes (2016) IV presents the state of the art in this research field. Published in 2016, the volume does not include recent years' innovations. In a rapidly growing sector like that of regenerative medicine, four years are significant amount of time. However, in this context, we are mainly interested in the basic idea. The authors start from the principle that the skin is the largest system of human organs and that "loss of skin integrity due to injury or illness results in a substantial physiologic imbalance and ultimately in severe disability or death". Let us think about victims of burns or cut injuries. Tissue engineering and regenerative medicine are born with a therapeutic purpose. They are designed to deal with diseases, illnesses, and disabilities. However, it is not difficult to understand that these treatments, once perfected, would not remain confined to the clinical and therapeutic field. In principle, they can be used to reverse the senescence of epithelial tissues, to regenerate the skin and epidermis of elderly individuals, or to prevent aging in healthy individuals. The senescence process itself could eventually be seen as a disease.

10.4 The Frontier of Parabiosis

A second promising strand in the field of anti-aging therapies is that of heterochronic parabiosis. It is a field of study that has an ancient history, which had momentum in the second half of the twentieth century and then went out of fashion, to finally return to being explored again today. The reason for the discontinuity of this research topic could be found in the sinister aspects of the basic idea and of the experimental methods. To get straight to the point, if one injects the blood of a young individual into the body of an elderly one, the latter rejuvenates. This fact is proved by experiments on laboratory mice which undoubtedly look gruesome to animal rights activists. A certain portion of skin located on the lateral back of two rats, one young and one elderly, is connected through surgery. From that moment, the two animals form a single body. They must walk together, eat together, and sleep together. The capillaries of the skin are also connected so that the blood of the two animals is continuously mixed. After a period of symbiotic life, the rats are disconnected and the biological parameters re-measured. After the "treatment," the elderly individual is significantly rejuvenated.

As a team of Italian researchers explain, "transfusion (or drinking) of blood or of its components has been thought as a rejuvenation method since ancient times. Parabiosis, the procedure of joining two animals so that they share each others blood circulation, has revitalized the concept of blood as a putative drug. Since 2005, a number of papers have reported the anti-ageing effect of heterochronic parabiosis, which is joining an aged mouse to a young partner" (Conese et al. 2017). The body's ability to regenerate most tissues reaches its apical moment in youth. One of the reasons why we age is the gradual loss of this ability. The decline can be partially traced to the impairment of the function of stem and progenitor cells. Parabiosis experiments demonstrate "that factors derived from the young systemic environment are able to activate molecular signaling pathways in hepatic, muscle or neural stem cells of the old parabiont leading to increased tissue regeneration." Going into technical details, the Italian researchers stress that "further studies have brought to identify some soluble factors in part responsible for these rejuvenating effects, including the chemokine CCL11, the growth differentiation factor 11, a member of the TGF-β superfamily, and oxytocin" (Conese et al. 2017). It should be, however, remarked that the mechanisms that produce the rejuvenation effects are still largely unknown. The explanation crafted by Loffredo, Wagers, and others, back in 2013, is under debate. For instance, it seems to be conflicting with the results published in *Cell Metabolism* by Egerman, Glass, and others, in 2015. This controversy has been briefly summarized by Reardon on *Nature* (2015). Still, in spite of the fact that there is no unanimously accepted theory explaining the phenomenon, there is little doubt that heterochronic parabiosis empirically works quite well.

If it works with rats, it could or should work with humans too. Experiments can hardly be performed on humans in the same way as they are on rats. However, since the trials do not involve drugs, but transfusions, any private clinic can put the idea into practice, once it finds young volunteers willing to donate blood and old patients wishing to have it injected. In the popular press, we already read of wealthy Silicon Valley managers who have no intention to wait for the conclusion of parabiosis clinical testing and are ready to buy young blood transfusions to reactivate the regeneration of their bodies (Dancyger 2018).

Conspiracy literature already speaks of "class vampirism". We can certainly agree that a world in which everything is for sale, included blood, appears to be quite sinister, but we must not forget that research on artificial blood creation is quite advanced. In 2017, a group of Bristol scientists published an article in *Nature Communication* that shows how artificial blood can be produced in large quantities.

In the incipit of the article, Trakarnsanga et al. (2017) note that the lack of blood is an important global health problem that risks becoming even more problematic in the future because, in the face of the extension of people's average life span, there is a decrease in the number of blood donors. That is why it is vital to find alternative methods to produce red blood cells.

Cultured red blood cells provide such an alternative and have potential advantages over donor blood, such as a reduced risk of infectious disease transmission, and as the cells are all nascent, the volume and number of transfusions administered to patients requiring regular transfusions (sickle cell disease, thalassemia myelodysplasia, certain cancers) could be reduced, ameliorating the consequences of organ damage from iron overload (Trakarnsanga et al. 2017).

As regards this challenge, many research teams are active and looking for the most effective and safe procedure. The mechanisms involved in the production of blood are still not fully understood, and different techniques under investigation have pro and contra. The British team recalls that "pluripotent stem cells (PSCs) provide a potentially unlimited progenitor source; however, there are substantial hurdles to overcome before these cells can be considered for manufacture of red cells". One of the problems is "the small number of erythroid progenitors generated to date and severely impaired enucleation of the resultant erythroid cells". The method presented in *Nature Communications* seems to overcome these problems. The strategy consists in generating immortalized adult erythroid progenitor cell lines. As the researchers emphasize, "such lines are capable of providing an unlimited supply of red cells and need only minimal culture to generate the final product".

We will not discuss the details of the technique, as the matter goes beyond our scientific competence. What we mainly want to emphasize is that the idea of blood as a rejuvenation resource must be understood within the framework of an almost unlimited availability of artificial blood. Certainly, ethical problems will be raised also as regards this scenario. Such is the axiological dissonance within the Western society, especially in the bioethical field, that there is virtually no idea immune from controversy. To provide just an example, in a 2018 article published in the journal *Transfusion Medicine and Emotherapy*, Hofmann (2018) hastens to affirm that, "even if rejuvenating substances from blood may be artificially and cheaply produced and justly distributed, problems arise: survival may have to be balanced against reproduction, as reproductive age increases". Hofmann admits that endless bliss and eternal youth have always been vital human dreams and that "parabiosis may bring us closer to the fountain of youth than ever," but at the same time, he stresses that "it is still too early to provide full-fledged assessments of its implications or to foresee how it will change health, aging, medicine, and society". Therefore, he invites us to discuss the possible scenarios opened up by this research line.

These observations are relevant, but it is also true that a pragmatic approach gives priority to the solution of real current problems, rather than hypothetical future ones. If parabiosis understood as class vampirism rightly arouses a feeling of rejection, once it is possible to produce young blood in unlimited quantities in the laboratory, we can expect that the treatment will remain ethically problematic only for a marginal fringe of the population. After all, there is already a percentage of people who, for religious reasons, refuse blood transfusions even for therapeutic purposes.

10.5 Growth Hormone, or the Holy Grail

Growth hormone (GH), or somatotropin, has been known for more than half a century as a potential elixir of eternal youth. It is well known that this peptide produced by the pituitary gland stimulates statural growth during childhood and adolescence. Somatotropin continues its action during adulthood, favoring the oxidation of lipid stocks and regulating various metabolic processes. In simple words, it helps to keep the body strong, fit, healthy, and young. The problem is that the amount of hormone secreted by the gland decreases over time. During childhood, the secretion of somatotropin grows and, not surprisingly, during puberty reaches its maximum, to remain constant during the young age. After the thirtieth year of age, the body begins instead to reduce the secretion of growth hormone. In a fifty-year-old individual of the human species, the secretion is halved compared to twenty years before, while in a seventy-year-old individual, the secretion of somatotropin is reduced to a third if compared to that measured when he was a young adult. The decline in production is dramatic, with all the consequences that each of us can ascertain.

There is evidence that this process is not inevitable. Lifestyle greatly affects the secretion of growth hormone. For example, a rich protein diet and heavy physical exercise cause the pituitary gland to secrete a greater amount of somatotropin, counteracting the physiological decline and aging of the body. However, the increase in secretion is greater in untrained subjects than in trained ones, because it is activated in the presence of a considerable and painful effort, capable of stimulating the production of lactic acid. Put it differently, once a certain degree of physical fitness is reached, training does not produce appreciable rejuvenating effects. Therefore, alternative solutions to endogenous production have always been sought, such as the intake of anabolic substances capable of stimulating the secretion of the hormone, or the intravenous injection of the hormone itself. However, the latter solution also presents complications.

To show what the problems are, let us briefly review the history of the discovery. In the 1950s, in Montreal, John C. Beck and some collaborators carried out clinical studies on the growth hormone of a thirteen-year-old male specimen of the Macaca mulatta species, or rhesus macaque. At a time when experimental protocols and examinations of ethical committees were less rigid and rigorous than those of today, the step from animal testing to human experimentation was relatively short. A first study, published in *Science* in 1957, showed that the administration of human and monkey growth hormone leads to a significant improvement in the conservation of nitrogen, to a retention of potassium, phosphorus, calcium, and sodium, to an increase in body weight and, ultimately, a significant increase in urinary aldosterone excretion. The effects were more pronounced when the human growth hormone was administered (Beck et al. 1957).

In 1958, on the *Annals of Internal Medicine*, the same research team published a similar study conducted on six people: an eighteen-year-old male with a statural and sexual delay and a skeletal age between thirteen and fourteen; a fifteen-year-old female with well-documented hypopituitarism secondary to craniopharyngioma; a fifty-three-year-old woman with breast cancer and widespread skeletal metastases; a sixty-eight-year-old woman with advanced postmenopausal osteoporosis; and a healthy twenty-four-year-old medical student. Once again, the ability of the substance to visibly affect metabolic processes was proven (Beck et al. 1958).

Since the first attempts to synthesize the growth hormone in the laboratory failed, the substance was initially extracted from the pituitary glands of corpses. The difficulty in finding the peptide limited its use for therapeutic purposes. It was mainly administered to patients suffering from idiopathic short stature. However, in 1985, it was discovered that subjects who had received the growth hormone extracted from corpses ten or fifteen years earlier often suffered unexpected pathologies. Unusual cases of Creutzfeldt–Jakob disease, better known as "mad cow disease," were diagnosed. The cadaver-derived growth hormone was therefore removed from the market. The hypothesis is that the infectious prions that cause the disease can be transferred together with the somatotropin extracted from the gland.

This gruesome discovery, however, did not stop the use of the substance, because just in those years, the procedure to produce biosynthetic growth hormones was eventually devised, through the recombinant DNA technique. The discovery is credited to Choh Hao Li, a Chinese biochemist, emigrated to the USA in 1935. In the mid-1960s, the Chinese scientist discovered that somatotropin is composed of a chain of amino acids. In the following decade, Choh Hao Li managed to identify, purify, and synthesize the hormone (cf. Cole 1996).

The pharmaceutical industry immediately comes into action. The Californian company Genentech paves the way to the first use of recombinant human growth hormone for therapeutic purposes. In 1981, Robert A. Swanson, President and Chief Executive Officer of the company, announces that three products—human insulin, human growth hormone, and interferon—all produced by microbes developed at Genentech, undergo human clinical trials (Genentech 1981). A little later, Nutropin is put on the market.

One of the practical problems related to this treatment is the need to administer the substance through injections. Oral administration would be ineffective because the substance would simply be digested. Injecting it daily, however, requires a continuous rotation of the areas in which to insert the needle or the insertion of a subcutaneous cannula fixed with a patch that can be used several times. In any case, the treatment can cause discomfort. Attempts to solve the problem fail. In 1999, the FDA approves Nutropin Depot, a drug patented by Genentech and Alkermes which provides for the prolonged release of the substance. This product requires an injection every two weeks, if not every month, instead of the daily treatment. However, Nutropin Depot is not successful, and in 2004, its sale is suspended. The production cost is too high and the price non-competitive, if compared to that of other products on the market.

Since 2005, many other recombinant DNA growth hormones have been synthesized. The list of approved somatropin-based products includes Pfizer's Genotropin, Eli Lilly's Humatrope, Novo Nordisk's Norditropin SimpleXx, Ipsen's NutropinAq, Merck Serono's Saizen, Ferring's Zomacton, and Sandoz Omnitrope. In addition to medical treatments, the potential use of GH for sports doping purposes was immediately glimpsed. The peptide hormone is now easier to find than when it was extracted from cadavers. Besides, the substance naturally produced by the adenohypophysis is difficult to distinguish from that surreptitiously introduced into the body. Although it remains an expensive treatment, the sport industry can sustain the expense and is constantly exposed to the temptation of doping, as the profitability of this business is mostly dependent on athletes' performance. As was the case with testosterone and anabolic steroids, the issue is constantly being examined by scientific literature and new strategies are continuously studied to avoid abuse (Holt 2011; Baumann 2012).

Abuses can also occur outside the sports world. Since it is known that this hormone has anti-aging effects, whoever has the necessary financial resources tries to obtain it from compliant doctors or from patients with medical prescription willing to resell it, with all the risks of the case, being the clinical tests still in progress.

The idea of pharmaceutical companies is to finally arrive at a safe product that can be legally sold also to healthy individuals. This path is leading to surprising results. On September 8, 2019, we learned that a group of Californian researchers has succeeded not only in stopping aging in adult human beings but even in reversing the natural process of senescence. The subjects of the experiment have been rejuvenated. The news quickly bounced from *Aging Cell*, the scientific journal in which it appeared, to traditional media and social networks.

In short, it seems possible to turn back the clock on biological age. The treatment had already been tested with encouraging results on laboratory animals, but it was still not known whether it was possible to reduce the biological age of adult human organisms in safe conditions. The study reports a clinical trial of Thymus Regeneration, Immunorestoration, and Insulin Mitigation (TRIIM), involving nine male individuals of Caucasian ethnicity, aged between fifty-one and sixty-five. The experiment was conceived and led by the immunologist Gregory Fahy, scientific director and co-founder of Intervene Immune, a private company operating in the field of biotechnology and health services based in the Los Angeles area. The trial took place at the Stanford Medical Center in Palo Alto, California.

For a year, nine healthy volunteers took a "cocktail" of three substances: growth hormone and two common diabetes drugs. At the end of the trial, the biological age of the patients was measured by determining the state of genome DNA methylation. On average, the volunteers resulted to be two and a half years younger. Their immune system also benefited from the therapy, showing signs of rejuvenation (Fahy et al. 2019).

Of course, a certain degree of skepticism is always advisable. For decades, we have been hearing that we are on the threshold of the lost paradise, one step away from achieving eternal youth or unlimited life. News on the amazing results of revolutionary anti-aging therapies circulate all the time, even though, unfortunately, not a few of the treatments on the market are scams. As is well known, private companies are interested in patenting treatments, to profit from them. As a consequence, they sometimes emphasize positive results and minimize negative ones, to attract investments or get approvals from drug agencies.

Caution is a must even when faced with serious research, as the one published in *Aging Cell*. The Californian researchers warned that those achieved are only preliminary results, obtained by a trial on a small sample of volunteers and in the presence of a conflict of interests. Therefore, further studies on more representative samples, including subjects of different age, sex, and ethnic groups, are needed. Studies performed by independent researchers would also be needed.

Still, there are reasons for moderate optimism. The project was screened by ethics committees. The trial was approved, back in 2015, by the FDA. The study passed the double-blind assessment process and was published in a prestigious journal. Solid foundations are laid for further experimentation. The news was also taken up by *Nature*. The methods and results of the experiment were summarized as follows:

The latest trial was designed mainly to test whether growth hormone could be used safely in humans to restore tissue in the thymus gland. The gland, which is in the chest between the lungs and the breastbone, is crucial for efficient immune function. White blood cells are produced in bone marrow and then mature inside the thymus, where they become specialized T cells that help the body to fight infections and cancers. But the gland starts to shrink after puberty and increasingly becomes clogged with fat. (Abbott 2019).

Growth hormone stimulates thymus regeneration but in the presence of unwanted side effects, such as the contraction of diabetes. Therefore, "the trial included two widely used anti-diabetic drugs, dehydroepiandrosterone (DHEA) and metformin, in the treatment cocktail." The results were a surprise for the researchers involved in the experimentation. Epigenetic analysis was conducted by geneticist Steve Horvath of the University of California, Los Angeles, who developed a multi-tissue predictor of age that allows us to estimate the age of DNA methylation of most tissues and cell types (Horvath 2013). His statement, also reported by *Nature*, ignites more than one hope on the effectiveness of this therapy: "I'd expected to see slowing down of the clock, but not a reversal". And he concluded: "That felt kind of futuristic".

10.6 Lengthening of Telomeres

In the scientific community, strong is the conviction that we get old because the cells that make up our body lose the ability to divide over time. One of the mechanisms identified at the base of this phenomenon is the shortening of telomeres, which are protective caps at the ends of chromosomes. Telomeres shorten with each cell division, and the time comes when they are so short, they damage the same genes they were supposed to protect. At that point, the cell stops dividing.

This mechanism works at different speeds in different individuals, not only for genetic reasons but also as a consequence of lifestyle or psychological stress level. Hence, it is useful to distinguish between chronological and biological age of a person. Telomere length is considered a marker of biological age. People having longer telomeres have a longer life expectancy and run less risk of developing age-related illnesses, such as Alzheimer's and Parkinson's disease, or cancer. From these considerations, the idea was born that "eternal youth" can be achieved by finding a way to slow down or stop the shortening of telomeres, or even cause their elongation. There is evidence that telomeres can change length faster than previously thought. Lengthening can also be achieved through specific mental or physical training and can be observed after a few months of treatment. A team of researchers has indeed discovered that the change in telomere length is associated with structural changes in the brain. The experiment was performed on 298 healthy adults, stimulated through mental activities and constantly monitored. When telomeres lengthen, with a certain degree of probability, the cerebral cortex also thickens, while telomere shortening is associated with a reduction in gray matter. As the researchers claim, their study "provides the first evidence to date for an association between short-term change in leukocyte telomere length and brain structure, suggesting that these processes may be mechanistically linked; the mental training used did not influence leukocyte telomere length of healthy, middle-aged adults" (Puhlmann et al. 2019).

The researchers also wanted to verify if the natural association between phenomena could be triggered by a mental training based on the growth of awareness and empathy, for a duration of nine months. As explained by the Max Planck Institute for Human Cognitive and Brain Sciences, "previous data from the ReSource Project, which was supported by the European Research Council (ERC), had already shown that certain regions of the cortex can be thickened by training, depending on the respective mental training contents of three distinct modules, each lasting for three months" (Max Planck 2019).

However, the researchers admitted that the mental training used did not affect the length of leukocyte telomeres in healthy, middle-aged adults, and therefore, other studies are needed to fully understand the mechanism and eventually control it.

Anyway, mental training is just one of many different strategies to elongate telomeres. A pharmacological treatment is also on the agenda of biologists and physicians. For instance, several studies have shown that there is a correlation between lithium intake and longevity, even if the mechanisms underlying this phenomenon have not yet been perfectly understood. Lithium is naturally found, in different concentrations, in tap water and in the food we eat, particularly in vegetables and derived food. The concentration of lithium in drinking water varies in different regions of the planet. In 2011, a team of Japanese researchers used "weighted regression analysis to identify putative effects of tap water-derived lithium uptake on overall mortality" (Zarse et al. 2011).

Independently, the same team of scientists has exposed some specimens of *C. elegans*, a small nematode commonly used for anti-aging studies, to comparable concentrations of lithium, quantifying their mortality. The results appeared encouraging. As for humans, "an inverse correlation between drinking water lithium concentrations and all-cause mortality in 18 neighboring Japanese municipalities with a total of 1,206,174 individuals ($\beta = -0.661, p = 0.003$)" was found. Consistently, it was also found that "exposure to a comparably low concentration of lithium chloride extends life span of *C. elegans* ($p = 0.047$)".

In the wake of this and other similar studies, the idea was born to verify if there is a correlation between lithium intake and telomere lengthening. A team of researchers led by Coutts (2019) made it clear that telomere length is both "a promising biomarker for age-related disease and a potential anti-ageing drug target," and provided evidence that lithium may "confer its anti-ageing effects by moderating the expression of genes responsible for normal telomere length regulation".

As regards the theoretical basis of the study, it is worth recalling that gene-enrichment analysis has identified thirteen genes associated with telomere length. Furthermore, the association between lithium and longevity can be measured not only on healthy subjects taking moderate amounts of lithium—through tap water, fruit, or vegetables—but also on subjects exposed to significant amounts of this substance, such as individuals subject to bipolar disorder. The depression that occurs during bipolar disorder is often treated with lithium. In this type of patient, antidepressants can cause manic episodes, while the response to lithium-based treatments is in most cases positive. The researchers confirmed that, concerning lithium, "chronic use in a sample of 384 bipolar disorder patients is associated with longer telomeres ($p = 0.03$)".

To corroborate the theory that lithium in our tap water has beneficial effects on health and longevity, and that lithium's effect on telomere length may be one the mechanisms by which it confers its anti-aging properties, the researchers "tested whether lithium affects the expression of genes responsible for telomere length maintenance". These were identified, from their gene-enrichment analyses, "in a relevant model system that recapitulates the drug's anti-ageing effects". Coutts and her collaborators found "that 3 out of the 13 genes identified from the gene-enrichment analysis had an assayed ortholog in a *C. elegans* model of lithium-induced extended longevity" and that "lithium had an effect on all three genes".

Other studies show how lithium exposure can have anti-aging effects and reduce mortality in distinct species, by acting on the length of telomeres, but the examples just provided may suffice for our purposes.

The question that arises is whether one can elongate telomeres by directly modifying the genetic makeup of individuals, employing the new gene-editing techniques. It has been proved that the length of telomeres is also linked to hereditary characteristics, and not only to lifestyle or nutrition. For example, it is known that children's organisms of different sex or ethnicity show different behaviors in this respect (Ly et al. 2019). Apart from all the ethical questions that can be raised regarding a modification of the human species with the means offered by genetic engineering, it has also been asked if the game is worth the candle. As we said, the telomeres gradually shorten to the point that the cell is no longer able to divide, but, at the same time, the cells accumulate mutations and anomalies over time. If genetically modified telomeres were created that do not shorten, the cells could multiply to infinity even when they are mutated or diseased. In other words, we may not grow old but may be more prone to tumors or other diseases. The "expiration date" that hangs over all of us could serve to avoid this scenario.

It is worth noting that the scientific literature has already given answers to this possible objection. A recent study in rats published in *Nature Communications* shows that an over-extension of telomeres does not lead to any undesirable side effects. Muñoz-Lorente et al. (2019) explain their experiment as follows:

> We generated mouse embryonic (ES) cells with longer telomeres than normal (hyper-long telomeres) in the absence of genetic manipulations, which contributed to all mouse tissues. To address whether hyper-long telomeres have deleterious effects, we generated mice in which 100% of their cells are derived from hyper-long telomere ES cells.

The results show no ambiguity whatsoever. It has been observed that "these mice have longer telomeres and less DNA damage with aging". Mice with hyper-long telomeres are slim and have low cholesterol and LDL levels, as well as better glucose and insulin tolerance. Besides, they also have "less incidence of cancer and an increased longevity". Otherwise stated, the study reports that having telomeres much longer than normal, in addition to slowing aging, not only does not increase the probability of contracting tumors but also, on the contrary, makes the onset of the disease less likely, at least in mice.

10.7 Concluding Remarks

In light of the medical studies carried out on astronauts during and after space missions, ethical concerns and doubts were rightly raised regarding plans of long-term space travel and colonization of other celestial bodies, starting with the Moon and Mars (Szocik et al. 2019b). Is it legitimate to send astronauts to space for long periods when the chance that their health condition could be hopelessly compromised is high? The issue of security is fundamental. We often forget, however, that medical research progresses very quickly, and as we believe we have adequately demonstrated, anti-aging medicine seems to be on the verge of a scientific revolution. If it were possible to regenerate and rejuvenate an entire human organism, many doubts about the safety of long-lasting space missions would *ipso facto* vanish. What we have presented here, both from a quantitative and qualitative point of view, are just some of the research lines in the field of anti-aging medicine. Our goal was to clarify that when we speak of "eternal youth," we are not talking about the regurgitation of the alchemical dream of a few visionaries, but of a huge scientific and industrial machine that has been set in motion in recent decades and is unceasingly producing new studies and discoveries. This is not to deny that there are still technical and ethical obstacles on the way. These scientific results could be questioned by new studies, and ethical or legal limitations to this field of inquiry could be imposed by governmental agencies in the future (Campa 2019, 160–173). However, there are good reasons to believe that this massive deployment of public and private economic resources and intellectual efforts will lead to important advances in these fields.

As we said, some scholars think that the very idea of rejuvenation, and of human enhancement in general, is morally wrong. They object that an expiration date of the human organism is necessary to maintain a social and natural equilibrium. It is our opinion that this objection can be rejected not only on scientific grounds but also on philosophical ones. It is evident that, for the maintenance and natural evolution of the species, the individuals serve as long as they are fertile and can take care of the offspring. Then, they become useless. If we were to decide to go along with nature, assuming it to be wiser than human rationality, we would have to question the whole of medical science and the very idea of health and social security services. It is not clear, indeed, why antibiotics and surgery are acceptable, not to say retirement pensions and health insurances, while hormones and telomeres should never be touched.

Since humans first began using tools, our evolution ceased to be purely natural and became self-directed, although not always knowingly. In a sense, our species is artificial by nature (Campa 2013). And if this is so, it is advisable to keep the pragmatic principle firm. First, we solve the ascertained problems, and then, we discuss the hypothetical ones. There is no doubt that aging, aging-related diseases, and death are problems. For those who accept this pragmatic perspective, the priority is to find effective anti-aging therapies. Then, if other health pathologies or social problems should arise in the lifetime gained, every effort will be made to find a remedy.

References

Abbott, A. (2019). First hint that body's 'biological age' can be reversed. In a small trial, drugs seemed to rejuvenate the body's 'epigenetic clock', which tracks a person's biological age. *Nature, 573*(173). https://doi.org/10.1038/d41586-019-02638-w.

Albanna, M. Z., & Holmes, J. H., IV. (2016). *Skin tissue engineering and regenerative medicine.* Cambridge, MA: Academic Press.

Baumann, G. P. (2012). Growth hormone doping in sports: A critical review of use and detection strategies. *Endocrine Reviews, 33*(2), 155–186. https://doi.org/10.1210/er.2011-1035.

Beck, J. C., Mcgarry, E. E., Dyrenfurth, I., & Venning, E. H. (1957). Metabolic effects of human and monkey growth hormone in man. *Science, 125*(3253), 884–885. https://doi.org/10.1126/science.125.3253.884.

Beck, J. C., McGarry, E. E., Dyrenfurth, I., & Venning, E. H. (1958). The metabolic effects of human and monkey growth hormone in man. *Annals of Internal Medicine, 49*(5), 1090–1105. https://doi.org/10.7326/0003-4819-49-5-1090.

Beyret, E., Liao, H. K., Yamamoto, M., Hernandez-Benitez, R., Fu, Y., Erikson, G., et al. (2019). Single-dose CRISPR–Cas9 therapy extends lifespan of mice with Hutchinson-Gilford progeria syndrome. *Nature Medicine, 25,* 419–422. https://doi.org/10.1038/s41591-019-0343-4.

Braddock, M., Campa, R., & Szocik, K. (2019). Ergonomic constraints for astronauts: Challenges and opportunities today and for the future. In R. Charles & D. Golightly (Eds.), *Contemporary*

Ergonomics & Human Factors 2019 (pp. 293–301). Chartered Institute of Ergonomics and Human Factors Publishers: Birmingham.

Campa, R. (2013). *La specie artificiale*. Saggio di bioetica evolutiva, Monza: Deleyva Editore.

Campa, R. (2019). *Eterna giovinezza. Il ritorno di un sogno antico tra studi scientifici e dilemmi etici*, Krakow: Orbis Idearum Press.

Campa, R., Szocik, K., & Braddock, M. (2019). Why space colonization will be fully automated. *Technological Forecasting and Social Change, 143,* 162–171.

Cappellesso, R., Nicole, L., Guido, A., & Pizzol, D. (2015). Spaceflight osteoporosis: Current state and future perspective. *Endocrine Regulations, 49,* 231–239. https://doi.org/10.4149/endo_2015_04_231.

Cavanagh, P. R., Licata, A. A., & Rice, A. J. (2005). Exercise and pharmacological countermeasures for bone loss during long-duration space flight. *Gravitational and Space Biology Bulletin, 18,* 39–58.

Cole, R. D. (1996). Choh Hao Li: April 21, 1913–November [illegible], [illegible]. *Biographical Memoirs of the National Academy of Sciences,* [illegible].

[illegible], M., [illegible], A., [illegible], & Angiolillo, A. (2017). The fountain of youth: A tale of parabiosis. *Stem Cells, and Rejuvenation, Open Medicine, 28*(12), 376–383.

Coutts, F., Palmos, A. B., Duarte, R. R. R., et al. (2019). The polygenic nature of telomere length and the anti-ageing properties of lithium. *Neuropsychopharmacology, 44,* 757–765. https://doi.org/10.1038/s41386-018-0289-0.

Crucian, B. E., Choukèr, A., Simpson, R. J., Mehta, S., Marshall, G., Smith, S. M., et al. (2018). Immune system dysregulation during spaceflight: Potential countermeasures for deep space exploration missions. *Frontiers in Immunology, 28,* 1437. https://doi.org/10.3389/fimmu.2018.01437.

Dancyger, L. (2018). Clinic to offer 'young blood' transfusions. *RollingStone*, Sept 27th.

De Santo, N. G., Cirillo, M., Kirsch, K. A., Correale, G., Drummer, C., Frassl, W., et al. (2005). Anemia and erythropoietin in space flights. *Seminars in Nephrology, 25,* 379–387. https://doi.org/10.1016/j.semnephrol.2005.05.006.

Egerman, M. A. (2015). GDF11 increases with age and inhibits skeletal muscle regeneration. *Cell Metabolism, 22*(1). https://doi.org/10.1016/j.cmet.2015.05.010.

Evans, M., & Kaufman, M. (1981). Establishment in culture of pluripotential cells from mouse embryos. *Nature, 292*(5819), 154–156. https://doi.org/10.1038/292154a0.

Fahy, G. M., Brooke, R. T., Watson, J. P., Good, Z., Vasanawala, S. S., Maecker, H., et al. (2019). Reversal of epigenetic aging and immunosenescent trends in humans. *Aging Cell, 18*(6), e13028. https://doi.org/10.1111/acel.13028.

Ferrucci, L., & Balducci, L. (2008). Anemia of aging: the role of chronic inflammation and cancer. *Seminars in Hematology, 45*(4), 242–249. https://doi.org/10.1053/j.seminhematol.2008.06.001.

Fitzgerald, J. (2017). Cartilage breakdown in microgravity-a problem for long-term spaceflight? *NPJ Regenerative Medicine, 2,* 10. https://doi.org/10.1038/s41536-017-0016-1.

Genentech. (1981). Genentech Reports Annual Revenues of $9.0 Million in 1980 versus $3.4 Million in 1979. www.gene.com/media/press-releases/4167/1981-02-18/genentech-reports-annual-revenues-of-90-.

Hargens, A. R., Bhattacharya, R., & Schneider, S. M. (2013). Space physiology VI: Exercise, artificial gravity, and countermeasure development for prolonged space flight. *European Journal of Applied Physiology, 13,* 2183–2192. https://doi.org/10.1007/s00421-012-2523-5.

Hofmann, B. (2018). Young blood rejuvenates old bodies: A call for reflection when moving from mice to men. *Transfusion Medicine and Emotherapy, 45*(1), 67–71.

Holt, R. I. (2011). Detecting growth hormone abuse in athletes. *Analytical and Bioanalytical Chemistry, 401*(2), 449–462. https://doi.org/10.1007/s00216-011-5068-2.

Honda, Y., Higashibata, A., Matsunaga, Y., et al. (2012). Genes down-regulated in spaceflight are involved in the control of longevity in *Caenorhabditis elegans. Scientific Reports, 2,* 487. https://doi.org/10.1038/srep00487.

Horvath, S. (2013). DNA methylation age of human tissues and cell types, *Genome Biology, 14*(10), Article number: 3156.

Iwamoto, J., Takeda, T., & Sato, Y. (2005). Interventions to prevent bone loss in astronauts during space flight. *Keio Journal of Medicine, 54,* 55–59. https://doi.org/10.2302/kjm.54.55.

Li, X.-L., Li, G.-H., Fu, J., Fu, Y., Zhang, L., Chen, W., et al. (2018). Highly efficient genome editing via CRISPR–Cas9 in human pluripotent stem cells is achieved by transient BCL-XL overexpression. *Nucleic Acids Research, 46*(19), 10195–10215. https://doi.org/10.1093/nar/gky804.

Loffredo F. S., et al. 2013. Growth Differentiation Factor 11 Is a Circulating Factor that Reverses Age-Related Cardiac Hypertrophy, *Cell,* 153 (4). https://doi.org/10.1016/j.cell.2013.04.015.

Lotz, M., & Loeser, R. F. (2012). Effects of aging on articular cartilage homeostasis. *Bone, 51*(2), 241–248. https://doi.org/10.1016/j.bone.2012.03.023.

Ly, K., Walker, C., Berry, S., et al. (2019). Telomere length in early childhood is associated with sex and ethnicity. *Scientific Reports, 9,* 10359. https://doi.org/10.1038/s41598-019-46338-x.

Mader, T. H., Gibson, C. R., Pass, A. F., Kramer, L. A., Lee, A. G., Fogarty, J., et al. (2011). Optic disc edema, globe flattening, choroidal folds, and hyperopic shifts observed in astronauts after long-duration space flight. *Ophthalmology, 118*(10), 2058–2069.

Martin, G. R. (1981). Isolation of a pluripotent cell line from early mouse embryos cultured in medium conditioned by teratocarcinoma stem cells. *Proceedings of the National Academy of Sciences of the United States of America, 78*(12), 7634–7638. https://doi.org/10.1073/pnas.78.12.7634.

Max Planck Institute for Human Cognitive and Brain Sciences. (2019). Cellular aging is linked to structural changes in the brain: Telomeres on human chromosomes change together with brain structure, *ScienceDaily*, 27 Sept 2019. www.sciencedaily.com/releases/2019/09/190927103248.htm.

Muñoz-Lorente, M. A., Cano-Martin, A. C. & Blasco, M. A. (2019) Mice with hyper-long telomeres show less metabolic aging and longer lifespans, *Nature Communications, 10*, 4723. https://doi.org/10.1038/s41467-019-12664-x.

Otsuka, K., Cornelissen, G., Kubo, Y., et al. (2019). Anti-aging effects of long-term space missions, estimated by heart rate variability. *Scientific Reports, 9,* 8995. https://doi.org/10.1038/s41598-019-45387-6

Payne, M. W., Williams, D. R., & Trudel, G. (2007). Space flight rehabilitation. *American Journal of Physical Medicine & Rehabilitation, 86*(7), 583–591.

Puhlmann, L. M. C., Valk, S. L., Engert, V., bernhardt, B. C., Lin, J., Epel, E. S., et al. (2019). Association of short-term change in leukocyte telomere length with cortical thickness and outcomes of mental training among healthy adults. *JAMA Network Open, 2*(9), e199687. https://doi.org/10.1001/jamanetworkopen.2019.9687

Reardon, S. (2015). 'Young blood' anti-ageing mechanism called into question. *Nature.*

Shuai, Y., Ma, Y., Guo, T., Zhang, L., Yang, R., Qi, M., et al. (2018). Dental stem cells and tooth regeneration. *Advances in Experimental Medicine and Biology, 1107,* 41–52. https://doi.org/10.1007/5584_2018_252

Sonnenfeld, G., Butel, J. S., & Shearer, W. T. (2003). Effects of the space flight environment on the immune system. *Reviews on Environmental Health, 18,* 1–17. https://doi.org/10.1515/REVEH.2003.18.1.1

Strollo, F., Gentile, S., Strollo, G., Mambro, A., & Vernikos, J. (2018). Recent progress in space physiology and aging. *Frontiers in physiology, 9,* 1551. https://doi.org/10.3389/fphys.2018.01551.

Szocik, K., Campa, R., Rappaport, M. B., & Corbally, C. (2019a). Changing the paradigm on human enhancements. The special case of modifications to counter bone loss for manned mars missions. *Space Policy, 48,* 68–75.

Szocik, K., Norman, Z., & Reiss, M. J. (2019b). Ethical challenges in human space missions: A space refuge, scientific value, and human gene editing for space. *Science and Engineering Ethics.* https://doi.org/10.1007/s11948-019-00131-1.

Taibbi, G., Cromwell, R. L., Kapoor, K. G., Godley, B. F., & Vizzeri, G. (2013). The effect of microgravity on ocular structures and visual function: A review. *Survey of Ophthalmology, 58,* 155–163. https://doi.org/10.1016/j.survophthal.2012.04.002.

Trakarnsanga, K., Griffiths, R., Wilson, M., et al. (2017). An immortalized adult human erythroid line facilitates sustainable and scalable generation of functional red cells. *Nature Communications, 8,* 14750. https://doi.org/10.1038/ncomms14750.

Truong, V. A., Hsu, M.-N., Nguyen, N. T. K., Lin, M.-W., Shen, C.-C., Lin, C.-Y., & Hu, Y.-C. (2019). CRISPRai for simultaneous gene activation and inhibition to promote stem cell chondrogenesis and calvarial bone regeneration. *Nucleic Acids Research, 47*(13), e74. https://doi.org/10.1093/nar/gkz267.

Wnorowski, A., Sharma, A., Chen, H., Wu, H., Shao, N.-Y., Sayed, N., Liu, C., et al. (2019). Effects of spaceflight on human induced pluripotent stem cell-derived cardiomyocyte structure and function. *Stem Cell Reports* (in Press).

Zarse, K., Terao, T., Tian, J., et al. (2011). Low-dose lithium uptake promotes longevity in humans and [illegible] https://doi.org/10.1007/s00394-011-0171-[illegible]

Part II
Human Enhancements: Philosophical and Moral Perspectives

Chapter 11
Two Planets, One Species: Does a Mission to Mars Alter the Balance in Favour of Human Enhancement?

Ziba Norman and Michael J. Reiss

Abstract In this chapter we examine the implications of a crewed mission to Mars, possible colonisation of the planet, and the wider implications this may have on genetic enhancement in both a terrestrial and space context. We consider the usage of both somatic and germ-line genetic engineering, and its potential impact on the evolution of *Homo sapiens*. We acknowledge that a mission to Mars may require the usage of such technologies if it is to be successful. Our investigation suggests that the use of such technologies might ultimately be linked with the transformation of our own species. We also consider projected timescales for the development of these genetic enhancements and the ethical questions raised by the possibility of speciation. Cooperation among spacefaring nations in this context and the development of norms for the use of such technologies is desirable.

11.1 Introduction

Although the initial exploration of Mars will involve only a very small number of humans, it seems very possible that the act of making the journey will alter how we view ourselves as humans. This is particularly likely if the journey is successfully undertaken so as to create a small colony to act as a 'reservoir' of humans in the event of a disaster wiping out humanity on Earth.

A precedent for this change in how we see ourselves is *Earthrise* (Fig. 11.1), the photograph taken by Apollo 8 crewmember Bill Anders on 24 December 1968. Sometimes described as the most influential environmental photograph ever taken, the iconic photograph of Earth as a fragile ball spinning in space gave us a sense of interconnectedness and a deeper understanding of humanity itself, and contributed to fostering a conception of responsibility for the stewardship of Earth's delicate

Z. Norman (✉) · M. J. Reiss
UCL Institute of Education, London, UK
e-mail: z.norman@ucl.ac.uk

M. J. Reiss
e-mail: m.reiss@ucl.ac.uk

K. Szocik (ed.), *Human Enhancements for Space Missions*, Space and Society,
https://doi.org/10.1007/978-3-030-42036-9_11

Fig. 11.1 Earthrise. https://commons.wikimedia.org/wiki/File:NASA-Apollo8-Dec24-Earthrise.jpg (Bill Anders/Public Domain)

ecological systems. It may even be that an awareness of this interconnectedness helped prepare us, at least in part, for the challenges of globalisation which followed. It provoked the realisation that we are one human family, sharing a common homeland we call Earth. A mission to Mars will similarly challenge our understanding of ourselves, not least because some form of genetic enhancement may be necessary to ensure the success of such a mission. In addition, if colonisation becomes a reality, even if it involves only the smallest group of humans, then our concept of interconnectedness is likely to become an extended one as we are represented both by humans on the Earth and by humans on another planet.

All this is taking place against a backdrop of our increasing awareness of the effects of deep space weather, of the danger of asteroid strikes[1] and of the place we occupy in the wider cosmos. Indeed, space science and the many technologies developed in this context have both increased our understanding of our place in the cosmos and provided access to information that may be vital for better modelling

[1] International Asteroid Day was established in 2016 by United Nations resolution (A/RES/71/90), to be held yearly on 30 June, the date of the Tunguska asteroid impact in Siberia, Russia, in 1908, to highlight the risk to life on Earth posed by asteroids.

of Earth's weather systems, which may prove pivotal as we look for solutions to the increasingly serious threats presented by anthropogenic climate change.

Research into deep space weather could be equally important if a manned mission to Mars is to be successful (Hapgood 2019), and it is entirely possible that we may gain valuable information as a result, which could make all the difference to understanding the rapid climate changes we are experiencing on Earth. In addition, the geological history of Mars itself may hold clues: research suggests that Mars was once Earth-like, with an ocean in its northern hemisphere, perhaps covering roughly one fifth of the planet's surface with water. Human exploration of Mars might allow for far more detailed studies than robotic studies offer, and may therefore play an important role in understanding climate change on Earth.

Of course, space exploration has adverse consequences—and not just the opportunity costs (think of the money). Although we have been a spacefaring species for only a few decades, we have already begun to pollute space: debris now orbits our planet and traces of our presence can be found even in remote parts of our solar system, as well as on our own Moon. A human presence on Mars will add to this. And the traces of our presence are not limited to inorganic materials, but include microbial matter that we have left behind.

Coupled with these concerns is an emergent astropolitics, running parallel to the geopolitical tensions that too often dominate relationships between nations on Earth. There are two major factors we can identify in this: (a) hardware in space is part of an essential infrastructure to support our globally integrated economies (e.g. satellite communications); this is naturally a concern; (b) a more general tendency to see space as an extension of territory, with a colonial approach being adopted in respect of resources that space may offer. China has suggested that a base on the Moon could be used as a centre for mineral extraction from asteroids, and that the Moon might be considered a staging post for missions to Mars as well. Does this mean militarisation of space is likely, if not inevitable? In August 2019, President Trump authorised the creation of a US Space Command,[2] while announcing the intention to press on with the creation of a Space Force, a sixth branch of the military, describing space as "the next war-fighting domain" (Rogers and Cooper 2019). President Trump duly signed the 2020 National Defense Authorization Act, which established the Space Force, in Hangar 6 at Joint Base Andrews, on 20 December 2019.[3] There are, of course, vital national interests that any country may indeed feel it needs to protect, especially as so much of the hardware which supports our globalised world is in orbit above us. Even so, at the very least, the rhetoric is regrettable and potentially inflammatory.

[2]A similar body was created during the Cold War, established in 1985 and disbanded in 2002, and was closely linked to the SDI (Strategic Defence Initiative, colloquially referred to as 'Star Wars'). The latest incarnation is charged with the development of Space Force Operations (see www.atomicheritage.org).

[3]https://www.whitehouse.gov/briefings-statements/remarks-president-trump-signing-ceremony-s-1790-national-defense-authorization-act-fiscal-year-2020/.

The language of colonisation is being increasingly employed. Ye Peijian, the head of China's lunar programme, when asked in 2018 why China is intending to send a *taikonaut* (Chinese astronaut) to the Moon by the 2030s replied:

The universe is an ocean, the moon is the Diaoyu Islands, Mars is Huangyan Island. If we don't go there now even though we're capable of doing so, then we will be blamed by our descendants. If others go there, then they will take over, and you won't be able to go even if you want to. This is reason enough (Hong 2018).

President Xi Jinping has identified 'leading in outer space' as one of China's goals for the 100th Anniversary of the foundation of the People's Republic of China. The driving principle behind this is wealth creation and the harvesting of mineral resources to support China's growth. Namrata Goswami, an analyst specialising in the geopolitics of space, remarks "Unlike NASA, which is aimed at space exploration and space science missions, China's space programme is aimed at long term wealth creation for the Chinese nation" (Goswami 2019). It is projected that the space economy will be worth some 2.7 trillion dollars by 2040, currently estimated to be 350 billion dollars, a sevenfold increase (Morgan Stanley 2019). Of course, the history of the European Age of Discovery suggests that governments do not always find it possible to draw a clear line between economic and military interests, especially when their vessels are some distance from home.

All this suggests that the astropolitics of the twenty-first century may indeed come to mirror the geopolitical tensions we have struggled to keep in balance on Earth since the 1950s. The Cold War itself was a catalyst for space exploration—the Apollo missions that brought humanity to the Moon an offshoot of the competition between the world's then two superpowers. And yet, despite this, space has also been a place which has allowed us to reflect on our humanity, on what it means to be human, as well as a place for healing tensions. The 1967 Outer Space Treaty (current signatories include the USA, China and Russia, 109 nations in total) established, inter alia, that space must remain free of nuclear weaponry and not be used for testing of such weapons. Though no mention is made of conventional weaponry, the Treaty did represent an awareness of the need for cooperation in space at a time of extreme global tensions. Under article II of the Treaty, neither the Moon nor any other 'celestial body' can be claimed as part of the sovereign territory of any country (United Nations Office for Outer Space Affairs 1967).

In this chapter, our focus is on the question of whether a mission to Mars would alter the balance of arguments in favour of human enhancement, not least because of the possibility that such enhancement will be necessary if the colonisation of Mars is to result in a self-sustaining community. We begin by setting out some of the scientific and cultural significance of a journey to Mars.

11.2 Journeying to Mars

There has been a long fictional history about humans journeying to Mars and inhabitants on Mars journeying here (May 2017). H. G. Wells' *The War of the Worlds* (1898) provided a dystopian portrayal, in which humanity is only saved from obliteration by the susceptibility of the Martian invaders to Earthly pathogens. While it is possible that Mars does harbour life, it won't be anything like that. More recent literature, with more of a semblance of reality, tends to concentrate on issues to do with us journeying to and surviving on Mars (e.g. Weir 2011).

The attraction of Mars in fiction and in reality is similar. From an astronomical perspective, Mars is close, though its distance from us varies greatly as a result of its orbital eccentricity and the fact that its orbit around the Sun is not synchronised with ours. Mars is also not too dissimilar to Earth in a number of important respects (size, length of day, typical light levels, force of gravity, length of year, axial tilt, the presence of ice, the presence of 'soil'—albeit with no good evidence, as yet, of any life therein). Of course, there are important differences. In particular, Mars has virtually no atmosphere, is usually much colder (average surface temperature of −46 °C with a range of −143 to 35 °C) and has substantially higher levels of damaging radiation.

A mission to Mars is probably still some time off though it remains unclear whether it would result from private or government funding. A lot of work is being undertaken to determine the practical and psychological consequences of such a mission (Messeri 2016). For example, the Mars Desert Research Station in Utah is a simulated Mars analogue habitat (there are plenty of digital Mars simulations too). Visitors or potential astronauts typically stay there for between one week and three months and engage in activities to help prepare for a journey to Mars or time on the planet (Fig. 11.2). Equally, a huge amount of work is being undertaken, mostly using rovers, to better understand the Martian landscape (Vertesi 2015).

11.3 Ethical Arguments Concerning a Mission to Mars

There are important arguments about whether a human mission to Mars is even needed. Ongoing advances in robotics are such that it seems virtually certain that there will at some point be no scientific arguments in favour of humans, as opposed to robots, going to Mars that are strong enough to outweigh the arguments against such a mission (Campa et al. 2019). These arguments are primarily about human safety, though there is also the argument that humans would risk introducing Earth life to Mars (widely agreed to be undesirable) to a far greater extent than would robots (which are much easier to sterilise).

By far, the strongest argument in favour of a human mission to Mars is to establish a human colony to serve as a space refuge for post-catastrophic Earth (Szocik et al. 2019). It can even be argued that humans have a *duty* to colonise Mars (or another

Fig. 11.2 A rover (foreground) participating in a Mars Society University Rover Challenge Hill Climb with Mars Desert Research Station in the midground. Available at https://commons.wikimedia.org/wiki/File:Mars_Society_University_Rover_Challenge_Hill_Climb.jpg. Mars Society/CC BY-SA (https://creativecommons.org/licenses/by-sa/4.0)

space body) to increase the chances of survival for the human species. Of course, it might be objected that the very substantial financial cost of such a colonisation programme means that the money would be better spent here on Earth reducing the chances of such a catastrophe (by dealing with climate change and environmental degradation, improving incoming asteroid detection and ways of dealing with these and possibly also by investing in peace education and demilitarisation). To this, we can respond that while the two of us are all for dealing with climate change and environmental degradation, investing in peace education and demilitarisation (a) we would not want to bet the survival of humanity on these coming about and proving to be of lasting success and (b) it might be wise both to strive for Earth's survival (for humans) *and* to plan for a refuge. As any conservation biologist knows, it is risky to have only one place for a species to live.

11.4 Should We Enhance Humans?

What is potentially wrong with enhancing humans? Isn't that something we do all the time when we help people to learn and when we improve technologies such as those used in communication, transport and clothing? We can set aside arguments to do with cheating in competitive sports. Those arguments are all about trying to get a level playing field. There is nothing immoral about using a particular kind of swimwear, designed to reduce friction, just because it isn't allowed in competition. In 2009, FINA, the Fédération Internationale de Natation, also known as the International Swimming Federation, decided to reverse its existing policy and ban all body-length swimsuits. These swimsuits allow for increased blood flow to muscles, hold the body in a more hydrodynamic position (so reducing drag) and increase flexibility. A partnership between Speedo and NASA had led to the LZR Racer, which has been reported as reducing competition times by between 1.9 and 2.2% (Anderson 2008), a huge amount at elite levels.

Rivals referred to the LZR Racer as 'technological doping'. Doping is, of course, rigorously prohibited in competitive sports but, for those of us who do not so compete, is it safety considerations alone that should caution us against taking such substances? If anabolic steroids help me to get the body shape, I want or erythropoietin helps me to run faster, which I find satisfying, why should I not be able to use such drugs? After all, they are prescribed by doctors in certain situations.

In what follows we concentrate on genetic engineering, but it is important for ethical analysis to consider whether it is enhancement per se, gaining an unfair advantage in competition, risking one's health or something else that is objected to. Ethical arguments about enhancement often rely, even if implicitly, on a concept of 'naturalness'. We tend to presume, for example, that it's fine to use medicines to restore someone to normality but hesitate or object to the same medicine being used to exceed normality. Consider, for instance, human growth hormone. Human growth hormone deficiency can result in children being of short stature. Treatment with the hormone, whether obtained 'naturally' from pituitary glands or made synthetically, can lead to increased growth. Indeed, Genentech's recombinant (genetically engineered) human growth hormone was approved for clinical trials use back in 1981 (Genentech 1981). It has since been widely employed for medical uses and is generally reckoned to be safer than the 'natural' version (which is more likely to be contaminated).

11.5 Should We Enhance Humans for a Mission to Mars?

At present, the genetic engineering of humans is in its early stages, unlike the genetic engineering of micro-organisms and plants. Many different plant species have been genetically engineered for such features as resistance to certain diseases, while microorganisms have been genetically engineered for a wide range of purposes including the production of such human proteins as insulin, clotting factors and (as mentioned above) human growth hormone.

Instances where genetic engineering of humans is taking place are in the treatment of diseases that have a genetic component. Examples are still infrequent but include the treatment of certain immune disorders, a type of heart disease and a type of blindness. Scientists and regulators have been quite cautious about allowing such work, no doubt fearing a public backlash and mindful of earlier trials of genetic engineering in humans that occasionally had unexpected and even fatal consequences—the most famous of which was probably the death in 1999 of Jesse Gelsinger in a gene therapy trial. Notoriously, the work in China in 2018 by He Jiankui (who was responsible for the birth of the world's first known gene-edited human babies, twins, to try to make them resistant to their father's HIV infection) proceeded without regulatory approval and has been widely condemned in his home country and internationally. He Jiankui has now been sent to prison for three years and given a heavy fine (equivalent to about USD 435,000).

A standard and important distinction that is often made in human genetic engineering is between somatic and germ-line genetic engineering. The somatic cells are the ones that make up most of our body (muscles, skin, nerves, bone, etc.); they are responsible for everything *except* producing our gametes. The germ-line cells are the ones responsible for producing our eggs and sperm. Even a change to the DNA in all of a person's somatic cells therefore does not pass to any offspring they have. However, if some of a person's germ cells are genetically engineered, that change may pass to their descendants. Accordingly, most countries with the available technology do not allow germ-line gene editing. It was largely the fact that He Jiankui engaged in germ-line gene editing that caused such a furore.

If it turns out that genetic engineering of humans would have benefits for astronauts, then it is perfectly possible that somatic gene editing would be used. However, this would only be likely to suffice if the astronaut returned to Earth. For colonisers—given the low likelihood, at least initially, of their having advanced laboratory facilities on Mars—it would make much more sense for germ-line genetic engineering to be employed. Techniques for gene editing, both somatic and germ-line, are advancing rapidly (CRISPR, prime editing, etc.), and it seems possible that regulatory authorities may indeed conclude at some point that the confluence of (a) genetic engineering of humans being safe enough; (b) there being a sufficiently pressing need for humans to colonise Mars; and (c) there being limitations in alternatives to genetic engineering mean that the genetic engineering of humans for the purpose of a mission to Mars is indeed permitted.

What sort of genetic engineering are we talking about? George Church, co-founder of Harvard Medical School's Consortium for Space Genetics, has at the time of writing identified some 55 genes that might be advantageous for long-term spaceflight (Pontin 2018; Church 2019). The list includes:

- CTNNBI—radiation resistance
- LRP5—extra-strong bones
- ESPA1—allows people to live with lower oxygen levels
- MSTN—reduced incidence of atherosclerosis

- ABCC11—endows its possessors with low BO (body odour)—useful on a space craft
- as well as a host of genes that might enhance our memories or make us smarter, less anxious or less likely to develop cancers (which often result from radiation damage).

At present, our detailed knowledge of how genes function and of the likely consequences of undertaking genetic engineering of humans is not robust. The geneticist Chris Mason at Weill Cornell has proposed a 500-year plan for space colonisation (iGEM 2011). Phase 1 is currently underway, indeed is intended for completion in 2020. It entails a base-by-base examination of the human genome to determine which parts are resistant to mutation and which tolerant of it. In Phase 2 (2021–2040), it is presumed that whole-genome sequencing and molecular characterisation are common, cheap and accurate. Work is focused on methods for contextualising variation and its effects. Efforts begin on integrating new elements into mammalian genomes. In Phase 3 (2041–2050), long-term human trials on genome engineering are begun. In Phase 4 (2051–2060), tests are begun in space environments. Thus far, many will consider that both the scope and timing are realistic, conservative even. Subsequent phases seem more optimistic—and the reader is not encouraged that there seems to be no Phase 5 … Phase 6 (2060–2100) entails the beginnings of settlements on other planets and genesis of synthetic genomes. In Phase 7 (2101–2150), new genomes allow toleration of extremely cold/hot and acidic/basic environments. In Phase 8 (2151–2300), the longest phase, these new genomes are sent off to begin seeding of Earth-like planets. In Phase 9 (2301–2400), humans are shipped off to these new worlds. Finally, in Phase 10 (2401–2500), there is human settlement of a new solar system, used as a model for future systems.

Futurology is never a straightforward science, but advances in genetic technologies are happening at an increasing rate. Various researchers have inserted a gene called Dsup (found in tardigrades, Fig. 11.3—notoriously hardy creatures) which seems to protect cells against radiation damage (Bittel 2016) into humans. More futuristic possibilities include enabling our kidneys to make the nine so-called essential amino acids and engineering the personalities of long-range astronauts so that they enjoy the journey more (cf. the dairy animal in Adams' *The restaurant at the end of the universe* (1980) that has been bred to want to be eaten).

So far, we have rather taken it for granted that the sorts of genetic engineering that we have been envisaging (radiation resistance, extra-strong bones, etc.) are indeed enhancements. And so, they are from the perspective of us on Earth. However, it could be argued that from the perspective of someone on Mars, they are more akin to medical treatments in much the same way that we see genetic engineering on Earth to tackle sickle cell anaemia, cystic fibrosis, heart disease and cancers as restorative treatment rather than enhancement. If this argument is to be accepted, it weakens ethical objections against the use of genetic engineering to facilitate human missions to Mars.

Fig. 11.3 A tardigrade (water bear). Tardigrades are notoriously hardy; inserting one of their genes into human cells makes them more resistant to radiation damage. https://commons.wikimedia.org/wiki/File:SEM_image_of_Milnesium_tardigradum_in_active_state_-_journal.pone.0045682.g001-2.png. Chokraie E, Warnken U, Hotz-Wagenblatt A, Grohme MA, Hengherr S, et al. (2012)/CC BY (https://creativecommons.org/licenses/by/2.5)

11.6 Consequences of the Genetic Engineering of Humans for a Mission to Mars

Ted Peters' piece in this volume discusses what he terms 'interplanetary sin transfer' and invites us to consider the theological questions that arise in the context of colonisation of Mars, and considers the possibility of developing a utopian community, and even creating a posthuman species that is morally superior to ours.

'Interplanetary sin transfer' sounds very grand, but all that is meant is that we humans would take to Mars our propensity to sin. Peters is acknowledging the theological origins of the term 'sin', the propensity to sin, in certain religious contexts, being considered an integral part of human nature. Even if one is not a theologian or religious believer, if we accept an everyday understanding of sin as both a capacity and a propensity to do things that are, in either a religious or legal context, undesirable, then the negative aspects of human nature will travel with us wherever we go. Indeed, the history of our migration as a species over the last 100,000 years or so has been one of carrying our sin with us into new places. We have continued with this 'intercontinental sin transfer' more recently as we have established stations in such places as Antarctica. Although crime rates are low on Antarctica, there have been instances of attempted murder and sexual harassment, inter alia (Rousseau 2016).

In this final section, while acknowledging these important issues, we concentrate on the possibility that a separate colony on Mars might indeed, accelerated by various sorts of genetic engineering to which we have already referred, result in the formation of a different species.

One response is to see the possibility of our species *Homo sapiens* dividing into two (eventually, possibly more if and when isolated outposts become established on other planets) as something that is undesirable and should be averted. On this reading, perhaps the most significant challenge raised by a long-term mission to Mars is the challenge to remain a unified species. This in itself addresses the problems of allegiance to Earth, raised in the context of conservation of our home planet, and presumes that, even if we are able to engineer our genes to make life on Mars and deep space travel possible, we have a responsibility, even a moral obligation, to ensure that we remain one species.

On this reading, there would need, even at this early stage in space travel, to be increased communication between spacefaring nations, with a view to establishing norms for the utilisation of human enhancement in space contexts. Envisaging the possibility of becoming a two-planet species may provide a suitable time for reflection and unification. The human element in this equation could provide a moment to pause and consider the effects of gene-editing biotechnologies in a terrestrial as well as a distant setting. In this sense, we should not become over-focused on the concept of space refuges and utopian dreams, nor should we limit ourselves for fear of the negative effects such a project might have on our allegiance to our own planet. Instead, we should consider the challenge as an opportunity to develop a wider understanding of stewardship both of our environment (on Earth) and ourselves as a species (wherever we are). Mars and its potential colonisation can then be viewed as a chance to create a larger, more adaptable, but unified human family.

However, before humanity speeds too far down this path, we should pause to ask two questions. First, how likely is it that space travel will lead to a new species? Secondly, if it did, would this necessarily be a bad thing? It is worth examining precisely what is meant by the term 'species'.

11.6.1 What Do We Mean by 'A Species'?

At one point, the dominant question in the embryonic field of history and philosophy of biology was what is meant by 'a species'. As so often is the case with controversies that are rooted in ideas rather than in data, with hindsight, the controversy seems overblown. The matter is still not decided—and the lack of agreement does have consequences for conservation biology (in deciding whether one species has gone extinct and how another might be preserved)—but this is simply because there are two main approaches to deciding what a species is.

The definition of which most people are probably aware is to do with successful interbreeding. Two individuals are said to belong to the same species if they can produce viable offspring. Of course, there are all sorts of caveats (individuals of the same sex don't produce offspring, juveniles can't either, some individuals are infertile, what is meant by viable—e.g. suppose the offspring live but are themselves infertile?, what do we do with species that reproduce asexually?, etc.), but overall, it's pretty clear what the criterion is and often it works well.

One problem with this definition is that it can be difficult to put into practice. In principle, one has to sit around and wait for individuals to mate and then see whether they produce healthy offspring. And what do we do about species scattered over very large areas? Do we need to set up a breeding programme to check that grey squirrels in Canada and grey squirrels in the UK belong to the same species?

The grey squirrel issue is exacerbated in the case of fossil species—and most species have long gone extinct. How do we decide where one species of ammonite begins and another ends? After all, there are no ammonites today, so we can't even extrapolate from today's to yesterday's ammonites.

The approach used by palaeontologists is to use the criterion of morphological similarity. Individuals are classified as belonging to the same species if they resemble each other sufficiently. Of course, this definition itself has problems. How dissimilar do two species have to be? And then consider dogs: the different breeds look very different yet are capable of interbreeding. Are we really saying that the domestic dog now contains dozens or even hundreds of species? Every time a new breed of dog is recognised, should it be recognised as a new species? This seems like *reductio ad absurdum.*

Despite these problems, biologist manage to agree in the great majority of case where one species, whether extinct or extant, ends and another begins—though there are always lumpers (who prefer there to be fewer named taxa such as species, genera and families) and splitters (who prefer tighter definitions, inexorably leading to more named taxa).

That this debate about the meaning of the term 'species' is not an arcane, angels-on-a-pin one is particularly evident when we move to consider our own species in the context of examining whether it is likely that space travel will lead to a new species of human.

11.6.2 Is It Likely that Space Travel Will Lead to a New Species of Human?

There is no doubt that all of today's humans belong to the one species, *H. sapiens.* Indeed, we are familiar with the phrase 'the human race' precisely because historical attempts to categorise humanity into different races were often tightly associated with racism. Careful work by geneticists and historians of science revealed the extent to which the previous biologists and anthropologists nearly always held views that would now be condemned as unacceptable—believing that discrete races could be identified and that such races differed in their mental capacities, with Caucasians almost always placed above other 'races' (Gould 1981; Lewontin et al. 1984). Indeed, in some cases, different sub-species were identified and named, in what, consciously or unconsciously, seems to have been a spurious attempt to provide a scientific veneer to unacceptable views. That such false science contributed to the perpetuation of slavery, inequalities in fields such as education, health and housing, and the Aryan views espoused in Nazi Germany only causes us to recoil from them still further.

And yet, to an evolutionary biologist, there is a different way of viewing biological variability. Under this perspective, the forces of evolution sometimes reduce variability among organisms and sometimes increase it. Reduction in variability principally occurs via the mechanism of stabilising selection when there is a single optimum in a population for a trait so that, for example, birds of a particular species that have bills either shorter or longer than this optimum do slightly less well in the struggle for existence, typically because these sorts of bills are slightly less adept at catching the birds' preferred prey. An increase in variability can have a number of causes. To continue with the bird example, it may be that, for whatever reason, some individuals begin to specialise on a slightly different sort of prey compared to the majority. Typically, such behaviour has no long-term evolutionary consequences. It is possible though, especially if these individuals become separated from the others (e.g. by reasons of geography), that such a simple behaviour is the very start of a long process that may eventually result in speciation—as is widely presumed to have happened in the iconic case of Charles Darwin's Galapagos finches.

So, how likely is it that comparable forces might lead to speciation in humans? For the foreseeable future, the simple answer is 'negligibly likely'. Even if genetic engineering of the type discussed above results in colonists with such characteristics as radiation resistance, stronger bones, reduced incidence of atherosclerosis and less BO (!), the genetic engineering being talked about is at the level of existing variants. Such individuals will still be as much members of *H. sapiens* as we are.

But this might change. If genetic engineering really does result in individuals whose genomes are very different, it is possible that on both the interbreeding and the morphological similarity criteria astronauts and their descendants could eventually belong to a different species. This is particularly likely to be the case if large changes are made to our genomes. For example, it is not impossible that astronauts heading to Mars might be given an additional chromosome, so that their 'normal' (diploid) cells would have 48 chromosomes and their eggs or sperm 24. This additional chromosome would be a convenient way of hosting major initiatives such as the ability, mentioned above, to synthesise all the amino acids.

Even without such step changes, the phenomena of natural selection and genetic drift will undoubtedly mean that if Martian colonies are sustainable (in the sense of producing offspring over successive generations without the arrival of new humans from Earth), the individuals in such colonies will slowly start to differ genetically from their Earth counterparts. Natural selection means that there will be additional strong selection for genetic variants better able to deal with high levels of radiation, low gravitational force and so on. Genetic drift is the phenomenon whereby isolated populations slowly diverge genetically as random forces lead to the disappearance of certain genetic variants in one population, whereas in another population, the same variants come to predominate.

The differences we see between indigenous people in very separate parts of the globe are precisely those that an evolutionary biologist would expect to see in any sexual species. Given reproductive isolation and the passage of sufficient generations—there are no hard and fast rules as to how many—geographically separated populations begin to diverge genetically. From an evolutionary perspective, there aren't absolute boundaries around species. In the early stages of such divergence, we see what we see today in humans, that there are recognisable morphological and physiological differences (cf. the Inuit and the Maasai women in Fig. 11.4), but these are insufficient to cause reproductive isolation. *H. sapiens* has only existed for about 200,000 years. As a rule of thumb (and such a rule is rough and ready), it takes about five times as long for speciation to occur in mammals.

But speciation can happen much more quickly. Scott Solomon has suggested that within a few hundred generations—perhaps just 6000 years—a new type (the context shows that he means 'species') of human might evolve on Mars, given that it is likely that selection pressures will be extreme (Solomon, 2016).

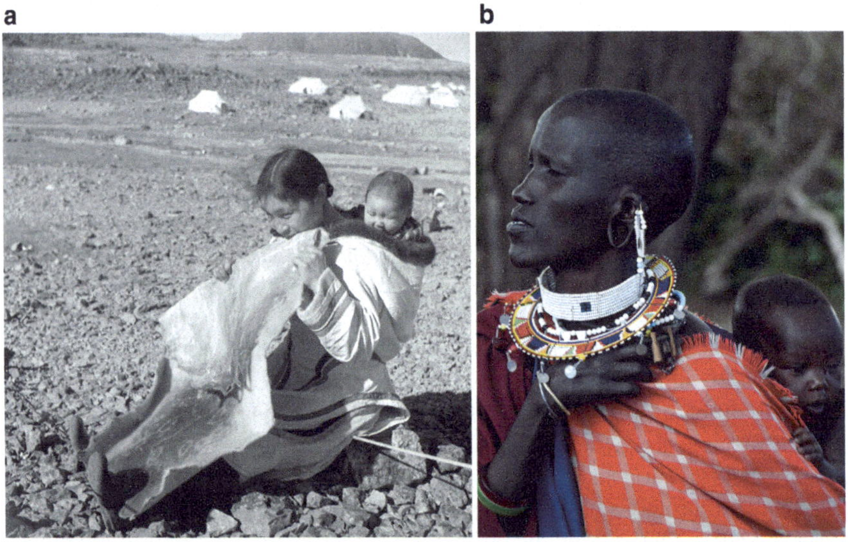

Fig. 11.4 **a** Inuit woman chewing sealskin to soften it for making boots, with child from Kinngait. https://commons.wikimedia.org/wiki/File:Inuit_woman_%E2%80%9CJosie%E2%80%9D_chewing_sealskin_to_soften_it_for_making_kamiits_(boots),_Kinngait,_Nunavut_(31497043966).jpg. BiblioArchives/LibraryArchives from Canada/CC BY (https://creativecommons.org/licenses/by/2.0). **b** Maasai woman carrying her baby in traditional clothing and jewellery in the Serengeti National Park, Tanzania. https://commons.wikimedia.org/wiki/File:Masai_woman-child.jpg. William Warby from London, England/CC BY (https://creativecommons.org/licenses/by/2.0)

11.6.3 Would It Be Undesirable for There to Be Two Species of Humans?

In principle, it would be possible to ensure that all the descendants of today's *H. sapiens* remain members of the same species—a judicious blend of mandated contraception and genetic tweaks should see to that. But why bother? We need to ask if there is anything problematic about there being more than two species of humans.

We note that this possibility is one that has been addressed in science fiction. In *The Time Machine* (Wells 1895), humanity has evolved (the year is 802,701) into the ineffectual Eloi and the thuggish Morlocks. Ursula Le Guin's writing features a number of humanoid species and explores the relationships between them—e.g. *The Left Hand of Darkness* (1969) has considerable interspecific altruism as Estraven (from Gethen) risks everything, and eventually loses his life, to save Ai (a Terran). In Mary Doria Russell's *The Sparrow* (1996), the story revolves around a Jesuit-funded trip in the year 2019 to Rakhat, which turns out to be inhabited by the peaceful Runa and the aggressive Jana'ata. Spoiler Alert: none of these books are optimistic about the coexistence of different humanoid species.

Precisely because of the history of racism and xenophobia, there may be much to be said for striving to avoid speciation in humans. As soon as such speciation happens (indeed, long before), there will be those who divide humanity into 'them' and 'us'—and such political divisions are likely to be exacerbated if we no longer belong to a single species, are separated by huge geographical distances and live in very different circumstances.

And yet, isn't this to buy into a presumption that our one species is superior to all others, that we have a right to existence that trumps the considerations of all other species? Is it possible that having two species of humans (*H. sapiens* and, say, *Homo astronauticus*) might cause us to re-examine presumptions of our superiority to other species? After all, since 1993, the Great Ape Project has striven to give rights to the non-human great apes: chimpanzees, bonobos, gorillas, and orangutans. We are more sensitive now to the accusations of 'speciesism', that it is wrong to favour our own species over others simply on the grounds that we belong to different species.

11.7 Conclusions

Mars has always held a special place in the human imagination. The technological capabilities we are now on the threshold of developing are likely to enable us to make a successful journey to Mars in the not too distant future, and perhaps even establish a colony there. In considering whether such a mission should be supported—the ethical and political implications it raises, specifically the need to focus on anthropogenic global climate change and concerns over the usage of genetic enhancement and even the possibility of speciation—the authors believe that by asking the right questions at this early stage, we have a genuine chance of ensuring that such a mission will not lead humanity into a dark corner, but rather has the potential to support the genuine flourishing of our species. We advocate for increased cooperation between spacefaring nations, and perhaps establishing guidelines for the utilisation of human enhancements for space travel.

If a mission to Mars does come to be the moment *H. sapiens* becomes a two-planet species, perhaps it will also mark the moment when we begin to transcend the violent tendencies of our terrestrial past. Indeed, a very high level of cooperation will be necessary if this, the most ambitious project we have ever undertaken, is to be successful. The dawn of the space age offered us an opportunity to see ourselves as one human family, and has made the technological integration of our planet possible. We hope that this next phase of space exploration will build on this, and may offer us an even greater chance to deepen our concept of interconnectedness and stewardship, both on our own home planet, Earth, and in the wider heavens which our descendants may inhabit in future; to become more than we are, to fully realise our human capacities, transcending the negative propensities we recognise in ourselves and perhaps even bringing life elsewhere in the universe.

References

Adams, D. (1980). *The restaurant at the end of the universe*. London: Pan.

Anderson, K. (2008). The war of the swimsuits. *Vault*, 23 June. Available at https://www.si.com/vault/2008/06/23/105705011/the-war-of-the-swimsuits.

Bittel, J. (2016). Tardigrade protein helps human DNA withstand radiation. *Nature*, 20 September. https://doi.org/10.1038/nature.2016.20648. Available at https://www.nature.com/news/tardigrade-protein-helps-human-dna-withstand-radiation-1.20648?utm_source=commission_junction&utm_medium=affiliate.

Campa, R., Szocik, K., & Braddock, M. (2019). Why space colonization will be fully automated. *Technological Forecasting and Social Change, 143,* 162–171.

Church, G. M. (2019). [No title] https://arep.med.harvard.edu/gmc/protect.html.

Genentech. (1981). *FDA-approved clinical tests on humans begin today with human growth hormone made by recombinant DNA*, 12 January Press Release. Available at https://www.gene.com/media/press-releases/4166/1981-01-12/fda-approved-clinical-tests-on-humans-be.

Goswami, N. (2019). China's get-rich space program. *The Diplomat*, 28 February. Available at https://thediplomat.com/2019/02/chinas-get-rich-space-program/.

Gould, S. J. (1981). *The mismeasure of man*. New York: WW Norton & Company.

Hapgood, M. (2019). The impact of space weather on human missions to Mars; TH need for good engineering and good forecasts. In K. Szocik (Ed.), *The human factor in a mission to Mars: An interdisciplinary approach* (pp. 69–91). Cham: Springer.

Hong, B. (2018). Starship troopers? China's looming land grab in outer space. *Daily Beast*, 22 June. Available at https://www.thedailybeast.com/chinas-looming-land-grab-in-outer-space?ref=scroll.

iGEM. (2011). *Colonization of Mars*. https://2011.igem.org/Team:NYC_Software/Tools/Colonization#.

Le Guin, U. K. (1969). *The left hand of darkness*. New York: Ace Books.

Lewontin, R. C., Rose, S., & Kamin, L. J. (1984). *Not in our genes: Biology, ideology and human nature*. New York: Pantheon Books.

May, A. (2017). *Destination Mars: The story of our quest to conquer the red planet*. London: Icon.

Messeri, L. (2016). *Placing outer space: An earthly ethnography of other worlds*. Durham: Duke University Press.

Morgan Stanley. (2019). Investing in space. *Research*, 2 July. Available at https://www.morganstanley.com/ideas/investing-in-space.

Pontin, J. (2018). The genetics (and ethics) of making humans fit for Mars. *Wired*, 7 August. Available at https://www.wired.com/story/ideas-jason-pontin-genetic-engineering-for-mars/.

Rogers, K., & Cooper, H. (2019). Trump authorizes a space command. Next, he wants a space force. *New York Times*, 29 August. Available at https://www.nytimes.com/2019/08/29/us/politics/trump-space-command-force.html.

Rousseau, B. (2016). Cold cases: Crime and punishment in Antarctica. *New York Times*, 28 September. Available at https://www.nytimes.com/2016/09/29/world/what-in-the-world/antarctica-crime.html.

Russell, M. D. (1996). *The sparrow (1996)*. New York: Villard.

Solomon, S. (2016). The Martians are coming—And they're human: How settling Mars could create a new human species. *Nautilus*, 27 October. Available at https://nautil.us/issue/41/selection/the-martians-are-comingand-theyre-human.

Szocik, K., Norman, Z., & Reiss, M. J. (2019). Ethical challenges in human space missions: A space refuge, scientific value, and human gene editing for space. *Science and Engineering Ethics*. https://doi.org/10.1007/s11948-019-00131-1.

United Nations Office for Outer Space Affairs. (1967). *United Nations treaties and principles on outer space, related general assembly resolutions and other documents*. New York: United Nations. Available at https://www.unoosa.org/pdf/publications/ST_SPACE_061Rev01E.pdf.

Vertesi, J. (2015). *Seeing like a rover: How robots, teams, and images craft knowledge of Mars*. Chicago: University of Chicago Press.

Weir, T. (2011). *The Martian*. Self-published.

Wells, H. G. (1895). *The time machine*. London: Heinemann.

Wells, H. G. (1898). *The war of the worlds*. London: Heinemann.

Chapter 12
Virtue Ethics and the Value of Saving Humanity

Koji Tachibana

Abstract Living in space will no longer be a science fiction in the twenty-first century. However, it remains arguable whether space exploration should be promoted for future space colonization. Some advocates claim that it is necessary because space colonization will be the only option for saving humanity. One plausible option for human settlement in space argues that genetic manipulations of the human mind and body might be required in order to dwell in the hostile environments of space. Such manipulations have various ethical issues such as it might harm several generations before the completion of colonization. Can space colonization under such circumstances be justified from ethical points of view? Utilitarianism and deontology can justify such a colonization because they endorse the assumption that saving humanity has the supreme value. In contrast, this paper argues that virtue ethics can reject the assumption. It concludes that, following a virtue-ethical point of view, ending our lives virtuously is more valuable than surviving even though humankind would cease to exist.

12.1 Pros and Cons of Space Colonization

In the twenty-first century, living in space will no longer be a science fiction. At present, astronauts and cosmonauts stay and work together at the International Space Station (ISS) for six months on average and up to a year at a time. In 2018, fourteen space agencies, including those who have collaboratively operated the ISS, reached an agreement on a new space policy to build a habitat in the orbit or on the surface

[1]Private enterprises such as the Mars One project and SpaceX have similar plans. Although the former has been said to being entering into bankruptcy and lacks technical feasibility (Do et al. 2016), the success of Crew Dragon Demo-2 by the latter in May 2020 will lead to the future space settlement (Carson 2018).

K. Tachibana (✉)
Kumamoto University, 2-40-1, Kurokami, Kumamoto 860-8555, Japan
e-mail: koji-t@kumamoto-u.ac.jp

K. Szocik (ed.), *Human Enhancements for Space Missions*, Space and Society,
https://doi.org/10.1007/978-3-030-42036-9_12

of the moon and to send humans to Mars (ISECG 2018).[1] Therefore, the idea of extended space sojourns is an actual concern of space policies and industries.

Colonization will be the next step in human space exploration. Although the possibility remains speculative at present, interdisciplinary academic discussions have been increasing year by year. For example, in a recent article, fifteen scholars including philosophers, anthropologists, theologians, and scientists discuss scientific, economical, ecological, ethical, and political issues of space colonization (Smith et al. 2019). Among these topics, moral feasibility is highlighted since it is distinct from other sorts of feasibilities, such as technical, economic, and political feasibility. Even when we have sufficient technology to colonize other planets and when it is endorsed by our economic and political institutions, we should not do it if we have a *moral* reason against colonization.[2] In short, moral feasibility is the last word on the topic of space colonization.

Philosophers have examined the moral feasibility of space colonization. Some deontic scholars argue that space colonization is our moral obligation because it can advance sciences and provide us with useful products, such as clean energy, which will help the Earth (Munevar 2019). Those who sympathize with utilitarianism also justify space colonization but for different reasons, such as that colonizing other planets will be a means of "increasing human welfare" (Szocik 2020).

In contrast, virtue ethicists tend to have negative opinions of space colonization. Since virtue ethics is a normative theory that prescribes pursuing virtues and avoiding vices, virtue ethicists claim that we should not colonize other planets if such behavior is vicious. For example, Sparrow (2015) argues that we should not colonize other planets because the advocates of space colonization exhibit two vices: hubris and insensitivity to beauty. For one thing, they exhibit hubris by acting as if they are gods who can do anything they want; for another, they are insensitive to beauty because they do not notice that colonizing other planets would ruin the beauty those planets originally have. He continues his argument as follows:

> To summarise: in terraforming Mars we would be drastically altering the character of a whole planet, a unique environment, which includes complex inorganic systems and possesses many features of striking natural beauty. Finally, of course, it must be pointed out that colonization (and thus terraforming) other worlds is by no means necessary for the survival of the human race. (Sparrow 2015, 163)

Sparrow's argument contains some truth regarding our virtuous way of living in the era of space colonization. However, the assumption his argument is based on—that space colonization is not necessary for the survival of humankind—makes it obscure what virtue ethics can prescribe on this issue. To clear away this obscurity, we have only to ask what if we were facing an existential risk because it was

---

[2]For example, there could be ethical criticisms of such a costly project. One could say that we should allocate our limited resources to solve problems on Earth, such as poverty and environmental affairs. I will not discuss these issues here. For a discussion of how some of these problems can be partially solved at once, see Tachibana et al. (2017).

determined that the Earth would be unable to feed us in the near future. Can virtue ethics continue to claim that we should not colonize other planets even if we would otherwise cease to exist? This new assumption—that space colonization is necessary for saving humanity—can reframe the position of virtue ethics about human virtues and vices in the era of space colonization.

12.2 Existential Risks and Two Options for Space Colonization

Some advocates of space colonization adopt the aforementioned assumption and claim that we should colonize other planets to save the human species.[3] Steven Hawking, one of this century's leading professors of astrophysics, was one such advocate. According to Hawking, "humanity needs to leave Earth and colonize the moon, Mars or other planets in order for our species to survive the impending doom of climate change" (Chow 2017). Another example is Elon Musk, an influential investor, who has said that we should promote space exploration because space colonization will preserve humanity (Carson 2018).

We should note that climate change, which Hawking emphasized, is not the only possible cause of human extinction. Extinction could also result from resource depletion due to rapid population growth, an enormous natural disaster, environmental pollution, nuclear war, a meteorite collision, a plague, or even the end of the Earth's lifespan (see Singer 2015, Chap. 15). Whatever the cause, no one can deny that the day of extinction will come. If we seriously account for the possibility that we might not be able to keep on living on Earth, then space colonization becomes an actual issue for consideration. This sense of crisis is something that the aforementioned advocates of space colonization seem to share. Let us schematize their thoughts as follows:

P1: The day will come when the Earth cannot feed us anymore.
P2: Space colonization is the only means to save the human species.
P3: The survival of humankind has the supreme value.
Conclusion: We should go to space and colonize other planets at any cost.

These three premises refer to different aspects of the world. The first premise (P1) expresses an astronomical fact, the second premise (P2) concerns a technological assumption, and the third premise (P3) expresses a moral belief. Although each fact can be doubted, we hereafter accept the first two premises, as they should be examined by empirical sciences. The questions we posed in the previous section will, then, be considered as the problem of whether virtue ethics can endorse P3 or whether it can posit something superior to (i.e., more virtuous than) our survival as a species.

---

[3]In this paragraph, I was helped by the survey in Traphagan (2019) to find such advocates.

To examine this question, let us specify two options that may realize space colonization. Needless to say, living in space is not an easy mission. Space colonization or even a long-duration sojourn in space has various risks for the human mind and body due to the hostile environments of space, including microgravity, changes in circadian rhythm, space radiation, isolation, confinement, and interpersonal issues (Kanas and Manzey 2008; Clément 2011; Hodkinson et al. 2017). The longer humans stay in space, the more severe these influences become. Therefore, to realize long-duration space sojourns and space colonization, we must address these problems.

The first possible solution to these problems involves advanced developments in space engineering and space medicine, which would allow us to build a space colony or habitat with a highly protective structure against the severity of the outer space environment and a highly livable housing standard (ISECG 2018). This option would solve many of the aforementioned issues because it addresses the hostile outer space environments and the limited ecosystem we would encounter. However, this option has a serious problem since building a closed colony or habitat would not to address the psychosocial problems that would arise from living in an isolated and confined environment. People living in such a colony for a long time may experience diminished behavioral health and may even develop various mental and psychiatric disorders. This might then become what I call a Hobbesian disastrous situation, in which it is difficult for people to keep their society sane, leading to "a war of all against all (Bellum omnium contra omnes)" (Tachibana 2019).

The second option approaches the problem differently by taking the stance that building a protective facility is not the only way to realize long-duration life in space. We can change not only our external environment, such as our accommodations, but also our internal environment, namely the human body itself. This option proposes that we should use biomedical technologies to genetically enhance humans so that we can survive, in the flesh, in outer space. For example, NASA refers to the possibility of genetic modification and "designer crews" as future biological countermeasures for space radiation (Allen et al. 2003). Although this is a speculative option at present, it is endorsed by some scholars; as Szocik (2020) put it, "[i]f human space missions are to be possible, human genome editing will not only be permissible, but required."[4] This option might solve the problem of the Hobbesian dystopia simply because genetically transformed humans could live in outer space without requiring an isolated and confined environment.

I have discussed the ethical problems of the first option elsewhere (Tachibana 2019, 2020). So, I will focus on the second one here and investigate its ethical implications in more detail. As I will explore, virtue ethics must weigh the relative merits of the value of our survival as a species and the ethical problems arising from genetic human modification.

[4]Although future deep space exploration will require some drastic countermeasures (Szocik and Tachibana 2019), it will also require cautious ethical, legal, and social consideration to such countermeasures (see Tachibana 2017).

12.3 Two Features of Genetic Manipulation

What sort of genetic manipulation would be required for humankind to survive in the flesh in outer space? Various means have been proposed, such as acquiring the ability of photosynthesis in order to survive with minimal food, developing an exoskeleton that may provide immunity to space radiation, and becoming a cybernetic organism that is united with a spaceship (Abney and Lin 2016; Kaku 2018, Chaps. 10 and 11). Let us assume that these sorts of changes will enable us to survive in outer space and, accordingly, that they can preserve humankind by reducing existential risks.

Such drastic transformations would have at least two features from an ethical point of view. First, genetic transformation would change our figures. By merging human and non-human DNA, we would become a sort of chimera or posthuman. With a photosynthetic body surface, our skin may come to look like and have the texture of a green plant. This sort of change might be acceptable to those who are familiar with green characters, such as the Jolly Green Giant or The Incredible Hulk. Meanwhile, with an exoskeletal body, we may look like insects or crabs. A possible imaginary depiction of this change would be something like a tardigrade that can survive even in the hostile environments of outer space. As for becoming cyborgs, our cybernetic body might be able to control our metabolism so that we could move to deep space planets as Helen America and Mr. Gray-no-more did in *The Lady Who Sailed the Soul* (Smith 1960). Or, like Tetsuo in *Akira*, we might experience a more drastic change in which we lose our physical structure and our souls merge with our spaceship (Otomo 1984–1991).

Second, such genetic modifications would require not only changing a large part of the human appearances because the ideal transgenic humans must not only be able to survive in outer space but also be able to reproduce offspring who can survive there. In other words, this type of genetic transformation would also require the manipulation of reproductive cells in order to transfer the target characteristics to future generations. Furthermore, if it takes several generations to reach a habitable planet, our descendants would be created on spaceships (Kaku 2018, Chap. 10). This means space colonization through genetic manipulation requires the transitional generations in two ways, namely, those who have been genetically modified or edited to some extent but not yet achieving its completion, and those who will be born in the spaceship and die there before humankind reach the targeted planet. Since the ethical problems of the second sense have been discussed (Milligan 2014, Chap. 10), let us focus on the first one and describe it in more detail.

Although drastic manipulation through the targeting of reproductive cells would change a large part of the human biological structure, we would not be able to manipulate all of the related genomes simultaneously. As in actual genomic experiments concerning, for example, fruit flies, several generations are needed to realize combinational or complex functions. The first "original" generation is classified into several groups, each of which is only partially manipulated. Then, individuals in each group that successfully express the target manipulations are chosen to be crossed with each other with the aim of creating a generation that expresses the combinational functions. This process lasts until generations that fully express the target combinational functions appear (Brand and Perrimon 1993; Bachmann and Knust 2008).

The same process would also be required in the case of human germline editing. At every stage, we would need to check whether the original and updated aspects of the transgenic humans satisfy the fitness requirements of outer space colonization and whether they are able to adapt to the hostile environmental factors. We must also check whether the manipulated characteristics are successfully transmitted to the next generation. The manipulations of both genetic modification and genome editing would require several generations before its completion. This means that many generations would live their lives as the transitional generations, having been genetically modified or edited to some extent but not yet achieving the ultimate goals of the process.

Furthermore, to hedge the risk of failure, not every person in a generation would receive the same genetic manipulation. Some would experience modification pattern A, some would receive modification pattern B, and others would not receive any manipulation and instead keep their "purity," just in case. Here, we should not overlook the fact that the life cycle of humans is much longer than that of laboratory animals. Fruit flies can give rise to progeny after only 10 days at room temperature, or 25 °C, whereas human beings usually require several decades for reproduction. Moreover, the partially modified humans would be raised with other children with the same types of modifications. They would not mingle with other types of modified children or non-manipulated children to avoid "genetic contamination," until the selected patterns of crossing are conducted (see Ishiguro 2005). They would live their entire lives on Earth in a strictly isolated environment.

12.4 Ethical Problems of Genetic Manipulation

These two features of the genetic manipulation of humankind have their respective ethical problems. Regarding the first feature, a drastic change in figure, which would lead humans to resemble other creatures, such as insects, may cause a personal identity crisis. It remains unclear which factors influence an individual's personal identity and to what extent. However, if physical identity contributes to personal identity, as Williams (1973) argues, a drastic change would likely cause mental and psychological stresses in the individual, even if the change progressed slowly. Furthermore, a drastic change in figure would not only lead to personal stress but also induce social and interpersonal frictions with other people, as Gregor Samsa experienced through his transformation into a beetle (Kafka 1915). As a natural consequence, transgenic humans may have difficulty living their lives with human dignity.

Furthermore, the crisis brought on by the first feature can imply a semantic problem; those who would survive in outer space may be called human in accordance with a *revised* notion of human but not with our existing notion of human. In this scenario, can we say that it is *our* humankind which survives as a species even if the humans are unlike us? What do we lose when this type of genetic transformation alters our original form? The actual effects of this transformation are currently almost impossible to know.[5]

The second feature causes a more serious problem because it seems to contradict our moral principles, which Kant (1785) clarifies as the principle that forbids us from treating a person merely as a means to an end. Following this principle, it is difficult to invent an ethically persuasive argument to justify using some generations for the sake of other generations. As Ishiguro (2005) vividly depicts the sorrow of those who are born only for the means of others, transitional generations would also feel sorrow because of the nature of their lives, as well as doom because it is pathetic to be born just for an unknown other's life, growing up in an isolated area and dying (or even being killed) for the person. Or, as Miyazaki (1982–1994, vol. 7) depicts, these individuals might even feel rage toward those who manipulated them and the space colonization policy that allowed it. They would not be pleased with their lives, and their dignity as humans would not be sufficiently respected. This would lead them to demand their dignity as humans and strive to act against the policy by claiming that they have the right to live their lives for themselves.

These defects will surely wound our morality. However, advocates of space colonization could still argue that their enterprise outweighs such ethical defects based on their doctrine (P3), namely, the supremacy of the survival of humanity. In this case, the problem is whether the doctrine is justifiable. Utilitarianism would support it because, compared with human extinction, the survival of the species would increase and would maximize the total amount of human utility (Abney 2019). Deontology may not endorse the P3 because of the aforementioned moral principle. However, this rejection could lead them into a dilemma because rejecting the doctrine (P3) might infringe on other aspects of moral duty, such as the prohibition of suicide. In fact, not a few philosophers assume that deontology may support the doctrine and accordingly endorse the colonization. For example, Green (2019) argues that space colonization is "a moral imperative" because "[c]ompared to ensuring human survival, all other moral values and actions are secondary." Likewise, Abney (2019) argues that human survival is "our absolute duty" because "if no morally responsible agents exist, then no morality exists," and the vanishment of morality is morally the worst situation. Szocik (2020) argues that "the moral duty to protect human survival as a species" permits us to enhance ourselves for settling in outer space (see also Schwartz 2011 and Klein 2007).

What, then, about the case for virtue ethics? Following Sparrow's (2015) course of the argument, the doctrine should be rejected as it contains various vices. In fact, Abney and Lin (2015, 252) argue that the likelihood of human extinction does not give us unconditional permission to colonize other planets:

---

[5]This may also lead us, or posthumans, to rethink the notion of being human and go beyond anthropocentrism (Ferrand 2016). However, this article considers such transformations from *our* virtue ethical point of view.

> Nonetheless, mitigating existential risk does not allow us to ignore more mundane ethical concerns. For instance, under virtue ethics, we would unacceptably demonstrate a vicious character if we were concerned with either individual or collective survival at the cost of the other things that make life worth living. So, it still seems implausible that a remote existential risk justifies pushing ahead with space exploration in whatever manner we want.

I agree with Abney and Lin in that we have *a* reason not to colonize other planets since doing so would demonstrate a vicious character. However, it remains uncertain whether, all things considered (ATC), we also have *the* reason not to do so because human extinction or choosing to die as a species also seems to be vicious. Although Abney and Lin (2015) kept silent on this point, I think this is the core problem for virtue ethics. If a virtue ethical theory can reject space colonization using the ATC judgment, the theory must provide a notion of virtue that would endorse the idea that spontaneous human extinction can be more virtuous than saving humanity in a certain situation. This is a very awkward notion of virtue because it implies that an individual not only chooses to die for him/herself but also calls upon all other human beings to choose to die no matter what they wish. Therefore, from the virtue ethical point of view, the central issue of the ethical validity of space colonization to save humanity is not in whether the enterprise can damage any particular human virtues, as Sparrow (2015) and Abney and Lin (2015) assume, but in whether we can have a notion of virtue in which spontaneous human extinction is a phenotype of human virtues.

12.5 Choosing to Die as a Virtuous Option[6]

It seems clear that choosing to survive is more desirable than choosing to die. In many cultures, this belief has been formulated as a moral doctrine called the prohibition of suicide. In the course of our lives, we face situations that are not easy to endure. However, those who commit suicide in such situations are usually the subject of disapproval, whereas those who persevere are praised. In this sense, the prohibition of suicide and the praise of survival have the same root; choosing to die has a negative and blameworthy value, whereas surviving has a positive and praiseworthy value. The positive value of survival is especially evident in the biological fact that all forms of life pursue self-preservation, reproduction, and the prosperity of their species. Based on this fact, some virtue ethicists define *good* as such that an individual that effectively preserves itself, reproduces, and contributes to the prosperity of its species is a good individual (Foot 2001). Following this course of the argument, choosing to survive will be virtuous, whereas choosing to die will be vicious.

These cultural and biological facts nourish the notion that any option that promotes human survival is desirable. Since this notion is so innate, we will never be confused to understand such a statement like this; "[a]lthough the end of humanity would

[6]This section requires a more cautious and comprehensive discussion, which is provided in my forthcoming paper (Tachibana, forthcoming).

greatly reduce the amount of harm, it would not end it all" (Benatar 2006, 224). As we cannot stop feeling that our survival is most desirable, we will choose to survive even if it might ruin the Earth. However, we should not forget that the notion itself does not imply that survival is most virtuous or that it has the supreme value.

How can virtue ethicists cancel this notion of supremacy while preserving the positive value of survival? In this section, I propose an argument to make the supremacy of survival defeasible by presenting three exceptions, which show, in a certain context, choosing to die can be acceptable or even more virtuous than choosing to survive.

The first example is euthanasia, or death with dignity. In this century, this notion is discussed in biomedical ethics to demarcate the border of ethically approvable ways of taking one's own life. Euthanasia involves a person suffering pain in his/her life and choosing to die instead of choosing to survive if some biomedical conditions are satisfied. For that person, dying can be preferable to surviving because death saves his/her human dignity. Insofar as preserving dignity is understood as a human virtue, such a death can be a virtuous option. Euthanasia is, in this sense, understood as a type of ethically acceptable death (see McMahan 2002, esp. 473ff.).

However, it would hardly be possible to use this notion to refuse space colonization in pursuit of the survival of the human race because it seems to lack normative power. Euthanasia concerns the life and death of an individual whose health is deteriorating due to disease, injury, or decrepitude. The decision to pursue euthanasia depends on this individual's thoughts about whether this deterioration is too severe for him/her to continue living. This means that two patients with similar medical conditions might reach different decisions about their lives; one can choose to die with dignity, while the other can choose to live, but both decisions are legitimately right. The point here is that these decisions are personal and, accordingly, do not have sufficient normative power to generalize one individual's ethically acceptable decision to that of another. Insofar as it is a matter of personal preference based on the sense of suffering, the euthanasia-based argument would not be able to be posed as a sufficient basis for canceling the notion of the supreme value of survival in space colonization because it does not have enough normative power to persuade those who agree with advancing space colonization not to do so.

The second example is martyrdom. Martyrdom, as a sort of religious belief, drives people to choose to die. However, such a death has a morally positive aspect, as it concerns, for example, the right, Christian way of living (Wendebourg 1987; Slusser 1992). Furthermore, the positive value of martyrdom has a certain normative power for other Christians (see Origenes 1899, 22.28–22.30), which naturally implies that apostasy is blameworthy. This can be observed from ancient times (Origenes 1899, 7–10, 35–44) to the missionary communities of the sixteenth and seventeenth centuries (Endo 1966). Therefore, martyrdom is a notion that endorses, in certain contexts, choosing to die as being more virtuous than choosing to survive, and this endorsement has normative power for others.

However, the normative ranges of this notion reach only those who belong to the same organization or profession, such as missionaries, but not to those who belong to other communities, such as those with different beliefs or in different organizations (Wendebourg 1987). Due to this limitation of normative power, this notion cannot cancel the supremacy of survival in the space colonization argument because space colonization is concerned not only with a certain group of human beings but with every human group. The notion required must be so normative that it reaches every human *qua* human and says that, in a certain situation, for every human being, it is more virtuous to die than to survive.

The third example, Socratic death, can have such normative power. Socrates is known as a philosopher who inquired about what constitutes the virtuous life for human beings. Characterizing the virtuous life as a just life that acquires the definition of virtue through dialogue, he argued that the good life requires living justly. In *Crito*, Socrates, who was sentenced to die in spite of his innocence, explains to his friend, Crito, why he would not break out of prison to survive even though he could do so if he wanted, stating that "the most important thing is not life, but the good life (ὅτι οὐ τὸ ζῆν περὶ πλείστου ποιητέον ἀλλὰ τὸ εὖ ζῆν)" (Plato 1900, 48b; Cooper 1997, 42). He argues that it is more virtuous to choose to die than to survive if we cannot survive without committing any wrongdoing, such as that of an innocent man breaking out of prison.

Socratic death is different from euthanasia and martyrdom in the range of its normative power. What Socrates talked about and showed in *Crito* is not his personal preference or a doctrine applicable only to an in-group people but a general prescription for living as human *qua* human—otherwise, his enquiry into the definition of virtue would be nonsense. Therefore, Socratic death proposes something radical here. It states that we should give up living and die willingly if we could not survive without committing wrongdoing. The life resulting from wrongdoing is not worth living. As Socrates ironically puts it: "is the life worth living for us with that part of us corrupted that unjust action harms and just action benefits? ('Αλλὰ μετ' ἐκείνου ἄρ' ἡμῖν βιωτὸν διεφθαρμένου, ᾧ τὸ ἄδικον μὲν λωβᾶται, τὸ δὲ δίκαιον ὀνίνησιν;)" (Plato 1900, 47e; Cooper 1997, 42). In short, the Socratic thesis is as follows: to die is more virtuous than to live unjustly.

The Socratic notion of virtuous life and death provides a good foundation for considering the supremacy of our survival as a species. This article has assumed that the time will come when the Earth will no longer be able to feed us. It has also assumed that the only option to save humanity in such a situation would be space colonization by a genetic manipulation of humankind. Given these assumptions, several ethical defects, such as identity crises and the exploitation of transitional generations, seem acceptable because saving our species may be worthy of pursuing at any cost. However, once we take the Socratic notion of virtue seriously, another possibility appears; we will be able to choose to end our lives as a species when the Earth does not feed us anymore because space colonization through human genetic enhancement, which this article has assumed to be the only option to save humankind, cannot be completed without engaging in unjust and vicious behaviors. From the virtue ethical point of view, only the Socratic notion of virtue provides a good argument to negate the supremacy of human survival as a species and justify that spontaneous human extinction can be virtuous.

12.6 Conclusion

Lucian of Samosata, an ancient Roman author in the second century, wrote a story of traveling to the Moon, describing it as "an utterly imaginary and impossible story (ἔτι δὲ μήτε ὅλως ὄντων μήτε τὴν ἀρχὴν γενέσθαι δυναμένων)" (Lucian of Samosata 1972, 1.4.9–1.4.13). He entitled this story, *A True Story* ('Αληθῆ διηγήματα), because he believed that he was telling a lie in the story. Maintaining this belief, he also told a lie in the title of the story. However, living in the twenty-first century, we know that space travel to the Moon is not imaginary or impossible at all. In this sense, Lucian's story is truly a *true* story for us.

What, then, is the true story for our century—to use Lucian's terminology? Space colonization through genetic enhancement could be such a story. However, just as space travel became a reality, contrary to Lucian's expectations in, so too can space colonization. Although I did not discuss the probability of this future in this article, I argued that this supposedly imaginary topic sheds new light on our existing notion of human virtue and the supremacy of survival. The concept of the ethical extinction of the humanity, as nourished by the Socratic notion of virtue, can exhibit a unique aspect of human virtue because it can doubt the validity of our biology-based notions of virtue, such as self-preservation, reproduction, and the prosperity of the species. Since any non-human animal would not choose to die either individually or as a species, these biology-based virtues are undoubtedly valid to them. In contrast, ending one's own life willingly can be a distinctive feature of human virtue because, in a certain context, human beings can and do choose to die as a result of their ethical decision-making. Incorporating this sort of choice into the notion of human virtue may lead us to fabricate a new story about what constitutes human virtue and what constitutes humanity. The imaginary topic of space colonization can provide such a *true* story.

References

Abney, K. (2019). Ethics of colonization: Arguments from existential risk. *Futures, 110,* 60–63.

Abney, K., & Lin, P. (2015). Enhancing astronauts: The ethical, legal, and social implications. In J. Galliot (Ed.), *Commercial space exploration: Ethics, policy, and governance* (pp. 245–257). Surrey: Ashgate.

Allen, S. C., Burnett, R., Charles, J., Cucinotta, F., Fullerton, R., Goodman, J. R., et al. (2003). *Guidelines and capabilities for designing human missions.* NASA Center for Aerospace Information. https://ntrs.nasa.gov/archive/nasa/casi.ntrs.nasa.gov/20030033919.pdf.

Bachmann, A., & Knust, E. (2008). The use of P-element transposons to generate transgenic flies. In C. Dahmann (Ed.), *Drosophila: Methods and protocols* (pp. 61–77). New York: Humana Press.

Benatar, D. (2006). *Better never to have been: The harm of coming into existence*. Oxford: Oxford University Press.

Brand, A., & Perrimon, N. (1993). Targeted gene expression as a means of altering cell fates and generating dominant phenotypes. *Development, 118*, 401–415.

Carson, E. (2018). Elon Musk wants to preserve humanity in space. https://www.cnet.com/news/elon-musk-wants-to-preserve-humanity-in-space/.

Chow, L. (2017). Stephen Hawking: 'I am convinced that humans need to leave earth'. https://www.ecowatch.com/stephen-hawking-leave-earth-2446264918.html.

Clément, G. (2011). *Fundamentals of space medicine* (2nd ed.). New York: Springer.

Cooper, J. (Ed.). (1997). *Plato: Complete works*. Indianapolis: Hackett Publishing Company.

Do, S., Owens, A., Ho, K., Schreiner, S., & de Weck, O. (2016). An independent assessment of the technical feasibility of the Mars One mission plan – Updated analysis. *Acta Astronautica, 120*, 192–228.

Endo, S. (1966). *Silence*. Tokyo: Shinchosha. (W. Johnston, Trans., published in 1969 by London: Peter Owen).

Ferrando, F. (2016). Why space migration should be posthuman. In J. Schwartz & T. Milligan (Eds.), *The ethics of space exploration* (pp. 137–152). New York: Springer.

Foot, F. (2001). *Natural goodness*. Oxford: Oxford University Press.

Green, B. P. (2019). Self-preservation should be humankind's first ethical priority and therefore rapid space settlement is necessary. *Futures, 110*, 35–37.

Hodkinson, P. D., Anderton, R. A., Posselt, B. N., & Fong, K. J. (2017). An overview of space medicine. *British Journal of Anaesthesia, 119*(S1), i143–i153.

International Space Exploration Coordination Group (ISECG). (2018). *The global exploration roadmap* (3rd ed.). https://www.nasa.gov/exploration/about/isecg.

Ishiguro, K. (2005). *Never let me go*. London: Faber and Faber.

Kafka, F. (1915). *Die Verwandlung*. Leipzig: Kurt Wolff Verlag.

Kaku, M. (2018). *The future of humanity: Terraforming Mars, interstellar travel, immortality, and our destiny beyond earth*. New York: Anchor.

Kanas, N., & Manzey, D. (2008). *Space psychology and psychiatry* (2nd ed.). Berlin: Springer.

Kant, I. (1785). *Grundlegung zur Metaphysik der Sitten*. Riga: Bey Johann Friedrich Hartknoch.

Klein, E. R. (2007). Space exploration: Humanity's single most important moral imperative. *Philosophy Now*, (61). https://philosophynow.org/issues/61/Space_Exploration_Humanitys_Single_Most_Important_Moral_Imperative.

Lucian of Samosata. (1972). Verae historiae. In M. D. Macleod (Ed.), *Luciani Opera* (Vol. 1, pp. 82–125). Oxford: Clarendon Press.

McMahan, J. (2002). *The ethics of killing: Problems at the margins of life*. Oxford: Oxford University Press.

Milligan, T. (2014). *Nobody owns the Moon: The ethics of space exploitation*. North Carolina: McFarland.

Miyazaki, H. (1982–1994). *Nausicaä of the Valley of the Wind* (7 Vols.). Tokyo: Tokuma-shoten. (in Japanese).

Munevar, G. (2019). An obligation to colonize outer planet. *Futures, 110*, 38–40.

Origenes. (1899). Exhortatio ad martyrium. In P. Koetschau (Ed.), *Origenes Werke* (vol. 1, pp. 3–47). *Die griechischen christlichen Schriftsteller* 2. Leipzig: Hinrichs.

Otomo, K. (1984–1991). *AKIRA* (4 Vols.). Tokyo: Kodansha. (in Japanese).

Plato. (1900). *Crito*. In J. Burnet (Ed.), *Platonis Opera* (Vol. 1, St I.43a–54e). Oxford: Clarendon Press, (repr. 1967).

Schwartz, J. S. J. (2011). Our moral obligation to support space exploration. *Environmental Ethics, 33*(1), 67–88.

Singer, P. (2015). *The most good you can do: How effective altruism is changing ideas about living ethically*. Connecticut: Yale University Press.

Slusser, M. (1992). Martyrium: III. Christentum. III/1. Neues Testament/Alte Kirche. *Theologische Realenzyklopädie, 22*, 207–212.

Smith, C. (1960). The lady who sailed the soul. A pseudonym for P. M. A. Linebarger. *Galaxy*, April, pp. 58–81.

Smith, K., Abney, K., Anderson, G., Billings, L., Devito, C. L., Green, B. P., et al. (2019). The great colonization debate. *Futures, 110*, 4–14.

Sparrow, R. (2015). Terraforming, vandalism and virtue ethics. In J. Galliot (Ed.), *Commercial space exploration: Ethics, policy, and governance* (pp. 161–178). Surrey: Ashgate.

Szocik, K. (2020). Is human enhancement in space a moral duty? Missions to Mars, advanced AI and genome editing in space. *Cambridge Quarterly of Healthcare Ethics, 29*(1), 122–130.

Szocik, K., & Tachibana, K. (2019). Research viewpoint: Human enhancement and artificial intelligence for space missions. *Astropolitics, 17*(3), 208–219.

Tachibana, K. (2017). Space neuroethics. *American Journal of Bioethics: Neuroscience, 8*(1), W6–W7.

Tachibana, K. (2019). A Hobbesian qualm with space settlement. *Futures, 110*, 28–30.

Tachibana, K. (2020). Workplace in space: Space neuroscience and performance management in terrestrial environments. In J. T. Martineau & E. Racine (Eds.), *Organizational neuroethics: Reflections on the contributions of neuroscience to management theories and business practice* (pp. 235–255). New York: Springer.

Tachibana, K. (forthcoming). Virtue in death: Euthanasia, martyrdom, and Socrates. In M. Gabriel & S. Breu (Eds.), *Nature, technology, metaphysics*.

Tachibana, K., Tachibana, S., & Inoue, N. (2017). From outer space to Earth—The social significance of isolated and confined environment research in human space exploration. *Acta Astronautica, 140*, 273–283.

Traphagan, J. W. (2019). Which humanity would space colonization save? *Futures, 110*, 47–49.

Wendebourg, D. (1987). Das Martyrium in der Alten Kirche als ethisches Problem. *Zeitschrift für Kirchengeschichte, 98*, 295–320.

Williams, B. (1973). Personal identity and individuation. In B. Williams, *Problems of the self: Philosophical papers* (pp. 1–18). Cambridge: Cambridge University Press.

Chapter 13
Ethical Problems of Life Extension for Space Exploration

Tony Milligan and Shin-ichiro Inaba

Abstract The possibility of living significantly longer than we currently live, and doing so in a more or less healthy condition, has repeatedly been bundled together with projects of space settlement. It is particularly strong within science fiction. It is there in works by Arthur C. Clarke among others. We may, admittedly, be sceptical about the classic reason once offered for thinking that longevity stretching into hundreds of years could be secured, i.e. the reduced physiological demands of microgravity. However, it is not entirely unreasonable to think that that a more modest extension of longevity could turn out to be an accidental side-effect of some other form of genetic modification geared to fit agents for the conditions of space. Should any modification with this effect be embraced or blocked? There are familiar and plausible ethical reasons to regard major changes in terrestrial longevity with caution, accepting them under some sociopolitical conditions, but not under others. For example, powerful elites living significantly extended lives while the mass of the Earth's population miss out would not be a particularly good model for social justice. This chapter will explore what happens when we attempt to transfer three of the more plausible ethical objections to life extension therapies from life on Earth to activity in space. The authors find that, just so long as genetic modification of humans is treated as permissible, some forms of such a longevity extending therapycould also be permissible. Nonetheless, the prospect does raise some deep concerns about the role of death within the 'meaning of life'.

T. Milligan (✉)
Department of Theology and Religious Studies, King's College London, London, UK
e-mail: anthony.milligan@kcl.ac.uk

S. Inaba
Department of Sociology and Social Welfare, Meiji Gakuin University, Tokyo and Yokohama, Japan
e-mail: inaba@soc.meijigakuin.ac.jp

K. Szocik (ed.), *Human Enhancements for Space Missions*, Space and Society,
https://doi.org/10.1007/978-3-030-42036-9_13

13.1 Introduction

The idea of going into space has often been a channel for utopian and transformational aspirations. From hopes for social progress (in Konstantin Tsoilkovsky's Russian science fiction of the 1920s) to aspirations for gender equality (from the 1960s onwards, in writers such as Ursula Le Guin); and on to social critique (both East and West, the brothers Strugatsky in the 1970s, and Kim Stanley Robinson since the 1990s). Sitting in the background of a good deal of this literature, we find the idea of personal survival across extended periods of time. It is there in much of Arthur C. Clarke's science fiction, with one of the most extreme versions being Frank Poole's disappearance in *2001: A Space Odyssey* (1968), and the subsequent revival of his freeze-dried body in *3001* (1997). Poole survives like [illegible] in open space. Survival by dehydration, along with survival by freezing in a box, is there again in Liu Cixin's first-contact trilogy, beginning with *The Three-Body Problem* (2007). It allows narratives to stretch across centuries from the moment of the first contact between humans and intelligent non-humans, while keeping central (also mortal) characters in place.

The mechanisms for extended survival vary, from going into stasis (Liu, and sometimes Clarke), through to the idea that space itself will somehow slow down the ageing process by placing less strain upon the human body. Unlike slower aging, stasis does not automatically result in a longer subjective experience of time. It simply distributes our limited experience of a life over a larger stretch of objective time. The subject is frozen and revived at a later date. They are not conscious of the centuries which pass, let alone the vast deserts of eternity which may lie beyond. In terms of awareness, they get what we get. The once-popular idea that space will reduce the strain, and allow some of us to live longer, is closer to the aspiration for a true extension of longevity. It could come complete with an awareness of survival across stretches of time which exceed the 120 year maximum for a human life, which seems to have been the norm throughout our recorded human history, even among the privileged sections of society who tend to leave more of a mark in the surviving records. This is as much as any of us get. Wealth and good living cannot buy more. Women approach a little closer to the maximum than men. The Gerontology Research Group lists dozens of women who have lived to between 114 and 119 years, and the single disputed case of Jeanne Calment who may have died at the age of 122. It also lists men who have lived to between 110 and 115, and the single case of Jiroemon Kimura who survived to 116 years of age.[1]

The aspiration to live longer in space has fed off of familiar concerns about human mortality, fear of death or at least the desire to defer it so that death does not come too soon for our hopes and plans to be realized. These fears and hopes range from the deep, the 'immortal longings' mentioned by Shakespeare's Cleopatra at the very point of death, through to the less deep. We already have corporations which catering for the later. The Alcor Life Extension Foundation, situated half an hour north of Phoenix, has almost 200 bodies and severed heads in cold storage. Cryogenically frozen in

[1] https://grg.org/Adams/B.HTM. Accessed March 9, 2020.

liquid nitrogen, in anticipation of a future revival, once science has finally caught up with the aspirations of wealth, mixed together with a good deal of what looks like wishful thinking (O'Connell 2017, 22–41). Stasis technologies may be possible. It is difficult to rule them out. They may even, at some point become available. But it seems very unlikely that we are already at or near to that point. Those who have their remains cryogenically frozen are, no doubt, aware of the limitations of the existing processes. They will have paid enough to research matters in more detail and to have realized that, unlike Frank Poole, and in the absence of improbable technologies, they are not coming back. Still, for some, the glimpse of a possibility is enough.

Hopes for an extension of life have also bled over into discussions about viable models of human expansion beyond the Solar System. Discussions which accept the improbability of warp drives, or the convenient discovery of nearby wormholes which might take us far away in minutes, hours, or at least less than a lifetime. Realistic attempts to set out a workable plan for interstellar travel have repeatedly crashed against the problem of the enormous periods of time which would be required, and the likelihood that future generations born on board any generation ship, any space ship built to cope with such a journey, would not come to share our goals. They might be expected to rebel against any settlement project that we happen devise. Their psychological responses to an unasked-for and confining predicament may be too unpredictable to be relied upon as a means of remote colonization or survival by proxy once life on Earth has come to an end. Cutting out these inconvenient *shipborn* intermediaries, and making at least some of the first generation *leavers* and the *arrivers* one and the same individuals, by putting humans into stasis, would seem to solve the problem (Milligan 2015, 146). It would, at least, go some way towards a solution if agents were successfully selected for their stability of aspiration, and unlikely to change their minds once the extent of their disconnect from all terrestrial authority is fully grasped.

Inconveniently, the actual mechanics of stasis for humans have remained elusive. The stuff of fiction or of a future and far more advanced science. The kind of thing that we may already research with a little help from DARPA funding. But the hoped-for return upon such investment in stasis research remains elusive. There have been some claims of progress (Roth and Nystul 2005). There are tales of mice and worms, frozen and unfrozen, and tales of chance human survival in extreme conditions. But they are never the real thing, never quite the capacity to freeze and unfreeze living humans with their capacities unimpaired (Shermer 2016). Interstellar space is out there, but the human longevity which would be required for direct human exploration of it remains elusive.

This is not to say that we cannot live longer. Life expectancy has risen on a global scale since the middle of the last century, and it currently sits at over 70 years almost everywhere except for parts of Africa (UN 2019). Matters can get worse, as well as better. Those born today may or may not be expected to live as long as members of any current generation. This too illustrates the variability of life expectancy. However, while average life expectancy has shifted, the maximum of longevity has remained stubbornly the same. Around 120 years is the most that we can ever expect to get, unless we are physically transformed in some way which reaches beyond changes brought about by diet, exercise, medication, and the welcome absence of excessive worries. Yet we know that this is far from maximum longevity for all creatures. Sponges and jellyfish can live much longer. It is even far from maximum longevity for intelligent creatures. Whales can live for centuries. Physicality, rather than routine, marks the difference between us and them. Physical transformation from *what we are* seems to be required if we are to match their extended lives. This may offer little hope to those of us who are already more or less 'fixed' in a ready-made state, but it need not push us towards the prospect of a posthuman future. That too may come, but genetic modification of more modest sorts will be in place before any prospect of that sort. Here, we assume that we are not, already, in a posthuman present.

Science fiction again offers pathways to cellular changes which would allow us to remain recognizably ourselves, at least on the surface. The French television series *Ad Vitam* (2018), set around the year 2070, allows indefinite regeneration of already-existing humans, through some sort of somatic cell therapy, inspired by the remarkable longevity of the *Turritopsis dohrnii* jellyfish (Kubota 2011; Miglietta et al. 2006). The creature is capable of reverting to a juvenile stage after reproduction. It sinks to the ocean floor, regenerates, and begins the process all over. The oceanic imagery, evident throughout the series, is both striking and familiar from correspondence between Romain Rolland and Freud on the subject of religious aspirations and the mystical dimension of religious attitudes towards *being* (Freud 1961). The Latinization of *L'Chaim/to life* in the title is also noteworthy and ironic. The kind of life made available is not a celebration. It is not even more of what we already have. Apparent immortality turns out to be unbearable for some. For the bravest, it is a desert. There are suicides. *Ad Vitam* raises a further ethical concern: the inseparability of life extension and acceptance of an exit route for those who do not like what they become. It is difficult to envisage any available process of life extension without the triggering of a change in Western attitudes towards suicide. Those who have a fixed opposition to the latter, and oppose any notion of a right to end life, should probably oppose life extension, even if it will help the space program. The ethics of these matters is not at all clear, irrespective of the level of complexity that we bring to the problem. If we focus upon consequences, and believe that ethics is grounded in clear-cut decision procedures, in the manner of a philosophical ethicist such as Peter Singer, it is still not obvious that we will arrive at one unique result. Singer's own ventures into this territory have produced an early article in which he was broadly opposed to life extension (Singer 1991), and a more recent article which is altogether more favourable (Singer 2012).

These comments survey some familiar concerns and attitudes. But they are intended as more than idle musings, or an obligatory opening review. Rather, they are deployed here as a way to shape a thought experiment about the ethics of life extension in space, in a way which lends plausibility, and helps to make it the description of a possible scenario rather than the kind of impoverished and unconstrained narrative that philosophical thought experiments about life and death so often collapse into (Milligan 2007, 2019). One obvious constraint upon such a thought experiment, if it is to be at all realistic, is that the initial and presupposed therapy will be germline, and *not* somatic cell. *Ad Vitam* is not our world. We are not jellyfish or sponges. Actual regeneration is a far more remote prospect than the genetic engineering of physiological differences which might extend the longevity of humans who have yet to be born, by some modest, but notable amount. The thought experiment below will be shaped, as far as possible, by what currently seems attainable, rather than indexed to any set of immortal longings, the needs of interstellar civilizations, or the requirements of narrative continuity in science fiction. To help avoid the imagined scenario answering to any of the latter, we will make the discovery of the technology of life extension accidental, a by-product.

13.2 A Thought Experiment

Let us assume that some germline therapy is devised in order to help future astronauts survive under the harsh conditions of space. Let us refer to it as a single therapy, for convenience, even if it is likely to involve several different things. The therapy will target problems such as reduced bone density, and some of the more familiar obstacles to putting human bodies into a situation which is very different from that of life on Earth. At some point, it is discovered that the therapy has the unintended side effect of slowing down the ageing process, opening up the prospect of pushing the maximum human age away from its current upper limit. Upon closer examination it turns out that the therapy, which involves genetic modification, alters patterns of cellular division and telomere shortening. Cells can keep on dividing for much longer, and in a more reliable way. Or, at least, some comparable change occurs, focusing upon the kinds of features that figure in the science of aging. It is not magic, nor intentional, but it is well within the bounds of what we may imagine as possible, and not mere logical possibility, but real material possibility. The convenient result has also been arrived at without having to go through any of the difficult arguments, and demanding procedures, that an actual proposal to research longevity extension might have to face.

Unlike various instance of life extension in science fiction, it is the therapy itself which produces the result. Space is not a place where one automatically tends to live longer as a result of microgravity, and its reduced pressures upon the human. Microgravity does not work that way. In our thought experiment, it is the genetic modification of human life for space which results in a higher maximum age than the historic human norm of around 120. The change also affects average life expectancy under reasonable conditions of health. Those who undergo it will tend to live longer, as well as enjoying a higher absolute maximum.

As it is a fortunate by-product, it need not match up with any estimate of the ideal extension of longevity which is desirable for the sake of some particular space-related aspiration. Science fiction, and thought experiments about what is ideal, may say that we need longevity to be extended over centuries. The fortunate accident, by contrast, results only in a modest extension of longevity. Those who undergo the therapy enjoy the possibility of an additional two or three decades more than the usual maximum, and no more. This is not, of course, the *guarantee* of a longer life. Accidents happen, and in space they will be frequent and often deadly. But let us allow that, if things go well, those who are born with the modifications (let us call them 'space candidates', abbreviated to 'candidates') will tend to live longer, and *could* live to around 150 years. If there were enough candidates, all of whom result from the relevant genetic engineering, someone would eventually get close to this age. Let us also match up the extension of actual life expectancy by a comparable length of time, say to 100, with survival to around 120 becoming statistically far more likely that it is for current humans. Survival to the current maximum will be less of an outlier and instead situated somewhere on the downside of the normal curve for 'years lived'. At the point when most present day humans are rapidly slowing down (at some point in their 70s and 80s), candidates will also still be in reasonably good working condition. They might not sign an extended deal for the New York Yankees, or win any medals for athletics, but they will be physically more robust than most of us would be if we reached a comparable age.

In our thought experiment, not all of these modified people will go into space. The terminology of 'candidates' captures the point that there is no guarantee that they will go. They will, however, belong to the pool of agents most likely to do so. Furthermore, they should remain more or less as well-adapted as the rest of us to life on Earth. Some things might be ruled out. It might even be the case that they take a little longer, on average, to recover from injury. But worries about condemning innocent future humans to an involuntary exile will not arise. The candidates will not be suited *only* to life elsewhere. And the therapy itself is not be the result of some grand eugenic scheme for breeding ultimate humans. It results accidentally, and only from a reasonable desire to avoid certain problems for those who go into space, a desire to create more of a level playing field for those who go, when compared to the rest of us.

Whether or not they go anywhere, our candidates will still die, like everyone else. But they will also tend to live longer and might even live longer than Jeanne Calment or Jiroemon Kimura, as well as enjoying better conditions of health than either of the latter experienced once they passed beyond their first century. Let us also recognize that there may be broader, and legitimate, social reasons for helping at least some human agents to live so long. Reasons which are entirely independent of our space aspirations, and independent of any quasi-religious hopes for survival. Agents who live longer, and in conditions of reasonably good health, could have special knowledge linked to their extended experience. Knowledge of a sort which cannot easily be acquired by the rest of us. Some of this knowledge may take the form of *knowing how*, rather than *knowing that* (Milligan 2019, 87–90). Or, it may involve a certain depth of understanding which grasps connections that the rest of us recognize only through a glass darkly, or in more shallow terms. Rather than old people, we might once again have elders who can act as a bridge between past and future, thereby allowing everyone to develop a better appreciation of time. This would be a good thing, when thought of from a social point of view, irrespective of whether or not anyone ever goes into space, and irrespective of whether or not we think of it as an individual good, a good for the individuals concerned. It might improve the quality of our thinking about multi-generational projects of various different sorts, and not just about space as a multi-generational project. We will take it that this is a change which would have some clear advantages, even if there was also a downside.

While broader social reasons exist for favouring an increase in longevity for some humans, and while our candidates may or may not actually go into space, in our thought experiment the genetic engineering is only carried out with a view towards space. It is not routinely available for those who wish to have children. Nor are the candidates drawn *in any unusual way* from otherwise privileged social groups. Privileged access will not be entirely avoided. This might be unrealistic, it could make the ethical requirements of the imagined scenario too demanding, and perhaps even utopian. However, such access will, at most, match the levels of privileged access which are socially typical for the time. Egalitarian societies will draw candidates in more egalitarian ways, less egalitarian societies may do so in less egalitarian ways. But, in each case, the bounds of privilege and inequality may be those which are consistent with liberal democracy (broadly understood) or with successor systems which have many or all of the same good features. The legitimacy of the space programmes which the candidates are part of will not be automatically compromised by questions concerning the legitimacy of the prospective but imperfect launch states.

The entire process might, however, be challenged on the grounds that all bioengineering of humans is ruled out. However, the authors suspect that this is a view which will not survive current processes of technological change. Once divorced from eugenic visions of a classic sort, there is also no obvious and overriding ethical objection to all germline therapies. There are only reasons for caution and restriction. Nothing else is likely to be politically viable in societies where biotechnology is viewed as normal, rather than with a special kind of suspicion. The process might also be challenged at a more philosophical level, on the grounds that it compromises the autonomy of the resulting candidates. The scenario does trades off plausibility against the acceptance of ethical worries of this sort. Plausibility is served by allowing that something more than somatic cell treatments will be required if longevity is to be extended, anytime soon, and with anything resembling current technologies. However, the agents most directly affected will not choose to have the required germline therapy because they will not yet be around to do so. They will be agents who are born with the modification(s), and they may then pass the modification(s) on to their children. No attempt at containment by rendering space candidates infertile will be made. First, because it would be deeply immoral and bring discredit upon any space program which was linked to such a punitive requirement. Second, because the advantages of longevity are seen in the light of sustained human activity in space, and the associated possibility of shifting from visitation to settlement. From the point of view of actual settlement, and especially that of eventual settlement autonomy, infertility would not be a good thing.

In order to avoid transposing into a familiar set of arguments, which are entirely independent of space, we will assume that worries about autonomy matter but that they need not dominate the conversation. Humans have brought other humans into being under a variety of different, and often adverse, circumstances of poverty and hardship. Each time, it has been without the consent of those who are born, irrespective of how they were born, and irrespective of their physical capabilities. If we say that adversity or significant difference from some norm requires consent, then we may be at risk of 'proving too much'. We will be in danger of making ordinary, current as well as historic, cases of bringing children into the world wrong. To insist upon strict conditions for the latter may be too demanding a requirement to be part of any workable social ethic for humans who are anything like ourselves. This does not mean 'anything goes', or that we can appeal to the idea that moral constraints apply only in relation to agents who currently enjoy the *terra firma* of existence (Salt 1914). With Bernard Rollin, we will assume that genetic modification of any creature (human or non-human) should not tend to make life worse for the resulting beings (Rollin 1998, 169). They should, then, enjoy the possibility of some regular manner of contentment. This minimal criterion ought to apply when bringing humans into being, with or without genetic modification. For present purposes, we will assume that if this limited requirement is met, then *all other things being equal*, humans can legitimately be brought into being, even though doing so will always be without their impossible-to-acquire consent. This could be reinforced by appeal to the familiar device of hypothetical consent, i.e., to *what the agents in question would say if asked, at a later point in time*. However, hypothetical moves look suspiciously like a different way of making the same point. We will also assume that broader 'anti-natalist' arguments against bringing *any* humans into being, under *any* actually available circumstance, will also fail (Benatar 1997). Human societies cannot live in accordance with such an ethic, and our concern here is with the kind of ethics which is socially viable.

As a qualification to our use of a thought experiment, we will allow that this kind of device may provide, at most, hints, clues and reminders about what is and is not ethically permissible. We will allow that no decision about ethically significant matters concerning life and death (e.g. abortion, euthanasia, embryo experimentation, and life extension) should be taken *solely* upon the basis of thought experiments, or by appeal to similar fictional scenarios. Nonetheless, thought experiments may show something, even if *what the experiment shows* can be a matter of debate. Our aim, in the present case, is not demonstration, but deliberation and the shedding of some light upon a difficult subject. Our question is the simple one of whether or not the thought experiment describes a permissible state of affairs. Put simply, 'Should we be allowed to do something of this sort?' It would take a good deal of argumentation, which has yet to be presented by anyone, to establish an actual *duty* to use life extension technology under conditions similar to those specified. And such a duty could not simply be underpinned by appeal to a duty to colonize space. It is, after all, far from clear that we have any such duty (Inaba 2016; Szocik 2019). Accordingly, we limit ourselves to the question of whether or not it would be *permissible* to modify humans for space exploration, in line with the scenario described.

13.3 Applying Three Classic Objections to Life Extension

We have set aside worries about autonomy, which we regard as manageable, in order to focus upon three of the classic objections to life extension, and to consider how they will carry over into the context of life extension for space. We have also considered the idea that life extension may be a social good only in general terms. However, we may need to be a little more precise about the social advantages which carry over into the context of space, if we are required to weigh them against the disadvantages. Two advantages, in particular, seem to carry over well. First, the epistemic/knowledge advantage. Even if exploration is limited to the Solar System, large engineering projects such as the terraforming of Mars (and Venus), or the building of a planet-size particle accelerator, are likely to require centuries. Success is far from guaranteed, but may be more likely if practical knowledge (*knowing how*, by contrast with *knowing that*) is sustained. There is a reason why the patriarchs of the Hebrew Bible are described as living to extreme old age: they needed to get the job done. Moses is not supposed to have led the same generation of Israelites into the promised land, but rather their children. The biblical narrative treats him as the bridge (epistemic as well as spiritual) between generations, even if not an ultimate beneficiary of the process. Continuity matters. Second, it would also help the multi-generational project of space expansion if our attitude towards time was to become less dominated by the short-term horizon which tends to shape budgetary appropriations. While a better attitude towards time, and the future, might come about through an improved sense of our obligations to future generations, we should allow for the possibility that good ethical arguments about duties to humanity, or to future generations, may fail to deliver the required change. The only way to finally break short-termism may well be for some humans to have longer lives and to be a bridge across the generations. We may need our Moses, or at least bridging agents of some sort. There are also epistemic (knowledge-related) dimensions to this claim, but it looks like a distinct consideration.

With this in mind, the three classic objections may now be specified: worries about overpopulation; worries about the unfair distribution of opportunities for a longer life; and worries about the erosion of life's meaning, which link the latter to a mortality which we should not seek to evade if we are to live fully human lives.

13.3.1 Overpopulation

The more overlap there is between the lives of generations, the more humans there will be at any given time. Everyone will still die, life extension is not magic. But if too many of us are alive *at the same time* we will create a greater environmental burden. Billions of humans spread across time, is a bearable burden, but if too many of our lives overlap, it will make matters worse. This is sometimes known as the 'Malthusian objection' to life extension (Davis 2005). It faces a number of difficulties, two of which will be highlighted here. *The first problem* is that it risks pushing us towards a life shortening or *Logan's Run* scenario. If it is better for lives to be shorter, they why not promote the latter as a social good? In Nolan and Johnson's *Logan's Run* (1967), published early in the awareness of emerging ecological problems, humans are allowed to live until the age of 21. Long enough to reproduce, but not long enough to become a burden upon the planet. Life is good, really good. Even hedonistic. But it is lived out under the shadow of a fiction of 'renewal', i.e. the chance of reincarnation following a deliberate end of life. This finely balanced society results from a youth-led revolt against environmental destruction, and a capacity for self-sacrifice in the interests of the greater good. Admittedly, this may also be too psychologically demanding for most of us. And we may wonder about just how good the resulting lives are. All options should, after all, allow for a reasonable chance of contentment and the authors do present the available lives as more hedonistic than they are rich and fulfilling. Yet, even if they were reasonably rich and fulfilling, we may still have strong intuitions that there is a significant difference between rejecting further life extension, and commitment to actual life reduction. The former is within the reasonable bounds of what humans could be persuaded to accept, within a liberal democracy. The latter may be outside of the bounds of what could be realized without excessive indoctrination and the promotion of a mythology. (Some variant of the 'renewal' story.)

The second problem is, perhaps, more telling. It turns upon an assumption that birth rates would be unaffected by life extension. The only thing that changes would be that some humans, the candidates, would live longer, and further into the lifetimes of the next generation. But once the numbers of space candidates increases to a noticeable level, their responses to increased lifespan would presumably tend to average out in ways which might reduce the urgency of reproduction. To the best of our knowledge, humans who live longer and well, in material terms, tend to have fewer children than those whose life expectancy is lower. They also tend to reproduce later in life. Some choose not to have children at all. (Another familiar contemporary trend.) This shift in childbirth is perhaps most evident in Japan since the economic boom of the mid-1950s to the 1970s. However, what initially appeared to be an atypical Japanese trend is now starting to look normal for advanced economies. The decline of the birth rate in Japan, from around two children per woman in the 1960s and 70s to an average of around 1.4 children today, is matched by an even greater decline in the levels of childbirth in the UK and the USA. If we chart the birth rate internationally, over recent decades, for country after country we will find it tracked by a line which is higher on the left than it is on the right. The rate has fallen, internationally and consistently.[2] While there might be a short term price to be paid for increased longevity, it could actually turn out to be a way to stabilize population, or at least to slow down its expansion after some initial brief increase. Increased longevity might carry ecological benefits.

---

[2]https://data.worldbank.org/indicator/SP.DYN.CBRT.IN?end=2014&locations=JP&start=1976. Accessed March 9, 2020.

And even if an increased population objection did work for the Earth, it would not automatically work for humans who spend a substantial amount of time in space. Allowing for the latter is the whole point of the genetic modification process set out above. It is unlikely that we would turn to genetic modification for brief visits to the International Space Station, or even for a round trip to the Moon or Mars. In our thought experiment, only those who were candidates for more extended activity in space, for long missions, or the descendants of such candidates, would benefit from life extension technology. The terrestrial population generally would not. The only circumstance under which candidates would account for a significant number of humans, in comparison with the overall human population on Earth, would be circumstances of settlement rather than visitation. In which case, the issue would be quite different because they would live *there* and leave their environmental footprint *there*, and not *here*. If there is a reasonable ground for worry, it might not arise from those who actually go into space, but from those candidates who carry the modification but do not go. The scenario allows for 'free riders', beneficiaries who do not contribute directly to the process and its benefits. Any risk of the exponential growth of this segment of the population would be something to monitor, in order to determine the optimal size of the next group of candidates. Yet even among this segment, longevity and greater life expectancy might have the same effect of reducing pressures to reproduce during the earlier stages of life. Agents with the modification might turn out to have fewer offspring. Whether they do or not is likely to depend upon extraneous, contingent matters of culture and available ways of living a life. At the very least, we may say that the scenario we describe would not *automatically* result in any increased environmental burden. There would, however, be the *risk* of such an additional burden, and such a risk would be worth taking into account in any overall assessment of the ethics of extending life in the way described. However, with a steadily mounting environmental crisis, to which there are no viable risk-free responses, the possibility of enabling humans to enjoy longer lives with fewer children may also be worth considering as a viable contributary response. Life extension for space might then function as a useful test case, before making any therapy more widely available, in a controlled manner.

13.3.2 Unfair Distribution of Opportunities

If worries about increasing our human burden upon the Earth turn out to be manageable, because increased longevity is not available to everyone, this itself could be a source of a second ethical objection: a concern about *unfair distribution*. Any therapy which avoided the risks of overpopulation because it was available only to a few, rather than the many, might well be distributed in an unfair way. An obvious example would involve cases where increased longevity and life expectancy were disproportionately weighted in favour of social elites. In any such case of injustice, especially if it was greater than the usual levels of injustice within society, increased longevity might still be a personal good for those who were able to live longer, but it would not necessarily be a social good. Whether it was a social good or not might, again, turn out to be a contingent matter, one which depends upon various other considerations.

The worry here is also not only about social justice, but also about the importance of political community, or social solidarity, i.e. the ability of citizens to see ourselves as sharing a common good for which sacrifices might be made, and from which duties might be determined. Given the vulnerability of modern liberal democracies to rapidly polarize, as witnessed with the emergence of populism and political agendas which focus upon appeals to special constituency interests rather than consensus, sustaining the sense of a common good is a matter of some importance. Of course, many therapies do start at the top, or at least among the reasonably well-off, and then become more widely affordable, HIV medication is a case in point. Therapies available at first only to the rich often do become more widely available over time but they might never become so widely available, or available as quickly, if a strict requirement for an initial just distribution was in place. A possible analogue here would be space tourism, which we expect to be restricted at first to the wealthiest, but then to spread down at least as far as the middle classes once prices come down, with some availability spreading across society as a whole.

However, in the present case, the scenario requires the distribution of increased longevity and life expectancy to *remain* limited, and not to spread quickly through the entire population. We want to suggest that, unlike the overpopulation objection, a version of this familiar concern would clearly still apply. Specifically, the candidates *would* have an apparent advantage over everyone else. There are, however, two important countervailing considerations. First, at least for those candidates who actually go into space, the benefits of increased longevity would come at the expense of the increased risks of attempting to live for extended periods of time in dangerous conditions. Space is not a place of security. Benefits would accrue to risk-takers and, in our thought experiment, only to risk-takers. This matters, at least in terms of familiar ethical intuitions that many of us share about risks. Consider, for example, a classic science fiction scenario reproduced in the American TV series, *The Expanse* (2015): some or other asteroid mining company draws down the rewards of risks which are taken by others who actually mine in space. In the case of the series, the latter role is occupied by the colourful 'belters' who hold much of the plotline together. From an ethical point of view, this question of distribution can be just as important as our assessment of probabilities of harm. When risks are significant, but beneficiaries are different people from the risk-takers, processes look unfair.

This is not an attempt to translate nineteenth- and twentieth-century labour relations in the context of space. The division between beneficiaries and risk-takers applies just as much among corporations and entrepreneurs. It is a reason why companies which invest in asteroid tracking, and take the initial but significant commercial risks involved in participation within the expensive first wave of actual asteroid mining, may have a special claim upon tenure, and upon entitlement to later protection from claim jumping by others who want to share the benefits, but have not taken these risks. If a process is socially worthwhile, but risky, the risk-takers should generally be among the beneficiaries, even if this does result in a distribution of benefits which is not even across society as a whole. Indeed, a completely even distribution of benefits might also be unfair given that some have taken the risks while others have not. As a result, it is far from obvious that the privileged position of those who actually go into space *and* enjoy the benefits of life extension, would involve any overall injustice. The main problem case would, as with the population problem, rest with space candidates who enjoy the benefits of a longer life but never leave the Earth for any significant period of time. However, the injustice of advantages enjoyed by these 'free riders' would be mitigated by the fact that they have in no way chosen their predicament, and unlike an unfair distribution of material goods, they cannot do anything to rectify it. (Other than going into space, and too much socially applied pressure to do this would probably introduce an unfairness of a different sort.)

Assuming that the population at large can appreciate that those who go into space in some sense 'earn' the advantage of a longer life, the problem of political community/social solidarity would also only apply if there was some mechanism whereby the numbers of Earthbound candidates became too large, or if candidates were drawn excessively from already privileged groups. While the scenario described has tried to rule this out, as a basic part of its description, nothing about the process of genetic modification actually ensures that its availability will be fair. The scenario specifies that the problem remains limited, but whether or not it could be kept limited in a real case is difficult to say. We can imagine things working out, but conceivability is not the same as practical possibility. The scenario might come unravelled, at this point. However, it is noticeable that actual astronaut selection in recent decades *has* been shaped by a growing level of concern for *inclusion*, and by a realization that sustained public funding depends upon the sustained legitimacy of space exploration. Gil Scott-Heron's infamous poem, *Whitey on the Moon* (1970), which struck against many things, including the narrow background of the Apollo program's astronaut selection, might fail to hit the mark today. The 'right stuff' *then* turned out to be white, military, and with a Harvard, Princeton, or more often MIT, background. This has changed, significantly, although we cannot expect that social skewing in astronaut, and prospective astronaut, selection can be entirely avoided. It is a microcosm of the society within which it occurs.

13.3.3 Life's Meaning

Finally, even if the overpopulation and unfair distribution objections are at least manageable, we may still have a concern of a deeper sort: a concern about life's meaning. This is somewhat harder to articulate. The clearest version has come from Leon Kass in 'L'Chaim and its limits: Why not immortality?' (Kass 2001). For Kass, there is something integral to *our kind of being* which involves acceptance of mortality, an acceptance which is certainly at odds with radical life extension technologies. Nick Bostrom has an amusing fable in which acceptance of death is like the sacrificing of a section of the population to a dragon tyrant, rather than attempting to slay it. While some accept the tyranny of the dragon as inevitable, others appeal to apparently more important social problems which might be tacked. Yet, all the while, the dragon is there, and it is *the* thing which needs to be urgently tackled (Bostrom 2005). For Kass, by contrast, death ought not to be seen as a tyranny or a disease, or as any sort of social or personal problem. It is not something to be fixed or cured in the way that a range of advocates of radical forms of life extension suggest. The wrongness of their position, from the Kass point of view, is not necessarily the belief that longevity *can* be extended, and extended even in radical ways, but the belief that this would be a good thing, or something consistent with living our kind of life.

Even though it might be somewhat elusive, we may accept the claim that discussions of life extension engender questions of 'life's meaning'. But what might we be talking about when we speak of 'life's meaning' and hold that it may be spoiled by living too long, or by not dying in the regular timely way? What is meant may perhaps just as readily be called 'death's meaning' because only living things die. And death may be at least as important and indispensable to 'life's meaning' as being born. But what exactly does that mean? Kass appeals to the idea that death might be needed to free up space for those who will be born later. This may be thought of both in terms of resources, and in the broader terms of giving others their own space. Theoretically, if the resource constraints could be avoided through space expansion or through some form of posthumanization, there might be no need to worry about extending the life of any particular generation, even to immortality, on the ground of resources. It would not exhaust resources for succeeding generations. And if this is all that is at stake then it is difficult to deny the simple additionism which says that 'it is better to live longer, all other things being equal' and to embrace life extension in all its forms (Gems 2003). It is the fact that all other things are not equal, that opportunities *are* limited, which then seems to introduce problems. But even this sets aside the idea of giving each generation their own space, and the real opportunities for social renewal that this allows.

There are other reasons to think that this particular question ought not to be reduced to one of resources. In a more holistic view that denies the simple additionism of 'life's meaning' and happiness (more life is better, and more lives better), the emphasis is placed instead on the narrative unity of human life. If having an end is necessary for such narrative unity, death may also be essential for the meaning of life. However, whether 'ending' is essential for the narrative unity is a question worthy of consideration, and the answer is not always obvious. Heidegger said something of this sort, and so too did Hegel. Both allowed death to shape the meaningfulness of life, bringing it into fullness. Heidegger, in *Being and Time* (1927); Hegel in the *Phenomenology of Spirit* (1807). But we cannot expect agreement among reasonable and well-informed agents. (Not all philosophers are Heideggerians or Hegelians.) These matters run deep. Perhaps a life that is *too* long, even if not immortal, might make it difficult to provide anything beyond its early stages with narrative unity, but the length of time required for large-scale projects in space could weigh against the importance of such narrative unity, and it is again difficult to see why one consideration or the other should automatically be trumped. We may think that there is a correct answer to this question, and there may be one, but it is not obvious that a compelling argument could ever be set out, one that others *ought* to accept if they are as rational as humans are expected to be.

If the Kass line of reasoning has any force, it is not obviously the force of a compelling formal argument, but (as with our approach) it may be the force of hints, clues, and reminders. It may, however, carry this over from the case of radical life extension into the more limited and modest scenario described in ourthought experiment. The latter does not describe radical life extension, but it *does* describe significant life extension, and so it too might be thought of as *tending towards* an evasion of our mortality. Or, at least, it might open up opportunities for such evasion. Even more so, if we take seriously a core idea from Aubrey De Grey's case for the practicality of radical life extension. For de Grey, the search for our best approximation to immortality does not require some magical single therapy or treatment (de Grey 2009). All it may require is the ability to survive into the next window of life-extending technological innovation. Life extension from that point onwards need only take us into yet another window of innovation, and so on. We need only keep advancing in this field, as we do in so many others. A difficult presupposition here is that modification can be carried out upon an already living human, an assumption which we have set aside. But that might turn out to be a limitation of the science that we know, rather than a limitation of science a century from now. And the timescale presupposed by the thought experiment does take us into this territory. Any germline therapy of the sort described would take *at least* a couple of decades to set up, and to clear the political obstacles which any good society wouldput in its way. Good societies take reasonable precautions. We would expect space candidates to go on to live until they are at least around 100, and life to 120 would not be unusual. Their later years would occur at least a century or more from now, and perhaps a good deal later depending upon the actual start point.

There may be something to de Grey's claim, although it looks like an unreliable process for anyone actually seeking immortality. Let us at least allow that if life-extending technology was discovered, as it has been in the thought experiment, then we cannot reasonably assume that it will then remain static afterwards. Perhaps an eternal jellyfish/*turritopsis dohrnii* scenario of rejuvenation, rather than the slowing of telomere shortening, might still be difficult to imagine during this time. But, even if workable therapies consisted only of the latter, and a slowing down of the aging process, they need only emerge early enough for the space candidates to still be in reasonably good health in order to make it worthwhile for them to consider a further extension. Whether or not they would do so would (again) depend upon a range of factors, including the likelihood that they would be a good match for the therapies in question. The *Ad Vitam* scenario considers that only some of us would be a match, while others would not. But it also assumes that those who *are* a match would also remain so after each treatment. Given the need to work at cellular level, this might not be the case. One therapy might rule out subsequent therapies. And, more in line with *Ad Vitam*, life might simply become less attractive to those who have lived so long. At some point, the psychological burden of having seen a previous generation pass away may become too great. Arthur C. Clarke, and othershave built in a recognition of this burden at the start of the discussion. It seems wise to keep it in play.

What the de Grey approach does suggest is that we cannot assume the existence of a firm line between radical life extension and more modest life extensions of the sort described in the scenario. The latter may result in the former, once life extension technology gets up and running. This being the case, modest life extension may well inherit the critiques of radical life extension, and especially the charge of an evasion of mortality, and/or a tendency to undermine life's meaningfulness. Continuity of life may still be regarded as an individual good, as a social good or as a human good. We may still say *L'Chaim*, with Leon Kass, but only if the toast to life takes the form of being fruitful, with one generation giving way to the next, and each one accepting when their time is up. This elusive objection may also turn out to be the strongest objection of all. At least, this may be the case just so long as the objection is not simply a disguised form of ethical conservatism and suspicion of biotechnology in general, but is framed in a way which touches upon something deep within a human life. It is, however, difficult to frame the argument in terms which avoid precisely such a collapse into suspicion of something new and untried, especially in the case of a sufficiently modest extension of longevity which carries no immediate threat of any more radical extension. Perhaps, at some level, an evasion of our mortality would be in play. But it is not obvious that this would actually be a sufficient reason to prohibit such modification, any more than it is a reason to prohibit cosmetic surgery for reasons of fashion or art or as lifestyle choice, or the heavy use of makeup in middle age or in later life.

This goes to the heart of the difficulty facing the deep problem about life's meaning. There may be a correct answer to the metaphysical question of 'What is the meaning of life?' once we have fully grasped what it is that the question asks. But when it comes to legislative restriction, this is metaphysics, and not territory where universal or even sufficiently broad agreement can be expected between all reasonable agents. As far as regulation goes, we may not get to decide upon the answer for others, at least not within broadly liberal societies. By contrast, the overpopulation objection might not be as philosophically deep, but it could give adequate grounds for restriction, if it worked. After all, we may have no option but to take measures for population control. The injustice objection might also give adequate grounds for restriction, if it worked. Functional social organization provides the required context for lives of individual freedoms and choice. Any excess of injustice can make such a life impossible. This too is a plausible candidate for grounds for regulation. Even if the appeal to life's meaning has a better chance of touching upon a deeper worry, and even if it this worry does apply to our modest scenario, and not only to radical life extension, the appeal to life's meaning will not easily translate into similar grounds for actual restriction. It might tell us that we should not attempt to realize the scenario, or that it should fall low in the list of human priorities. But it would not tell us that it describes either an ethically or politically impermissible state of affairs, one that a good society cannot allow and should legislate against.

13.4 Conclusion

There are, no doubt, many circumstances under which it would be ill-advised to engage in the genetic engineering of humans. These include more or less all circumstances in which such engineering is carried out in pursuit of an imaginary human perfection. Above, we have assumed that any general prohibition of genetic engineering of humans will not hold in the face of social change. At some point the questions we face will concern regulation and restriction rather than outright prohibition. Acknowledging the presence of risks is also not the same as saying that we should fail to take advantage of *any* possibility of life extension which falls into our hands, especially if it does so more or less by accident. Particularly so when the potential societal advantages of life extension are real, and when the special advantages of life extension for space exploration are clear. It is also noteworthy that, in the scenario described by our thought experiment, some of the problems are not associated with those who are modified and go into space, but with those who are modified, enjoy greater longevity and life expectancy, but do not go. A recurring acknowledgement of this is among our hints, clues and reminders. This particular concernis, however, some what indirect and does not strike in any clear way at the permissibility of modification for those who do go into space. Nothing above is inconsistent with a recognition that it could be a bad thing if the relevant technologies emerged within the wrong kinds of political system. But these risks too, are not obviously greater than those associated with other technologies which could be used for bad ends. After all, the brevity of an ordinary human life can also be used for bad ends, even if the risks of brevity often escape our attention.

Acknowledgements The publication was supported within the project of Operational Programme Research, Development and Education (OP VVV/OP RDE), 'Centre for Ethics as Study in Human Value', registration No. CZ.02.1.01/0.0/0.0/15_003/0000425, co-financed by the European Regional Development Fund and the state budget of the Czech Republic.

References

Ad Vitam. (2018). [Film]. ARTE, November 2, 2018.

Benatar, D. (1997). Why it is better never to come into existence. *American Philosophical Quarterly, 34*(3), 345–355.

Bostrom, N. (2005). The fable of the dragon tyrant. *Journal of Medical Ethics, 31*(5), 273–277.

Clarke, A. C. (1968). *2001: A space odyssey*. New York, NY: New American Library.

Clarke, A. C. (1997). *3001: The final odyssey*. New York, NY: Ballantine Books.

Davis, J. K. (2005). Life-extension and the Malthusian objection. *Journal of Medicine and Philosophy., 30*(1), 27–44.

de Grey, A. (2009). Radical life extension: Technological aspects. In D. F. Maher & C. Mercer (Eds.), *Religion and the implications of radical life extension* (pp. 13–24). New York, NY: Palgrave Macmillan.

Freud, S. [1930] (1961). *Civilization and its discontents* (J. Strachey, Trans.). New York: W.W. Norton & Company.

Gems, D. (2003). Is more life always better? The new biology of aging and the meaning of life. *The Hastings Centre Report, 33*(4), 31–39.

Inaba, S. (Ed.). (2016). *Uchu Rinrigaku Nyumon: jinko chino ha supesu coroni no yume wo miruka? (An introduction to space ethics: Do AIs dream of space colonies?)* Nakanishiya Shuppan.

Kass, L. R. (2001). L'Chaim and its limits: Why not immortality? *First Things*, May 2001. https://www.firstthings.com/article/2001/05/lchaim-and-its-limits-why-not-immortality. Accessed March 10, 2020.

Kubota, S. (2011). Repeating rejuvenation in *Turritopsis*, an immortal hydrozoan (Cnidaria, Hydrozoa). *Biogeography, 13,* 101–103.

Liu, C. [2007] (2016). *The three-body problem*. London: Head of Zeus.

Miglietta, M. P., Piraino, S., Kubota, S., & Schuchert, P. (2006). Species in the genus *Turritopsis* (Cnidaria, Hydrozoa): A molecular evaluation. *Journal of Zoological Systematics and Evolutionary Research, 45*(1), 11–19.

Milligan, T. (2007). Lockean puzzles. *Journal of Philosophy of Education, 41*(3), 351–361.

Milligan, T. (2015). [illegible]. Jefferson, NC: McFarland.

Milligan, T. (2019). Thought experiments and novels. *Studia Humana, 8*(1), 82–92. https://doi.org/10.2478/sh-2019-0006

Nolan, W. F., & Johnson, G. C. (1967). *Logan's run*. New York, NY: Dial Press.

O'Connell, M. (2017). *To be a machine: Adventures among cyborgs, utopians, hackers, and the futurists solving the modest problem of death*. New York, NY: Doubleday.

Rollin, B. (1998). On *Telos* and genetic engineering. In A. Holland & A. Johnson (Eds.), *Animal biotechnology and ethics* (pp. 156–171). New York, NY: Chapman and Hall.

Roth, S. M., & Nystul, T. (2005). Buying time in suspended animation. *Scientific American, 29*(6), 48–55.

Salt, H. [1914] (1976). The logic of the larder. In T. Regan and P. Singer (Eds.), *Animal rights and human obligations*. Upper Saddle River, NJ: Prentice-Hall.

Scott-Heron, G. (1970). [Record]. Whitey on the Moon, *Small Talk at 125th and Lenox*, Flying Dutchman Records.

Shermer, M. (2016). Methusalah's moon shot. *Scientific American, 315*(4), 84.

Singer, P. (1991). Research into aging: Should it be guided by the interests of present individuals, future individuals, or the species? In F. C. Ludwig (Ed.), *Life span extension: Consequences and open questions* (pp. 132–145). New York, NY: Springer.

Singer, P. (2012). Should we live to 1,000? *Project Syndicate*, https://www.project-syndicate.org/commentary/the-ethics-of-anti-aging-by-peter-singer?barrier=accesspaylog. Accessed March 10, 2020.

Szocik, K. (2019). Human space in the outer space: Skeptical remarks. In Szocik (Ed.), *The human factor in a mission to Mars* (pp. 233–252). Cham: Springer.

The Expanse. (2015). [Film]. Alcon Entertainment, November 23, 2015.

United Nations (UN). (2019). *Human Development Report*. United Nations Development Programme, New York, 2019, https://hdr.undp.org/en/content/2019-human-development-index-ranking. Accessed March 9, 2020.

Chapter 14
The Accessible Universe: On the *Choice* to Require Bodily Modification for Space Exploration

James S. J. Schwartz

Abstract Humanity should not attempt to establish space societies that would not be open to "baseline" humans (e.g., those with species-typical oxygen or radiation protection needs). Two arguments are provided for this conclusion: The first argument is via analogy with disability and accessibility. Just as it would be impermissible today to mandate disability-removing medical procedures, so too would it be impermissible to require bodily modification in order to participate in a space society. We should only create space societies that are accessible to humanity, broadly speaking. The second argument is that the requirement of bodily modification would pervert the vision of the human expansion into space. The future we ought to strive for is one in which all humans can travel easily between different destinations in space, in much the way that humans can travel easily between different destinations on Earth.

14.1 Introduction

Discussions of space settlement often have a conspicuously utilitarian flavor. Given the great expenses associated with space missions, proposals for settlement initiatives tend to devolve into analyses of the cheapest possible means for establishing a permanent human presence in space. The challenges here are legion: An inveterate feature of space environments is their unrelenting hostility to human life. Nowhere save for Earth is there air ready for breathing, water ready for drinking, food ready for eating, or temperatures suitable for living. Thus, human inhabitation of the space environment requires highly engineered habitats and in situ resource exploitation machinery, among other technologically demanding systems. The need for these systems contributes greatly to the cost of initiating and supporting human space settlement. On a much grander scale—both in terms of cost and time—we might seek to terraform entire planetary environments in order to make them more suitable for human life.

J. S. J. Schwartz (✉)
Department of Philosophy, Wichita State University, Wichita, Kansas, USA
e-mail: james.schwartz@wichita.edu

K. Szocik (ed.), *Human Enhancements for Space Missions*, Space and Society,
https://doi.org/10.1007/978-3-030-42036-9_14

The general tenor of these proposals is that the hostile space environment must be modified in various ways to permit a sustained human presence. However, another genus of proposals holds that none of these costly measures is necessary, because we can establish a permanent human presence in space by *modifying*[1] *humans* so they are better adapted to space environments. In the words of space visionary Olaf Stapeldon,

> …the colonization of some of the planets may in time become practicable, if terrestrial man continues to develop his control of the physical environment through atomic power, and if he attains sufficient biological knowledge and eugenical art to breed, or otherwise construct, human or quasi-human races adapted to strange environments. (1948; reprinted in 2012, 33)

More recently, as Athena Andreadis argues, "[i]f we truly wish to be an integral part of life on the new planets, rather than tourists [illegible] behind the glass of air conditioned [illegible] genetic engineering of the prospective colonists" (2013, 272). Possible modifications include the ability to hibernate; the ability to absorb solar energy; the ability to survive in low-oxygen environments; increased tolerance for radiation; etc.[2]

The question I shall consider here is: Should we pursue space settlement initiatives that *require* each settler to be modified in some or all of these ways? In other words, should we seek to create a society that bars the participation of all "baseline" or unmodified humans? My answer is that we should not. As I shall argue, a necessary (but not sufficient) condition for space settlement is that it remains available to human persons *generally*, which means it should be open to persons lacking artificial tolerances for high radiation environments, low-oxygen environments, etc. I lay out two arguments in support of my answer.

The first argument is via analogy with disability. An important lesson from disability scholarship is that disability is not (or is not purely) a medical phenomenon. Disability does not reside (solely) in the body, but instead, as Melinda Hall argues, in complex interrelationships "between the body, social structures, and social norms" (2017, 133). On this picture, stigmatization is the antecedent of disability; disabled persons exist wherever there are cultures that stigmatize forms of embodiment regarded as "abnormal," and that discriminate on many fronts against these individuals (with one example being the way that physically disabled persons are discriminated against through architecture which prefers stairs to ramps and elevators). In Sect. 14.2, I argue that a consequence of this understanding of disability is that we should not deliberately create environments that are generative of cultures that stigmatize forms of embodiment, and thus in turn, we should not deliberately create environments accessible only to "space modified" humans.

---

[1]Here the term 'modification' is preferred to terms such as 'treatment' and 'enhancement.' In this context, 'modification' has a value-neutrality not shared by either of the other terms: Those in need of *treatments* have something wrong with them that needs treating; those who seek *enhancements* seek to become better version of themselves. Meanwhile, those who seek *modifications* are merely seeking changes or alterations to themselves—with no tacit presumption or connotation about the value (positive or negative) of these changes or the persons seeking them.

[2]See ibid. as well as Abney and Lin (2015).

The second argument is aspirational, and derives from a kind of "ideal picture" of the human future in space. On this picture, which I describe in Sect. 14.3, space travel is taken for granted in the same way that we take air travel for granted. Humans are able to travel to, visit, and live freely throughout the solar system. This attractive vision of our future is not possible if our strategy for human expansion into space relies substantially on modifying humans, as opposed to modifying the space environment.

I close the chapter in Sect. 14.4 by responding to the objection that my criticism of bodily modification is inconsistent with my views on the need to preserve the space environment for scientific study.

Two themes are important throughout: The first theme is that we have a duty to respect persons, which includes a duty to respect persons' self-concepts and the ways that persons identify (or not) with their bodies. To the greatest extent possible, we should promote systems that reinforce these duties; and we should oppose systems that violate these duties. Humans do not exist simply to be of service to space settlement—or to be of service to any other project, for that matter.

The second theme is that *choice* plays an important role in thinking about space settlement. The nature and scope of the human future in space is very much open, and very much within our power to shape. We are not predestined to witness a future in which space belongs exclusively to modified humans. Whether this or some other future comes to pass will be the result of collective *choices* we make about how to pursue space exploration and space settlement. Inclusive futures are possible because inclusive choices are currently open to us—but we must make these choices sooner rather than later.

14.2 Space Settlement and Models of Disability

Imagine the following—admittedly fanciful—situation:

> An eccentric billionaire has purchased an uninhabited island and plans to use all of their wealth create a self-sustaining utopia on this island – the kind of idyllic paradise that many humans dream of. They send an open invitation to the rest of humanity: Anyone is free to join this society (up to a certain maximum number of persons that can be sustained on the island). However, there are to be no accommodations for disabilities on the island: no ramps, no elevators, no braille signs, etc. This idyllic paradise will be designed exclusively for nondisabled persons.

A possible reaction here is that there is nothing untoward about this eccentric billionaire's plans. After all, according to pervasive ideas about personal autonomy, people should be free to spend their own money as they please and to associate with whomever they please—so long as they cause no harm to anyone else. But from a different perspective, there is something remarkably unfair about the creation of a highly attractive opportunity—life in an idyllic utopia—that is closed entirely to disabled persons[3]—the same individuals who already face pervasive discrimination and highly constrained opportunities for participating in society.

Suppose that this concern is brought to the attention of the eccentric billionaire who, graciously, takes a more compassionate perspective and agrees that their idyllic paradise ought to be open to all persons, including disabled persons. They announce their solution: Any disabled person is welcome to join the new society provided that they undergo medical procedures to remove their disability—procedures that will be paid for by the eccentric billionaire.

This offer is in keeping with the "medical model" of disability, which, according to Anita Silvers, "fixes on reducing the numbers of people with disabilities by preventative or curative medical technology" (1998, 59). And as Richard Hull explains, this understanding of disability assumes that disadvantages associated with disability "inhere, and unproblematically, originate in functional limitation," which "lends support to the view that disabled people should be grateful for any charitable measures granted them" (1998, 203). In our example, it is easy to imagine that many would expect disabled persons to express impassioned gratitude to the eccentric billionaire for the offer to "cure" them so that they can participate in the idyllic utopia.

As Silvers argues, the medical model fails to promote justice for disabled persons. Since the medical model encourages eliminating functional limitations, policies based around this model would direct resources to "developing and purchasing restorative and rehabilitative treatments" (Silvers 1998, 95). A significant problem here is that it may not be possible—either technically or economically—to remove every kind of functional impairment. Thus, focusing primarily or exclusively on medical treatments.

> …exacerbates the depreciation of those who cannot be restored, labeling them as failures, and authorizing their being at greater psychological risk than would be permitted for so-called normal people. Because it exaggerates rather than eliminates the probability that being impaired exposes people to isolation, loss of purpose, depreciation, and powerlessness, conceptual frameworks that assume the medical model do not offer an avenue for equalizing people with disabilities. (ibid.)

It is important to recognize that these concerns remain even in technologically advanced, post-scarcity milieus. Even in a world where technological developments allow for the removal of all "impairments," disabled individuals will continue to exist and will continue to come into existence (either through genetics, injury, aging, or other means). These individuals would very likely feel depreciated and powerless in the face of intense social pressure to receive treatments.[4] To be sure, some disabled persons might welcome such treatments. However, some may not desire to be changed or altered, and regard the prospect of being "cured" or "fixed" as deeply insulting and humiliating. A case often discussed in the disability scholarship is a Deaf couple[5] who deliberately conceived a deaf child so that they could have a child

---

[3]I will follow Tremain (2017) in eschewing "people-first" language.

[4]See, e.g., Bradshaw and Ter Meulen (2010) and Sparrow (2011).

[5]Here 'Deaf' (with a capital D) indicates membership in Deaf culture; whereas 'deaf' (with a lower-case d) indicates hearing impairment.

able to participate with them in Deaf culture (Mundy 2002). Some individuals, then, do not view their "impairments" as disabilities, but instead as sources of identity, community, and fulfillment.

For these reasons, we should be especially mindful of the importance of personal autonomy regarding decisions about who is to receive (or to be pressured or coerced into receiving) modifications, treatments, or enhancements. Consider David DeGrazia's comments on human enhancement:

> Enhancement projects, if successful, are likely to affect one's narrative identity by affecting one's self-conception. A person who succeeds in becoming physically fit, or better looking, or more masterful in some valued activity will probably perceive herself somewhat differently than she otherwise would have. (2005, 267–8)

Degrazia is ultimately unconvinced that enhancements will prove to violate "inviolable core characteristics" (ibid., 280). Whether he is right or not about the harmless nature of the consequences of the enhancements often promoted by posthumanists and transhumanists, the consequences may be far from harmless to disabled persons. For some disabled persons, procedures that dramatically changed their bodies would affect their self-conception in an unwelcome way. This points to a fundamental error promulgated by the medical model, as Silvers helpfully describes:

> Justice for people with physical, sensory, or cognitive impairments should not be couched in expectations that they be cured. How can laying claim to the right to be altered so as not to be an impaired individual be central to any person's self-respecting self-identity as an individual with a disability? (1998, 137–8)

Disabled persons are not inherently "broken" and should be taken at their word when they identify with and express satisfaction with the conditions of their bodies. The medical model, which encourages treatments which might be destructive of the identities of disabled persons, should therefore be rejected.

In its place, Silvers believes we should frame these issues through the "social model" of disability. Whereas the medical model conflates "impairment" with "disability," the social model distinguishes these ideas: "impairment" refers only to a (purportedly) value-neutral accounting of the status of one's body; "disability" marks the failure of society to create environments that are accessible to persons with impairments. As Silvers describes it:

> The social model traces the source of...disadvantage to a hostile environment and treats the dysfunction attendant on (certain kinds of) impairment as artificial and remediable, not natural and immutable. It transfigures individuals with disabilities from patients into persons with rights, which, when acknowledged, should eliminate the social disadvantages... Their environment is inimical to them because, in respect to almost all social venues and institutions, people with disabilities are neither numerous nor noticeable. (ibid., 75)

The social model calls for a dramatic shift in how we think about disabled persons. When we are confronted with the realization that "it is not the individual but the environment that is defective" (ibid., 94), it is no longer appropriate to promote medical procedures as the sole or primary means for accommodating disability. Instead, urges

Silvers, "the social model directs us to address the suffering of people with disabilities by eliminating or reforming the external circumstances that contribute to it rather than by eliminating or revising the people themselves" (ibid.).

However, as Melinda Hall argues, the social model overlooks the value-ladenness of the notion of impairment. Our understandings of sickness and health, of normal and abnormal, of functional and non-functional, are shaped by contingent cultural factors (Hall 2017, 48–9). There is no real, natural category of "impaired" that human bodies fall under; "impaired body" is not a natural kind. Rather, a person only becomes "impaired" in the context of a group, society, or culture that regards this person as unhealthy, abnormal, or under-functional. Thus, impairment and disability each is a social construct. Any differences between bodies of different persons represent merely different forms of embodiment.[6]

Hall proposes the cultural model as a replacement. This model "responds critically to the false choice of *either* the social world *or* the body as an explanatory mechanism," it does not assume that contexts and constructions of disability "are merely or only tragic or negative," and it "attempts to understand locations of disability as complex interplays between *both* embodiment *and* the social world" (ibid., 46). On this model, then, disability arises in the ways that different forms of embodiment navigate in the social world and in how the social world responds to different forms of embodiment. Since the cultural model denies genetic and biological norms of embodiment, there is no basis for identifying preferred or "normal" forms of embodiment. Such preferences and norms only arise in cultural contexts, and *stigma* plays an important role in the explanation of why disabilities are regarded negatively in many cultures. Cultural factors determine which traits and abilities are valued, and thus, which forms of embodiment are lauded as normal and which are stigmatized as abnormal and discriminated against. Crucially, "[t]he faintest whiff of deviation from the norm is enough for the pernicious effects of stigma to operate" (ibid., 55). Since any form of embodiment is potentially a deviation from a culturally induced norm, any form of embodiment is a potential victim of stigmatization. Meaningful progress—no matter the society nor the forms of embodiment stigmatized in this society—ought to focus on "the revision of political and social circumstances, seeking justice for those with disabilities and acceptance of diverse forms of embodiment" (ibid., 139). This adds to rather than negates the prescriptions of the social model: Not only should we revise inaccessible built environments; we should also revise ableist and stigmatizing cultural and political attitudes and practices.

At this point, the implications for the eccentric billionaire's idyllic utopia should be quite clear: Rather than requiring procedures to remove disabilities, the billionaire should opt instead to build an idyllic utopia that is accessible to persons regardless of their forms of embodiment. This would remove a significant structural barrier to the creation of a culture free from the kinds of norms and values that stigmatize certain forms of embodiment. Moreover, if this is done proactively rather than retroactively—if the built environment is designed from the outset as an accessible environment—then the overall costs should not greatly exceed the original construction costs.

---

[6] *Cf.* Tremain (2017), especially Chap. 3.

What does any of this have to do with bodily modification for space settlement? Consider a new scenario, not too dissimilar from our first example:

> An eccentric billionaire has developed the means to transport large groups of people to initiate the settlement of Mars – fulfilling a dream held by many humans. They send an open invitation to the rest of humanity: Anyone is free to join the settlement (up to a certain maximum number of persons that it can sustain). However, each settler will be required to be altered in ways that enable them to survive exposure to Martian conditions (through, e.g., procedures which increase their radiation, temperature, and pressure tolerances; and that enable them to breathe Martian air). There are to be no pressure-sealed, heated habitats to support "baseline" or unmodified humans. This settlement will be designed exclusively for persons modified to live on Mars.

In this case we have an interesting inversion of issues related to disability, for in this scenario, people which today are considered to be "able-bodied" persons—including the most physically fit and skilled among this population—would be lethally impaired and severely stigmatized in the eccentric billionaire's Martian settlement.[7] Although many individuals may rejoice at the opportunity, many also may become frustrated that a longstanding and deeply held dream—living on Mars—requires too great a sacrifice: modifications to their bodies that they believe would irreparably alter their identities and self-concepts. From a perspective informed by the social and cultural models of disability, this would be an inherently unjust way to initiate human space settlement. Rather than creating humans suited to Martian environmental conditions, we should instead create Martian habitats that are accessible to unmodified humans. And rather than creating a Martian culture that would stigmatize unmodified forms of embodiment, we should instead create a culture that does not stigmatize any forms of embodiment.

A likely rejoinder is that, given the vast expenses associated with space missions, and given the urgency behind instigating space settlement to ensure the long-term survival of the human race, we need to establish space settlements as quickly and as cheaply as possible. Therefore, since species survival is urgently on the line, we are licensed to do whatever is necessary for successfully establishing space settlements, up to and including modifying humans to be better suited to natural Martian conditions. As Andreadis notes, "[f]or space travellers and planetary settlers, such changes will be made to confer not the dubious enhancements touted by transhumanists but the ability to survive" (2013, 273).

There is much to say in response to this line of reasoning. To begin with, we should reject any view that reduces human persons to mere instruments to the uncertain goal of long-term species survival. Respect for persons places boundaries on what can be asked or compelled of persons, especially in support of goals that may not be their own. Of course, the mere existence of such boundaries does not tell us on which side compulsory modifications fall. The main point I wish to make here is that it is wrong to presume that we are obligated to do "whatever is necessary" for ensuring species

[7] *Cf.* Trijsje Franssen on enhancement more generally: "if the enhanced human becomes the norm…then the non-enhanced human becomes a dysfunctional being. One of the consequences of this would be that the vast majority of human beings alive today would become subhuman or even disabled…" (2014, 162).

survival. There are many things we are obligated not to do, even if failing to do them results in the expiration of humanity.[8]

Nevertheless, I agree that space settlement is ultimately necessary for ensuring long-term human survival, and correspondingly, that we have a duty to settle space *in principle*. Moreover, it is worth pointing out that the *urgency* of this need is open. Some view space settlement as an urgent duty (Green 2019; Abney 2019); others think we have time to spare (Schwartz 2019a; Milligan 2011). But however urgently we need to settle space, the need for compulsory modifications is entirely artificial, because these modifications are unnecessary in the first place. Space habitat engineering is much more mature than biomedical engineering of the kind alluded to above. If space settlement is something we need to initiate right now, then it will have to be accomplished by "baseline" humans living in artificial habitats—since we do not at present have the ability to implement modifications to humans that would enable them to survive exposure to the Martian environment. Perhaps we could develop such abilities more quickly should we choose to dramatically increase funding for biomedical engineering research and development. But we could also choose to dramatically increase funding for space habitat engineering research and development. Either way, it is our *choice*—neither option is forced on us.

Meanwhile, if space settlement can wait on however long it takes for the maturation of biomedical engineering, then it can also wait on however long it takes for the maturation of space habitat engineering. Whichever area matures first depends on many factors—science and technology development are often unpredictable. But a significant factor will always be *how we choose* to apportion funding for research and development. We are more likely to make discoveries and devise new technologies in areas that are well funded than in areas that are poorly funded. So again, *we have a choice*—no particular space settlement strategy is forced on us. So, it is incorrect to presume that bodily modifications will be necessary for space settlement. Thus, it is incorrect to presume that bodily modifications will be necessary for satisfying an obligation to ensure the long-term survival of the human species.

This element of choice points to an important disanalogy in the use of the social model terrestrially versus extraterrestrially—however, it is a disanalogy that strengthens considerably the case for applying the social model to space settlements. An incessant bugbear to proponents of the social model is the objection that it is too costly to remodel the built environment so that it is accessible to disabled persons. It is always easier to design accessible spaces from the outset than to renovate previously inaccessible spaces, and too many places remain inaccessible and without funding for adequate renovations. I do not wish to comment on the merit of this objection—I raise it merely to highlight the fact that the space environment is not yet a built environment. There are, as yet, no spaces outside of low-Earth orbit that are accessible to any human. *It is entirely within our power to determine how accessible (or not) the built environments will be in space. This, in turn, gives us a great deal of power to influence the cultural attitudes of future space-dwellers.* If we want space settlements to be places that are accessible to "baseline" humans, and if we want to reduce the

[8] See Chap. 6 of Schwartz (2020) for discussion.

probability that space cultures will stigmatize various forms of embodiment, then we need only plan for this from the outset—saving us from having to make costly and politically infeasible renovations down the line.

We have in the case of space settlement an unprecedented opportunity to determine the nature and scope of human participation through our selection of the "dominant cooperative framework" for space societies. According to Allen Buchanan et al., a dominant cooperative framework is a "set of basic institutions and practices that enable individuals and groups in a given society to engage in ongoing mutually beneficial cooperation" (2000, 79). A society's dominant cooperative framework determines, *inter alia*, which individual traits and abilities count as assets and which count as deficits. However, as Buchanan et al. observe, the "dominant cooperative frameworks for entire societies have never been chosen" and "have emerged not according to any overall conception or plan or as the result of collective deliberation concerning alternatives, but rather from the cumulative and largely unanticipated effects of many individuals' actions over many generations" (ibid., 290). While we are not powerless to revise existing frameworks (Silvers and Francis 2013), nevertheless these frameworks often place social, political, and economic boundaries on what sorts of revisions may realistically be accomplished. But no such frameworks bind us in space; and we are free to establish whichever dominant cooperative framework we choose.

Regarding the terrestrial situation, Buchanan, et al. are open to the possibility that the best cooperative framework may be one that includes some compulsory enhancements. They note that each individual has an "important and morally legitimate interest" in accessing a cooperative framework that is "the most productive and rewarding" for them (Buchanan et al. 2000, 292). They hold that this requires that we acknowledge enhanced persons' interests in a cooperative framework that may be too demanding for unenhanced or disabled persons, because "those who could participate in a more productive and rewarding scheme but are barred from doing so by restrictions designed to make the scheme more inclusive lose something of value" (ibid.). Although they acknowledge that excluding the unenhanced (and disabled) from participating in *any* cooperative framework may be a greater evil than denying enhanced persons the most productive and rewarding framework, they maintain that this does not imply that the interests of the enhanced count for naught. "In some instances," they claim, "a proper balancing of" of interests "may require changing individuals" (ibid., 294). The upshot is that.

> In a just society in which the powers of genetic intervention are highly developed, both society and individuals will be changed in order to reduce the incidence of disabilities, both for the sake of those who would have been disabled and for the sake of others. (ibid.)

Here Buchanan, et al. point to an oft-envisioned benefit of enhancement—that the enhanced will produce goods (through, e.g., their superior creative and intellectual skills) that will be accessible to and will greatly benefit the unenhanced.

I will not dwell on whether these benefits make the enhancement pill easy enough to swallow for individuals living in contemporary terrestrial societies,[9] but we should consider what judgment is appropriate for the creation of space societies. There is a perspective on which this issue is moot: If a space society is genuinely independent from all terrestrial societies, then it will have its own independent cooperative framework. If this society contains only modified persons, then there will be no unmodified persons around to experience any disadvantages or to suffer the effects of stigmatization. This suggests that there is nothing suspect about creating a space society exclusively for modified persons.

Meanwhile, if a space society engages in mutual cooperation with terrestrial societies, then there will be some overarching cooperative framework that covers space society and terrestrial society. On the one hand, this framework might accord a high value to universal participation and mandate that space settlements remain open to the unmodified. On the other hand, this framework might permit modified-only space settlements, if these provided substantial benefits to unmodified Earth-dwellers. I must admit that I find this latter possibility difficult to imagine. Whether or not they contain modified humans, it is not clear that space settlements will provide significant, tangible benefits to terrestrial humans.[10] But even if space settlements will provide significant benefits to terrestrial humans, there is no reason to suppose that these benefits have to be realized exclusively by modified space-dwellers. Moreover, if a cornucopia truly does await us in space, then there will be no resource limitations that prevent us from building accessible places in space.

Nevertheless, I think these reflections miss the point in a crucial way. There do not exist any space societies, and there do not exist any modified space-dwellers whose interests we are obliged to consider. Our question is, and has been all along, about whether and how to *create* a space society. This is a task that, should we pursue it, will be carried out by persons belonging to our existing cooperative frameworks, and that will be supported by the resources of persons belonging to our existing cooperative frameworks. This means that the interests of all persons belonging to contemporary terrestrial societies, regardless of their forms of embodiment, matter in the construction of a new cooperative framework that includes space settlement. In this respect, any decision to create a "modified-only" space society is no different, and no less objectionable, than a decision to create a "disability-free" terrestrial society. This point is reinforced, rather than undermined, when revisiting the oft-cited rationale of species survival. To hold up space settlements as an escape from urgent terrestrial catastrophes, and then to deny escape to all unmodified humans, would demonstrate a malicious repugnance for humanity.[11]

---

[9]Though I am sympathetic to a point Buchanan acknowledges elsewhere, which is that "getting more of the goods that can be made available to [the unenhanced] through the participation of others in the dominant cooperative scheme may be less important to the unenhanced than being able themselves to participate in the cooperative scheme" (2011, 231).

[10]See Chap. 5 of Schwartz (2020) for my reasoning.

[11]If a terrestrial analogue is needed to help motivate this point, then consider a modification to the original example of the eccentric billionaire: Instead of an "idyllic utopia", this time the billionaire wants to create an enclave that will enable its denizens to survive a civilization-ending sea level rise.

As Hall implores, a more productive vision of enhancement, both here and elsewhere, should be "an expression of care, care of existing individuals—not idealized future subjects that cannot, and will not exist" (2017, 139). This is a point often lost in technology-focused discussions of space settlement, which view humans more as vehicles for accomplishing grandiose plans than as persons deserving of respect and care. The social and cultural models of disability, then, enable a more productive framing of questions about bodily modification for space settlement. Feasibility objections—which might *appear* persuasive in application to terrestrial societies—are comparatively toothless in space contexts. In space, there are no already-existing inaccessible built environments requiring costly renovations. In space, there are no exclusionary cooperative frameworks. In space, there are no stigmatizing cultures. None of these things will exist unless we create them deliberately. *Space societies will be exclusionary, discriminatory, and stigmatizing societies only if we choose to create them this way, only if we choose not to make space inclusive.*

We must begin making inclusive choices today. A crucial step toward realizing an inclusive future in space is the creation of an inclusive present in space. Decisions made early on in any effort can have a dramatic impact on what is possible in the future. Decisions made now and in the near future—especially those concerning space mission infrastructure—will constrain future possibilities for spaceflight and will in effect determine who is and who is not able to participate in spaceflight and (eventually) space settlement. If accessible, inclusive choices are not selected early on, it may become prohibitively difficult to pursue these choices later, as renovating for accessibility is often more costly than designing for accessibility from the outset. Moreover, as Wells-Jensen et al. (2019) argue, crewed space missions of all kinds will be more successful if they employ principles of Universal Design, and if their crews include disabled persons. Further still, the inclusion of disabled persons throughout the design and construction of space settlements may be necessary for successfully designing and constructing accessible places in space. This is because nondisabled mission planners are more likely to make incorrect presumptions, predictions, and judgments, both about the abilities and needs of disabled persons and about how to design space habitats to accommodate these abilities and needs.

14.3 The Human Future in Space

Like many space exploration researchers, I grew up with the *Star Trek* vision of the human future in space. The salient aspect of this vision is that space travel and

Rather than designing this as an accessible space (which the billionaire could do easily), they will only save disabled persons if these persons agree to disability-removing modifications. Or suppose instead that the eccentric billionaire opts against building water reclamation and desalinization equipment (which they could do easily) and instead requires all of the saved to be modified so that they are able to hydrate by drinking seawater. As the products of choice, rather than necessity, these exclusionary moves are all the more criticizable.

space settlement are the province of all of humanity. It is a future in which humans—humans with minds and bodies like those that exist today—are free to travel the galaxy, to visit other planets, and to call these places their homes. As the enduring international popularity of *Star Trek* demonstrates, this seductive vision is widely shared. Rather than viewing this as a puerile and naïve optimism to be cast aside the moment the first practical concern is raised, we should instead view it for what it is: a valued future, and one that we have every reason to work toward creating in space. This is not a merely fictional ideal valued only by fictional future persons—it is our ideal, valued here and now, and one that should play a prominent role in how we create and shape space societies. As Nicholas Agar insists, "human beings are good enough to travel to the stars" and we should not be "required to transform ourselves into different kinds of beings—robots or posthumans—to be worthy of that honor" (2014, 197–8).

Much more important than preserving the human genome, I insist, is the preservation of the continuity of human cultures and cultural processes (excised of their discriminatory features, ideally). While using modifications to enable space settlement may save "the species," these modifications will not save *us*—our cultures, our communities, our values, our hopes, our dreams. To borrow again from Agar,

> We can veridically project ourselves into the possible future populated by Captain Kirks and Lieutenant Uhuras. We expect that the worlds of these future humans will be very different from our own. But we can look upon their experiences as ones that we might have had, had our births been postdated by a few hundred years... We have an interest in the human story. We would like it continue. (ibid., 200)

The preservation of this "human story" is the true source of our obligation to preserve humanity.

Contrast this vision with that promulgated by William Sims Bainbridge, a transhumanism and enhancement advocate:

> We can agree that the planet Earth should remain a refuge for traditional humanity, living in a variety of low-tech societies in what technophiles would call a perpetual Dark Age. Those who wish to transform themselves into a very different kind of intelligent entity will need to leave the Earth... The original *Star Trek* motto – *to boldly go where no man has gone before* – has been criticized for splitting an infinitive and employing sexist language, and I now criticize it for implying that space travelers will be humans in the antique sense of the term. Another motto from the science-fiction subculture is better, leaving open the nature of spacefarers and playing nicely off an old religious motto: *The meek will inherit the Earth, but the bold will go elsewhere.* (2007, 212; emphasis in the original)

The advanced technological sheen of this vision of the human future in space only thinly veils its moral bankruptcy. It includes a humiliating and dehumanizing portrait of the unmodified as "meek" and confined to living in a "perpetual Dark Age." Space, according to this vision, is not a place for members of "traditional humanity," who are as mere wildlife that should be grateful they have been graciously granted Earth as a "refuge." (Perhaps the modified will equate visiting Earth with childhood fieldtrips to the zoo!) I can think of no more appropriate example of a proposal that justifiably invites the criticism that we need to "learn our lessons on Earth" before venturing into space. To endorse this vision is to fail spectacularly and entirely to recognize and seek to rectify the evils and divisiveness that derive from ableist thinking. As Hall appreciates, emerging technologies are of dubious value without "a revolution in values and concerted effort to dismantle the stigma surrounding disability" (2017, 108). Moreover, transhumanists "make a universalizing gesture when they categorize all humans as deficient, but this move serves to merely shift, rather than ameliorate, stigma connected to deficiency – thus maintaining and even strengthening ableism" (ibid., 133). Without a revolution in values and concerted effort to dismantle the stigma surrounding "traditional humanity," Bainbridge's future looks bleak indeed.[12]

There are, to be sure, pragmatic reasons to be wary of a vision like Bainbridge's. We cannot assume that a single set of modifications will be sufficient for surviving in every potentially inhabitable space environment. Instead, each planetary environment is likely to require its own unique set of modifications. The modifications needed to thrive on the Moon will be quite different from those needed to thrive on Mars, and each of these modifications will be quite different from those needed to thrive on Titan, Europa, Enceladus, or elsewhere. This is likely to generate nearly impermeable barriers to interplanetary travel, commerce, and social participation, and could lead to a great fracturing of human culture rife with conflict throughout the solar system. Human provincialism is problematic enough on Earth, where there are few biological boundaries to intra- and international travel. Provincialism will follow us into space, and it will only be intensified in a future where humans identify geographically *as well as biologically* with their home environments (as in, e.g., *The Expanse* with its depiction of persistently fractious Earthly, Martian, and "Belter" cultures). The *Star Trek* vision, in contrast, seeks to *minimize* barriers to interplanetary travel and settlement, providing a more stable foundation for interplanetary cooperation.

Which of these futures would you like to work toward? Which of these futures could you realistically be a part of? Which of these futures could you reasonably ask the rest of humanity to help you build? *For the moment we have a choice about which future to attempt to create in the space environment.* No external pressures will *force* us into a *Star Trek* future, or the future Bainbridge envisions, or any other future. Our decision could reflect very well on us, but also very poorly on us. And importantly, *this moment will some day pass*—but hopefully not before more inclusive attitudes prevail.

14.4 Modifying Space Environments?

Throughout this chapter, I have advocated for the position that, when it comes to space settlement, we should prefer environmental modifications to biological modifications. Some might view this as incompatible with my defense of the need to preserve space for scientific study.[13] This is because the environmental modifications needed to support human life in space will require extensive in situ resource

[12]Bleak to us unmodified folk, yes—but I don't see anyone else around…

[13]See Schwartz (2019a, b, c, 2020).

utilization, which will be quite destructive of any space environment in which it is conducted. Meanwhile, if humans are biologically modified, then they will be able to survive in space environments with a much smaller environmental footprint. Modified humans will also be able to explore more adeptly and in more places than unmodified humans. For these reasons, then, shouldn't I prefer biological modifications to environmental modifications?

First, we have no assurances that the environmental footprints will be small either way. Populations of biologically modified humans will expand and consume increasing quantities of resources, perhaps at a greater rate than non-modified populations, whose growth will be constrained by the rates at which habitats can be constructed and resources provisioned.

Second, the efficiency point is granted. Modified humans would be able to conduct more effective research of the space environment. But from a planetary protection perspective, they may be just as contaminating as unmodified humans. So as far as the exploration of *pristine* space environments is concerned, there may be few significant differences in the contamination risks and scientific rewards of modified-versus-unmodified explorers; since robotic explorers may be preferable to both.

Third, while I maintain that there is significant value—instrumental as well as intrinsic—in the scientific knowledge and understanding to be gained through space research, nevertheless I have never claimed that this value is overriding. In general, our obligations to respect bodily autonomy are stronger than our obligations to carry out scientific research. I regard it as unlikely that the scientific gains associated with a modified human presence in space would outweigh the losses associated with the creation of conditions generative of discriminatory, stigmatizing cultures.

Finally, as I have made clear elsewhere, I do not believe that space settlement should take place in a space environment until after this environment has been subjected to extensive scientific study. This chapter should be read under the assumption that we are only seeking to establish space settlements in well-explored environments. In such cases, there are negligible scientific tradeoffs between settlements with and without modified humans. Thus, every case in which I view space settlement as permissible is a case in which I view environmental modifications (e.g., the construction and provisioning of habitats capable of supporting unmodified humans) as permissible.

Acknowledgements Thanks to Susan Sterrett, Sheri Wells-Jensen, Melinda Hall, and anonymous referee for discussion and comments.

References

Abney, K. (2019). Ethics of Colonization: Arguments from Existential Risk. *Futures, 110,* 60–63.

Abney, K., & Lin, P. (2015). Enhancing astronauts: The ethical, legal, and social implications. In J. Galliott (Ed.), *Commercial space exploration: Ethics, policy and governance* (pp. 245–257). Dorchester: Ashgate.

Agar, N. (2014). *Truly human enhancement: A philosophical defense of limits*. Cambridge, MA: MIT Press.

Andreadis, A. (2013). Making aliens. *Journal of the British Interplanetary Society, 66*, 269–274.

Bainbridge, W. (2007). Converging technologies and human destiny. *Journal of Medicine and Philosophy, 32*, 197–216.

Bradshaw, H., & Ter Meulen, R. (2010). A transhumanist fault line around disability: Morphological freedom and the obligation to enhance. *Journal of Medicine and Philosophy, 35*, 670–684.

Buchanan, A., et al. (2000). *From chance to choice: Genetics and justice*. Cambridge: Cambridge University Press.

Buchanan, A. (2011). *Beyond humanity?* New York, NY: Oxford University Press.

DeGrazia, D. (2005). Enhancement technologies and human identity. *Journal of Medicine and Philosophy, 30*, 261–283.

Franssen, T. (2014). Prometheus descends: Disabled or enhanced? John Harris, human enhancement, and the creation of a new norm. In M. Eilers, K. Grüber, and C. Rehmann-Sutter (Eds.), *The human enhancement debate and disability* (pp. 161–182). London: Palgrave MacMillan.

Green, B. (2019). Self-preservation should be Humankind's first ethical priority and therefore rapid space settlement is necessary. *Futures, 110*, 35–37.

Hall, M. (2017). *The bioethics of enhancement: Transhumanism, disability, and biopolitics*. Lanham, MD: Lexington Books.

Hull, R. (1998). Defining disability—A philosophical approach. *Res Publica, 4*, 199–210.

Milligan, T. (2011). Property rights and the duty to extend human life. *Space Policy, 27*, 190–193.

Mundy, L. (2002). A world of their own, *The Washington Post*, March 31, 2002. Available at: https://www.washingtonpost.com/archive/lifestyle/magazine/2002/03/31/a-world-of-their-own/abba2bbf-af01-4b55-912c-85aa46e98c6b/?utm_term=.bcab9a6d0694. Accessed August 3, 2019.

Schwartz, J. (2019a). Space settlement: What's the rush? *Futures, 110*, 56–59.

Schwartz, J. (2019b). Mars: Science before settlement. *Theology and Science, 17*, 324–331.

Schwartz, J. (2019c). Where no planetary protection policy has gone before. *International Journal of Astrobiology, 18*, 353–361.

Schwartz, J. (2020). *The value of science in space exploration*. New York, NY: Oxford University Press.

Silvers, A. (1998). Formal justice. In A. Silvers, D. Wasserman, & M. Mahowald (Eds.), *Disability, difference, discrimination: Perspectives on justice in bioethics and public policy* (pp. 13–145). Lanham, MD: Rowman & Littlefield.

Silvers, A., & Francis, L. (2013). Human rights, civil rights: Prescribing disability discrimination prevention in packaging essential health benefits. *Journal of Law, Medicine & Ethics, 41*, 781–791.

Sparrow, R. (2011). A not-so-new eugenics: Harris and Savluescu on human enhancement. *Hastings Center Report, 41*, 32–42.

Stapeldon, O. (2012). Interplanetary man? *Journal of the British Interplanetary Society, 65*, 30–39. First published in 1948 in *Journal of the British Interplanetary Society, 7*, 213–223.

Tremain, S. (2017). *Foucault and feminist philosophy of disability*. Ann Arbor, MI: University of Michigan Press.

Wells-Jensen, S., Miele, J., & Bohney, B. (2019). An alternate vision for colonization. *Futures, 110*, 50–53.

Chapter 15
Who's Afraid of Little Green Men? Genetic Enhancement for Off-World Settlements

Kelly C. Smith and Caleb Hylkema

Abstract There are many threats facing the Earth that could put humanity (and indeed the entire terrestrial ecosystem) at existential risk. This is a brute fact we must face head on. And we should also be clear that this risk imposes a clear prima facie moral duty to establish successful settlements on other worlds sooner rather than later in order to ensure our survival. That will be no easy task, but one promising approach that could make it considerably easier is the use of genetic enhancement to give off-world settlers some of the biological adaptations they need to thrive in the harsh conditions they will encounter. Of course, there are many ethical concerns here that also should not be taken lightly, though it's our view that, in general, they can be handled if we approach them with care. We thus begin this chapter by responding to several of the more common objections to genetic enhancement. Some of these are well taken in a general sense but don't pose an insurmountable obstacle to using this technology for settlements, while others are based on unexamined terrestrial assumptions that do not apply in off-world contexts. Next, we discuss how this debate, as with so many debates about the future, tracks an unacknowledged fault line between idealists and pragmatists. Idealists tend to see an off-world settlement as an opportunity to start afresh and create a better world than we have ever experienced on Earth, while pragmatists tend to argue that tough choices will inevitably have to be made and, in the final analysis, it's better to have an acceptable settlement now than hope for a perfect one in the distant future. To illustrate our approach, we critique Schwartz's (The accessible universe. Present volume, 2020) recent argument for the creation of an off-world settlement accessible to all who wish to go. While this is certainly a laudable vision, for practical reasons it is simply not a feasible demand, at least initially. Finally, we offer a vision of a realistic use of genetic enhancement that might make settlements easier to establish and then discuss what sorts of unique problems these could create. In the final analysis, we believe it's possible to create a morally acceptable off-world settlement in the near future, though doing so may

K. C. Smith (✉) · C. Hylkema
Department of Philosophy & Religion, Clemson University, Clemson, SC 29630, USA
e-mail: kcs@clemson.edu

C. Hylkema
e-mail: chylkem@g.clemson.edu

K. Szocik (ed.), *Human Enhancements for Space Missions*, Space and Society,
https://doi.org/10.1007/978-3-030-42036-9_15

well involve controversial technologies like genetic enhancement. Given the stakes we face, we have a moral duty to forge ahead, though this does not absolve us of the responsibility to listen carefully to critics who seek to restrain some of our less thoughtful impulses. We will have to strike a complex balance, but it is ultimately not morally defensible to substitute our dreams of an ideal settlement for the possibility of a real one.

15.1 Introduction

The Earth may only be habitable for a short while—until some point in the future when the processes of environmental degradation, including climate change, pollution, and population growth resulting from human civilization render it unsuitable for human (and other) life. Even if we find a way to fix these problems and live sustainably on Earth, sudden and devastating catastrophes such as a nuclear war, meteor strike, or bio-disaster will still pose existential threats. Off-world settlements thus seem a prudent hedge, increasing the probability of our long-term survival. Given the utility of genetic enhancement in creating and sustaining successful off-world settlements, this technology suggests itself as a way to make such a settlement a reality in the near future. But are genetic manipulations ethical in this context?

We begin our answer to this question by discussing a few objections to genetic enhancement and show how off-world settlements present a unique set of circumstances within which many of these lose their force. First, we address the issue of obtaining the informed consent of settlers for genetic enhancement. Children pose a special challenge here but, we argue, not an insurmountable one. Second, we discuss the potential for genetic enhancement to exacerbate social inequality and argue that the unique circumstances of an off-world settlement defuse much of the force of these worries. Third, we respond to the question of whether and to what extent we should seek to accommodate disability in an off-world settlement. In particular, we consider to what extent we are morally required to expend capital (time and money) in order to accommodate the dreams of disabled would-be settlers. We also offer replies to some common objections to altering humanity that are based on implicit notions of "normality" which do not apply in a straightforward way within the off-world context. Finally, we provide a realistic sketch of what genetic enhancement in off-world settlements might look like and discuss the ethical issues unique to such a scenario.

15.2 Working Assumptions

There are two key assumptions underlying our position that we will not argue for in detail here, since that would take us far beyond the scope of this chapter. However, it's important to articulate them clearly at the outset to avoid confusion, so, for the purposes of this chapter, we will assume:

1. Off-world settlement may be essential to long-term human survival. As has been discussed extensively elsewhere (Bostrom and Ćirković 2008; Currie and Ó hÉigeartaigh 2018; Kareiva and Carranza 2018) there are many existential risks to humanity that an off-world settlement would significantly reduce. Although the exact level of any one of these risks is debatable, ignoring them entirely is unconscionable. Since the existence of humanity (not to mention other life we share the planet with) could be on the line, even a miniscule risk must be very seriously considered. It seems that opponents of settlements often simply ignore or dismiss these risks without careful consideration, but they *must* play a critical role in any discussion of the morality of such projects.
2. We will have the technology in the near future to undertake safe and effective genetic engineering of the sort that will make an off-world settlement much easier. At the moment, there are many objections to genetic engineering that are purely technical—for example, we just don't yet know how to modify human DNA without major risks, both to the modified individual and to the human germline. Until we can offer the kinds of basic safety assurances we expect of other morally defensible medical procedures, we should not attempt this kind of genetic engineering outside of a research context. However, it seems a safe bet that the current technical limitations are temporary, and thus, our chapter assumes a future where these problems have been adequately addressed and a safe and effective genetic enhancement technology is available.

15.3 Two Perspectives Concerning Off-World Settlement

Considering the implications for the future of humanity, we argue that the creation of a self-sustaining settlement should be the primary objective, with concerns about ideal social structure being secondary. To some this will seem patently obvious, while to others it will seem like an abandonment of moral principles. There is a dichotomy of perspective here that is rarely discussed explicitly but is important for understanding the dynamics of this debate. On the one hand, there are the idealists who see an off-world settlement as a unique opportunity to create something fundamentally different and better—a new type of society free from many of the deeply ingrained problems we face on Earth.[1] On the other hand, there are the pragmatists who argue that we have excellent reasons to establish an off-world settlement as soon as possible, even if it is far from perfect. If this distinction is not kept in mind, confusions can occur. For example, both sides might agree with a principle like "We should make an off-world settlement as inclusive as possible," but then disagree radically on what "possible"

[1]A variation on this basic theme claims we have no moral right to establish settlements elsewhere until we have first perfected society on Earth. For examples of both sides of this "Earth First" debate, see Smith and Abney (2019).

means, since many things are *logically* possible, but not practically so within a reasonable timeline. To make the situation even more complex, idealism tends to dominate among humanists and social scientists, while pragmatism tends to dominate among engineers and scientists. This disciplinary partitioning of perspective causes other sorts of problems: for example, making it much more difficult for those with a critical perspective to get a fair hearing at space science conferences where these issues are typically discussed, as they tend to be dominated by the scientists and engineers.[2]

At the end of the day we side with the pragmatists, but allow that there are still legitimate worries to entertain regarding the creation of an off-world settlement. Having productive debates with idealists is thus useful in tempering potentially extreme pragmatic dispositions, as unrestrained enthusiasm in favor of off-world settlement could lead to moral disaster. But we also feel that, given the existential importance of off-world settlement, those expressing idealistic concerns have an obligation to be realistic as well. Critiques of settlements often seek to hold settlements to a standard we have not successfully achieved *on Earth* or that do not take account of the unique conditions that would apply in the context of an off-world settlement. Given the practical impossibility of a perfect settlement, we will have to settle for imperfection if we are to settle at all. The real question is thus *how far* from ideal we should be willing to go in order to establish a settlement efficiently and effectively.

Given the harsh conditions of extraterrestrial environments, human genetic enhancement may have an important role to play in making off-world settlements a reality and thus should be carefully considered. Lower levels of oxygen and gravity, greater exposure to radiation, and the difficulty of securing sufficient food and water are only some of the challenges any near-term off-world settlement will face (Szocik and Braddock 2019). Genetic enhancement offers ways of dealing with these problems that seem likely to be cheap and efficient (Szocik et al. 2019). Off-world settlement could still be pursued without enhancement, but the costs (in both dollars and human lives) may be quite high, which would make it morally undesirable given a viable alternative. Unfortunately, there is a great deal of opposition to *any* discussion of genetic enhancement, and this is often informed by standard terrestrial objections which do not apply well to the context of an off-world settlement. In what follows, we offer pragmatic replies to some of these, beginning with two classic objections which are rarely thought of within the context of off-world settlements per se before proceeding to more targeted critiques.

[2]This is part of the reason one of us has spearheaded the creation of a new interdisciplinary group dedicated to the exploration of the social and conceptual, as opposed to empirical and practical, issues surrounding astrobiology and space exploration: the Society for Social and Conceptual Issues in Astrobiology (*SSoCIA*).

15.4 Informed Consent

The first classic objection has to do with the adequacy of informed consent. It's long been a mainstay of applied ethics that we should not expose people to risk without their informed consent (Beauchamp and Childress 2019). But that is easier said than done and one perennial problem is that securing robust informed consent is far more difficult than many people realize (Manson and O'Neill 2007). Securing the right kind of informed consent therefore poses serious issues that must be addressed carefully.

Let's begin with the low-hanging fruit: getting informed consent from adult settlers. This is not nearly as complicated as is sometimes implied. Yes, we are asking settlers to incur massive risks relative to what ordinary Earthbound people encounter, but as long as we take the time and effort to make sure they fully understand the risks involved,[3] then they should be allowed to make their own choices—indeed, the principle of autonomy is the primary moral justification for informed consent in the first place. Settlements would certainly be extremely risky enterprises but, to put it bluntly, if informed adult settlers die on a mission, they died doing what *they chose* to do. After all, if a highly intelligent, well-educated, and superbly trained settler can't make such a decision, it's unclear if *anyone* can, in which case our concept of informed consent is far too strong.

The truly knotty problem involves children. For purely technical reasons, it's likely that the kinds of genetic modifications needed to give settlers the adaptations they need would have to be performed before adulthood—on embryos or at least pre-adolescents (Walters and Palmer 1997). This is because any major bodily modifications will probably need to be programmed into the DNA *prior* to the development of the characteristics in question. Obviously, this is morally much more problematic than altering adult astronauts. We hesitate, and for good reason, to make decisions for our fellow humans before they are in a position to fully understand the potential impacts. But since children are not in a position to decide for themselves,[4] and failing to decide while they are young has morally significant consequences as well, the real question is who should get to decide for them, and under what circumstances.

15.4.1 Parental Consent

One of the most distressing aspects of becoming a parent is the realization that one is unavoidably forced to make decisions for one's child, even when the future impacts of those decisions are unpredictable. Worse, a child will often object to parental

[3]Which is certainly not always the case—the recent *Mars One* project offers a particularly nasty example of "consent" done wrong.

[4]We often do require children to give *assent* (which is not legally binding) as a kind of compromise. But even this requires a level of maturity which may not be reached in time for a decision about these kinds of genetic engineering.

decisions in ways that *might* be legitimate. Should I force my son to take years of grueling piano lessons, which he hates, or allow him to fill his free time with video games, which he enjoys? If I opt for lessons, he may thank me later for his musical abilities or it may be a source of ongoing resentment of the sort he will one day discuss with his therapist. I just don't know for sure which outcome will occur, but I can't simply wait until he comes of age to decide for himself, since by then the opportunity will have passed. This is a relatively trivial example, obviously, but it shows how parents must make controversial judgment calls about what is best for their children all the time.

And there are much more momentous parental decisions as well. A parent may have to decide whether to force their child to undergo a risky surgery, for example. Typically, this would only occur when the interests of the child are very obviously in favor of the surgery—for example, if it's clear that this is their best chance of survival. But the case is not always so clear. Consider, for example, a parent who wishes to move to a third world country with their child. This might be good for the child in some ways—for example, she will learn about another culture, etc. But it will surely have some major disadvantages as well—for example, by denying her the benefits of living in a first-world country, such as access to western standard medical care, a well-funded educational system, etc. In this case, it's entirely possible the child will end up *worse* for the move, all things considered—indeed, it's not inconceivable that she could *die* as a result of her parents' decision. Nevertheless, we tend to agree that parents have the right to make these calls, since they can typically be assumed to have the child's best interests at heart. Various permutations of this question have been extensively pursued in the literature (see, e.g., Parlett and Weston-Scheuber 2005).

Parents can and do make the wrong decisions, to be sure. But here we must distinguish what is ethically *ideal* and what is legally *permissible*. We may think the parent is not making the best ethical decision, but this does necessarily mean the government should infringe on their freedom to choose. In modern democracies, we tend to give the benefit of the doubt, whenever we can, to the parent. Only fairly egregious and obvious harms (e.g., refusing lifesaving medical treatment) are deemed worthy of limiting the parent's autonomy. Thus, the US government allows parents to refuse vaccinations for their children, even when doing so is potentially life-threatening to the child. Given all this, one non-negotiable requirement for any genetic enhancement of children must be obtaining fully informed consent from the parents, with all the inconvenient complications that entail.

15.4.2 Risk is Relative

The risks involved with off-world settlement are substantially greater than those present in most terrestrial situations, so it may seem unjustified to incur them if nothing important stands to be gained. But here we remind the reader of our initial assumption that the risks of genetic engineering per se are no higher than for other major medical procedures we would normally allow. Given that, this kind of objection is not really about genetic engineering so much as it's about the risks of putting settlers into dangerous circumstances of the sort any off-world settlement would surely impose.

This is tricky, in large part because we tend to judge risk in a *relative* fashion: This seems risky *relative to* that. Typically, we judge risks as acceptable once they are on a par with those we already consider "normal". For example, in vitro fertilization certainly has risks, but objections on these grounds largely faded away once the technology progressed to the point where the risks were roughly the same as those associated with a "normal" pregnancy. Many of our intuitions about what is morally permissible are thus based on implicit, and thus critically unexamined, ideas about normality. An off-world settlement changes these calculations in fundamental ways we have yet to really grapple with.

On the Earth of today, for example, we are highly skeptical of a parent who is willing to endanger their child's life, since the baseline expectation is that the child will likely do just fine without taking that risk. But this is a context-dependent judgment—today we would likely object to incurring a risk by genetically engineering children so they are resistant to smallpox, but would change our minds quickly should our old nemesis become commonplace again. Our intuitions are shaped by the fact that it has been a very long time since we had to think much about very high baseline risks.

But such risks are inherent in the settling of a new land. Consider that the first Pacific Islanders who set out in canoes for undiscovered territory took enormous risks, both for themselves and their children. And they surely weren't *unaware* of these risks, they simply made the calculation that, all things considered, this risky move offered the best future prospects for their family. No doubt even at the time most of their compatriots disagreed, and certainly the most common outcome of such adventures was a disaster, but some of these pioneers secured a bright future for their families, just as they had hoped. Again, our point is not that the moral calculation will necessarily endorse settlement in all contexts, just that those making an informed choice to take such risks should be afforded a certain benefit of the doubt under the circumstances.

15.4.3 Enhancement as Therapy

Reimagining "normal" applies to all human abilities. It's not normal on Earth to photosynthesize, and this is fine because there is no pressing need to do so (and because the technology hasn't been developed yet). But survival on another planet may well *require* adaptations that can be gained in no other way. One author in this volume (Schwartz) observes that disability is not a natural kind, which is to say there is no truly *objective* standard of ability. We agree, but add that *normality* isn't a natural kind either. To use a standard distinction in the literature (Harris and Chan 2008; Kilner et al. 1997; Kiuru and Crystal 2008) to make much the same point, what counts as a genetic "enhancement" as opposed to a "therapy" may vary with what background conditions we take for granted. If people simply cannot survive on a world without a given ability, should giving them that ability be considered enhancement *in that context*? Andreadis (2013) makes this point in the context of off-world settlement when he observes, "[f]or space travelers and planetary settlers, such changes will be made to confer not the dubious enhancements touted by transhumanists but the ability to survive." Settlers on a very different world should not be forced to maintain terrestrial standards of normality, especially when they are not at all practical.

It's worth noting that this is not merely a pragmatic point—it's not just about what we need to do to establish a *practical* off-world settlement, but about what is in the *best interests of the settlers*. This might seem strange, since no matter how good our technology, the first explorers will face a very high likelihood of death or disfigurement and certainly will suffer from extremely high levels of psychological stress. But that doesn't change the fact that genetic enhancement could significantly reduce their *relative* risk. Of course, if we are extremely risk-averse, we could ban off-world settlements outright, with or without genetic engineering, but such a move may have very severe negative consequences for all of humanity. If we thus accept that we are *obliged* to pursue off-world settlements in some fashion, then we have an attendant obligation to offer settlers the kinds of technology they will need to have the best chance of a decent life. Whether we agree with their decision or not, once they have made it, we should do all that we can to help them thrive, which may well entail genetic enhancement. Indeed, it could even be argued that a parent who has decided to become a settler but refuses to allow genetic enhancement of their child is guilty of a form of child abuse (e.g., by exposing their children to an unnecessarily extreme risk of cancer, etc.).

15.4.4 Benefits of Settlement

Of course, most parents will recoil in horror at the risks involved in becoming an off-world settler, much less a genetically enhanced one. That is certainly their right—taking risks that one can avoid is rarely popular. But the question is not whether most people would agree to these risks, much less whether these risks are ethically *ideal*, but simply whether a parent who chooses to take them is acting in such a blatantly immoral fashion that we have no choice but to restrict their freedom to choose. If there is a real benefit that may be gained by taking these risks, it's much easier to see them as rationally defensible, even if we personally would not take them. So, are there real potential benefits of settling on another world?

Of course there are. First, if the off-world settlement is successful, settlers and their families will become founding members of a completely new society on a new world. Many would argue (and we count ourselves among them) that this is a benefit of *incalculable value*—the kind of opportunity that comes along, not even once in a lifetime, but once in a millennium (if that often). To be sure, most parents will not want to expose their children to such risks, but this is not directly relevant to the question of whether parents who *do* want to take such a risk are acting *indefensibly* or *irrationally*. If there is a realistic chance of success, even if it's smaller than most of us are comfortable with, it does not seem at all obvious that the parents are misguided.

Second, there are the interests of humanity in general. If it's true, as we assume, that an off-world settlement may be necessary to ensure the long-term survival of humanity, then we should be willing to relax our normal expectations of acceptable risk. What seems an unnecessary risk from the point of view of an individual parent might well be a *necessary* risk for the broader society in which those individuals live. To go back our previous example, part of the reason Pacific Island culture thrived was that it spread itself so effectively through the efforts of generations of risk-takers. In other words, societies (or species) which refuse to take "unnecessary" risks may unwittingly put themselves at risk of an even worse long-term outcome.

15.4.5 Safeguards

Since there is no way to avoid all the problems inherent in parents taking major risks for their children, safeguards should be in place to ameliorate the problems as much as possible. For one thing, as we noted above, decisions on the behalf of children should be in the hands of fully informed parents, provided they are not abusive, rather than an external agency. We must be extremely careful to make sure those parents understand the risks—both to themselves and their children—as fully as possible and are acting out of what they see as the best interests of the children. We might even decide to screen out parents who seem to be taking on these risks for more "base" motives (e.g., fame and fortune), though we should be cautious here, as one person's base motive is another's noble one. Finally, if possible (though we expect it will not be possible in general), we should always seek *assent* (which is legally distinct from consent) from the children themselves.

Few parents will agree to expose their children to the risks of settlement, but it seems hard to argue that carefully prepared parents who do so are simply immoral, which is what would be required for us to infringe on their liberty by force of the law.[5] Of course, there is a sense in which taking these kinds of risks is never *ideal*, but then we don't live in an ideal world.

---

[5]Medical students are often willing to endorse overriding patients' wishes when they want to act against medical advice—until they are forced to consider what this will mean in practice (e.g., forcibly restraining someone in order to perform surgery).

15.5 Social Inequity

The second classic objection to genetic engineering is that it will exacerbate social inequalities (Mehlman and Botkin 1998; Baumann 1999; Fenton 2007). The argument tends to assume that (1) genetic engineering will confer some real advantage but (2) this advantage will be available only to those with money and influence. Thus, to allow genetic engineering is to further cement existing social inequalities by adding biological advantage to all the other advantages that accrue with wealth and power.

Even in terrestrial contexts, it's not obvious that genetic engineering must *necessarily* be limited to those with wealth and power, as many goods and services that were initially very expensive and difficult to acquire sparked just this sort of debate initially, only to became widely available in time (e.g., [illegible] computers). [illegible], this objection should give us pause when considering genetic enhancement in *terrestrial* societies, though it poses much less of an issue in the context of off-world settlements. The basic reason is that if genetic engineering is *required* to make a settlement possible, then (1) all the reasons provided previously about the necessity of off-world settlement establish, by extension, the necessity of genetic enhancement and (2) access to genetic enhancement would be universal *in the context of the settlement*.

We will leave the discussion of the first point for later, but the second is rarely appreciated adequately and deserves a brief discussion. Off-world settlers would be on perfectly equal footing, since they would all have to obtain the same genetic enhancements as a condition of their inclusion in the settlement.[6] Since all members of an off-world settlement would need the same enhancements to cope with the environmental conditions of the new world, equality of access *within the settlement* is guaranteed from the beginning. If there is no differential access to genetic enhancement among the settlers then, from *their* point of view, everyone is equal.

Of course, there might be a difference in access between settlers and terrestrial citizens.[7] Moreover, genetic modification would only bring settlers up to a performance level, in the context of the settlement, that would be considered normal for unmodified humans on Earth (if that). Thus, these modifications would only count as "enhancement" from a hypothetical, cross-context, perspective and wouldn't pose a threat to fairness and equality of the sort that generates a moral problem. In other words, nobody will be using these "enhancements" to get an unfair leg up, either on their off-world neighbors or their fellow humans back on Earth. And it's hardly the case that these would be the only differences between terrestrial humans and settlers.

---

[6] At least initially, acquiring the genetic enhancements would presumably be *required* to become a settler. However, it's worth noting that this need not continue indefinitely—once the settlement reaches a certain level of maturity, it might be able to sustain a more mixed population of settlers (e.g., by building domed areas to provide extra oxygen and radiation shielding to "wild type" humans).

[7] This may or may not be the case, since it's certainly possible (if difficult) to set up a system of universal access on Earth as well. Whether we do this is a question about our choice of socioeconomic systems that has nothing to do with genetic engineering per se, any more than the US' choice of a system restricting access to healthcare is a feature of healthcare per se.

Settlers would have routine access to many things that are hard to acquire on Earth (high tech equipment), and terrestrial residents would have access to many things that would be very difficult to acquire on the settlement (e.g., attending a Clemson football game). In other words, it's not possible to guarantee truly equal cross-context access to everything people might care about no matter what we do—the contexts are just too different. This is a pretty suspect frame of reference to adopt anyway, since moral theory has long defended different provisions of goods to differently situated persons.

15.6 A Utopian Vision

Now we will turn to a more targeted objection to genetic engineering in the context of an off-world settlement. As it happens, another article in this volume (Schwartz 2020) presents a concrete example of idealistic thinking that offers a useful contrast to our own approach. Schwartz argues that any settlement we create, since it would be something many would desire to be part of, must avoid unnecessary restrictions on who can be a settler, or on what kinds of interventions (such as undergoing genetic transformation) settlers must agree to. Instead, he argues, we should seek to make settlements maximally inclusive and, in particular, we should allow settlers with *disabilities* to participate in the great adventure. After all, we will have to develop many new technologies to make any off-world settlement a reality, and there is no reason to suppose that achieving accessibility through genetic modification will be inherently easier than doing so via physical structures.

To make his point, he offers an analogy of an eccentric billionaire who creates an island paradise that has no accommodations whatsoever for those with disabilities. The billionaire then provides free access to all abled persons who wish to enjoy his creation. Although people object that this is unfair to those with disabilities, the billionaire is unwilling to make accommodations to allow them equal access. Or rather, he is only willing to offer accommodations of a certain sort, since he does agree to provide free genetic modification that will make a previously disabled person normal should they wish to move to his island.

This scenario is meant to be analogous to a proposal for off-world settlement in which genetic enhancement is required as a condition of being a settler. Since we intuitively recoil at the hypothetical billionaire's restrictions, Schwartz thinks we should also recoil at such restrictions for an off-world settlement. Unfortunately, it's not a very useful or fair analogy, so we offer a more realistic modification of it below in the section on financial considerations. For now, we wish to note that Schwartz falls prey to a common failing among idealists: vagueness. Whether he is talking about off-world settlements or an island paradise, he is rather vague about precisely what sorts of disabilities he has in mind and how we should seek to accommodate them, which makes his position difficult to critique in detail. However, at times he seems close to suggesting the truly utopian position that (1) *no* disability should be a bar to serving as a settler and that (2) it is *entirely* up to us which sorts of technologies we develop to enable accommodation. This strikes us indefensibly optimistic, as do some of his other implicit assumptions. We will consider some of the more important of these in what follows.

15.6.1 Goals Versus Requirements

Who could argue with Schwartz's inclusive vision? In a sense, no one—which in itself is an indicator that something is amiss, since thoughtful people do in fact disagree about how inclusive off-world settlements ought to be. We freely admit that a fully inclusive settlement should be the *ultimate* goal we strive for, but this impacts shorter-term goals only to the extent it is actually possible to achieve within reasonable limits of time, resources, and risk. Unfortunately, the simple fact of the matter is that *any* settlement on another world will not be inclusive in all sorts of ways of necessity. The main reason for this is purely pragmatic: certainly initially, and likely for many years after it is established, a settlement will have a very small population. It is thus critical that each and every settler, transported off-world at massive expense into an incredibly precarious and dangerous situation, be able to contribute as much as possible to the settlement's success. *Some* disabilities would surely be a major problem in this regard, particularly in a situation where we don't have access to the kinds of elaborate support we can take for granted on Earth, and we will simply not have the luxury of including settlers who are less able to contribute to the success of the settlement in the fullest possible way.

That said, it's fair to point out that some restrictions that might be imposed on settlers would likely be arbitrary—a point often overlooked by zealous pragmatists. Thus, it's critical that any restrictions we impose be carefully considered in the context of an off-world settlement. As Schwartz correctly notes, "disability" is a relative concept, not a natural kind. Thus, it's entirely possible that some conditions which cause problems on Earth might well be neutral or even advantageous in space. For example, a blind colleague once suggested that blindness would be an *advantage* in space, since there would be no need to waste power on lighting.[8] Similarly, some types of mobility issues that would constitute a hard barrier to exploration on Earth might not be much of a problem in a low or zero-gravity environment. So, just as we must calibrate our *moral* intuitions to the appropriate context, we should avoid judgments about *physical* restrictions based on terrestrial circumstances that do not apply on the new world.

The legal concept of a *bona fide occupational qualification* (BFOQ) is useful in making sense of which restrictions we should and should not apply (29 CFR § 1625.6). US law allows discrimination in employment decisions (even when the

[8]Even if we choose to provide lighting for our light-challenged colleagues, blind settlers would have a major advantage during a power failure, since they would not be hindered by the removal of the crutch their sighted colleagues have come to rely on.

results seem to violate legal protections such as the Age Discrimination in Employment Act), provided that the feature in question is an *essential* attribute of the job (Reed 1983). Thus, we can and should require fighter pilots to be able to withstand high G loads, since this is critical to their job. And, insofar as youth correlates with the ability to withstand such forces, the military can justify a youth-biased distribution among its fighter pilots. In a similar vein, it also seems reasonable to require that pilots have adequate vision, not because this is necessary in a *logical* sense, as we could in principle redesign our air traffic control system to operate solely on auditory cues. However, such an accommodation would be massively difficult and expensive and therefore isn't considered morally *required*. On the other hand, it's much less clear that we need to require pilots' vision be 20/20 *without correction* or that they be able to do a specified number of pullups, though these are common standards in the military by tradition. When the fighter pilot retires from the military and becomes a civilian airline pilot, a new urgency is added, since any failure on her part will endanger hundreds of passengers—nobody will want to fly in a plane piloted by someone less than fully capable of handling all aspects of the job as well as we can reasonably expect.

Every settler in a newly established settlement will be in the position of an airline pilot, since any mistake could well result in the failure of the entire settlement and the death of all their colleagues. If a disability impairs a settler's abilities relative to their "normal" colleagues, or if they are at higher risk of becoming ill and depriving the settlement of their services,[9] then allowing them to participate will incur a critical risk we cannot simply ignore in the name of inclusion. How much of a risk of a debilitating medical problem in the middle of a complex and critical operation are we willing to risk? How much are we willing to spend to enable the dreams of a few would-be settlers? There are no easy solutions, but we have a responsibility to address these issues head on. The question we must focus on, given the imperfect world we inhabit, is unfortunately not *whether* we should exclude people, but *which* people we should exclude and why.

Our intuitions here are not usually well thought out and are often influenced by stilted portrayals in popular media. In the movie GAATACA, for example, the hero (Vincent) wants more than anything to be part of the crew of a space mission. He is a genetically unmodified human with a heart condition, however, so in the movie's futuristic society where genetically perfect humans get all the plum assignments, he would not normally even be considered. But Vincent is unwilling to accept this limitation and steals the identity of Jerome, a genetically perfect human, to make it past the screening committee. He then works extremely hard to excel in the training,[10] becoming an example to all his genetically perfect colleagues. The audience cheers his determination to achieve his dreams and delights in the thought that anyone can

[9]For example, they might require medication that cannot reliably be manufactured off-world. Interestingly, this is precisely the kind of problem that might be easily corrected with genetic engineering.

[10]Though he also cheats to some extent, for example faking the EKG's used to monitor his heart during extreme stress.

succeed if they try their utmost. What the movie does not consider, however, is *why* the screening committee would not want someone with a heart condition on a mission where every person is critical. In the movie, everything works out because, well, that's the story we like to hear and movies are about vicarious wish fulfillment, not education. But consider what would happen if Vincent were to suffer a fatal heart attack during launch? At the very least, this would be a massive challenge in a situation where every crew member plays a specific, and critical, role in the success of the mission. It seems more realistic, if less psychologically gratifying, to view Vincent as incredibly selfish for pursuing his dream at such a risk to others, especially when those others are not even aware of the risk, much less have consented to it.

There is, of course, an injustice inherent in the fact that some individuals may simply be better disposed to become settlers, but this is the case in virtually every situation where people operate on the edge. The alternative is having relatively ill-equipped settlers who are less likely to secure the sort of sustainable settlement that may be necessary for the long-term survival of humanity. This would endanger not only those individual settlers, but their fellow settlers and even the future of humanity. Universal inclusion is an excellent ultimate goal, but surely not at any price.

15.6.2 Technology of Inclusion

Schwartz is clearly right to note that some types of accommodation might be possible, even easier, than genetic engineering. For example, we *may* learn to build habitats that can efficiently provide food and oxygen before we learn how to engineer humans to live with less of these essentials. He is also correct to point out that, since we can (at least collectively) allocate research dollars as we wish, what technologies we develop is a factor we have some control over. But we have to consider the very real possibility that the future may not play out as we wish, no matter how much we wish it.

One rather obvious point to make here is that, even if we develop both types of technologies equally rapidly (which is highly debatable), it's almost certain that genetic modification will be much *cheaper* than building specialized physical structures. Once we know what genetic changes to make, it is an excellent bet we could do that cheaply and efficiently, since thousands of laboratories all over the world already have the basic technology—indeed, we already engage in genetic engineering on a daily basis.[11] In contrast, even if we knew how to build truly inclusive structures on another world, the additional complexity would likely require a significant increase in the amount of specialized materials that would have to be transported from Earth. This is no small problem, since launch costs at the moment are something on the

[11]This has been true for some time. One author ran an undergraduate cell biology laboratories 30 years ago in which each student created a new kind of microbe that likely had never existed before.

order of \$10,000*per pound* (Marshall Space Flight Center), so even a "mere" 10% increase in the weight of material we would need to lift out of Earth's gravity could increase project costs by billions of dollars. Of course, this may not be a problem if truly radical futuristic technologies (e.g., extremely efficient propulsion, space elevators, nanomachines constructing structures from local regolith, etc.) come to fruition (Naser 2019). But radical possibilities such as these are, well, radical. This means they are almost certainly much further down the road than safe and effective genetic enhancement, where we have understood the basic principles for decades. So, while it is certainly *possible* that we could build inclusive off-world structures in 50 or 100 years as cheaply and easily as we could genetically engineer humans to live without them, it's not *likely*, and we should bias our discussions in favor of what we think we will probably be able to do in the foreseeable future.

15.6.3 Financial Considerations

The discussion above hints at another kind of (usually implicit) objection, namely that worrying about the *cost* of a settlement is somehow grubby and inappropriate in a high-minded moral discussion. To some extent, we agree—money is not the only, and often not the most important, consideration. It is extremely common for ethicists[12] in particular to be suspicious of monetary considerations, since there is a long history of people doing immoral things in the name of economic efficiency. Thus, no thoughtful person would argue that *anything* we might do to build a settlement more cheaply is therefore a good idea. But that doesn't mean we can simply ignore such considerations either.

If nothing else, money represents opportunity—at least in our current socioeconomic system.[13] If we choose to spend \$100 *billion* to build inclusive structures on Mars, rather than genetically engineering settlers for \$100 *million*, there is a real opportunity cost associated with that choice. It's easy to say this is "just money" and shouldn't influence our moral calculations, but we invite the reader to pause for a moment and think of all the good that could be done (either for Earth or a settlement) with the \$99.9 billion difference between the two approaches. For example, we could decide to use the savings to address accessibility problems on Earth, in which case going with genetically engineering a settlement would provide a significant benefit *to many more people* (something Schwarz, oddly, finds hard to envision). On a related note, it would be quite difficult to get people in a democratic society to spend that kind of money in order to realize the full potential of a handful of settlers. Should

[12]And European ethicists in particular, since they are used to a more managed economy than their US counterparts.

[13]Of course, we can argue about the desirability of adopting a different type of socioeconomic system, but this is an extremely complicated debate that has no inherent connection to the question at hand.

this be the case? Not in an ideal world, of course—but neither Earth nor an off-world settlement is ideal, and will not be for a very long time indeed.

We would like to end this section with our own variation of Schwartz's hypothetical island example discussed previously—one which offers a more realistic portrayal of the choices we would face with an off-world settlement. Suppose an industrialized nation plans to build an undersea settlement on Earth to mine some newly discovered resource. The resource would benefit millions of people—perhaps, for example, it's an essential ingredient in a promising new treatment for cancer. As a consequence, those who participate will be considered heroes and there is thus no shortage of volunteers. However, building the infrastructure for this undersea settlement will force us to face tough choices. For example, living in this settlement for the extended periods of time the project requires will be extremely challenging, both physically and mentally—to the point where most people will not be able to cope without extensive and expensive physical accommodations. Suppose it's logically possible to do that—we *could* build an elaborate structure that would allow anyone, no matter their disability or psychological profile, to participate. The problem is that such a structure would cost \$100 billion. Alternately, we could build a more minimal structure that would accomplish the task, costing only \$100 million, but this would require settlers to come from a very small pool of people with rare physical and psychological profiles. The suggestion that we should automatically pick the more inclusive option, despite the massive cost differential, is simply not morally defensible.

15.7 An Alternate Vision

Discussions of genetic manipulation often suffer from a lack of detail, which lets the imagination run rampant. Thus, some critics may imagine truly horrific alterations of humans and oppose all genetic modification on these grounds, even if they are never explicit about their reasoning. It's worthwhile, therefore, to offer a realistic picture of what these genetically engineered inhabitants might look like. There are many known biological challenges an off-world settlement must overcome in order to be self-sustaining. For example, settlers will need to grow enough food, on a world where this is extremely difficult to do, to avoid starvation. On many worlds (including the most realistic near-term targets of Mars and the Moon) and in deep space, they will also need to resist much higher levels of radiation than on Earth. We could build vast farm domes and construct elaborate radiation shielding, of course, but this will be extremely expensive and limit the settlers to safe areas most of the time. So, what might genetic engineering have to offer?

One genetic modification that would be helpful is to make humans *smaller*. Since both the need for food and the chances of developing cancer scale with the absolute number of cells in the body, the fewer cells a human has, the less food they will need and the more resistant they will be to the adverse effects of radiation (Nunney 2018). And there is every reason to expect this would be a relatively simple genetic modification—not only are some humans already small, but we know of past human populations which have been smaller still.[14] Nor is this a monstrous modification, both because it's already present in the population and because smaller humans are fully human in every way that matters from a moral perspective (Randy Newman was just wrong on this score).

Another useful modification would be the ability to extract food from light. Humans cannot do this presently, but there are animals on Earth who do so by incorporating photosynthetic microorganisms within their membranes.[15] So, it might not be difficult to tweak the human immune system to allow such endosymbionts—after all, the human body already contains more endosymbiotic microbes than human cells. While we are at it, we could probably alter the pigment molecules expressed in our skin cells to better filter harmful radiation.

The result? Little green men who require much less food to survive and are much more resistant to radiation than "normal" Earthlings. We do not yet know how to do this, of course, but these modifications are not so radical from a biological point of view, and thus we can reasonably expect the technology to implement them to appear relatively soon, particularly if we make it a focus of our research efforts. If the price of a successful settlement on another world is that it be inhabited by little green people, is that really such a high price to pay? We think not.

15.7.1 Altering Humanity

If we truly think that modifying humans puts at risk what is essential to humanity, then clearly there is a potent objection here. To put it baldly, one might object that creating a successful settlement of engineered persons is not really preserving *humanity*, since they are not *humans*. This objection cannot be fully answered without a complete theory of what it means to be human, of course, which is a task far beyond the scope of a single chapter. However, to the extent this is a concern about an engineered person's *moral* standing, it is not that difficult to assuage, since none of these changes would alter what most moral theories have taken to be our moral essence.

To be sure, there is disagreement over exactly what that moral essence is, and thus humans' moral standing is variously thought to result from our rationality, capacity to feel pleasure and pain, possession of self-awareness, capacity to love, etc. Fortunately, we do not need to venture into this thorny patch of conceptual weeds, since *none* of these characteristics would be impacted by the genetic modifications discussed above. In other words, what clearly matters morally are not the details of biology in which unmodified humans and little green men would differ, but more fundamental attributes we would share. Of course, some people might still insist that alterations

[14]There were species closely related to modern humans that seem to have been extremely small, such as *Homo floresiensis* of the Indonesian islands (sometimes referred to as "hobbits"—see Aiello 2010).

[15]For example, in a classic example of mutualism, the cnidarians which build coral reefs incorporate photosynthetic dinoflagellates within their bodies (Fransolet et al. 2012).

to our biology somehow undercut what it means to be morally human. But to say that is to go against the grand sweep of ethical thinking over the last few hundred years, which has increasingly argued that biology is incidental to moral status. For example, Africans and Native Americans are (to some extent anyway) biologically distinct from Europeans, and we used to think this was a critical difference. But now we know better and realize these details have nothing to do with their moral status. Humanity, in the moral sense, is not about biological *structures* but instead the *abilities* these structures enable.[16] This is precisely why ethicists have recently adopted the term "person" instead of the biologically loaded "human" when debating the moral status of embryos, other animals, etc.

In short, if little green people can think about philosophy, establish loving relationships, create works of art, and dream about a better future, then what difference does it make what they look like? Our differences would surely be minor in the grand scheme of things—indeed, they are better conceived as grounds for a celebration of *diversity* as opposed to moral distinction.

15.7.2 Revulsion

Perhaps the most common reaction to proposals of genetic engineering is a powerful *revulsion* at the "unnaturalness" of such altered beings (Borenstein 2011). This feeling is based on deeply rooted intuitions as to what's "normal", "human", and "natural", as well as implicit belief in the *goodness* of these attributes. Genetic enhancement would produce people who are not, at least by commonplace definitions of those adjectives, any of these.

There is hardly sufficient space here to go into all the reasons why this objection is problematic, though we will note in passing that it's hard to see this as anything more than a particularly naked invocation of the appeal to nature, which boils down to the claim that what is "natural" is good, and what is "unnatural" is bad. Appealing to nature—in this case biology—as a basis for moral judgment has long been viewed by ethicists as extremely problematic. For one thing, claims that something is "normal", "human", or "natural" are *descriptive* in nature and fundamentally different from *normative* claims about moral goods. What counts as "normal", "human", and "natural" in the descriptive sense is largely the by-product of evolution by natural selection, which does not select those traits which are *morally* good, but rather those which confer the greatest fitness upon individuals. The progress of modern civilization has largely been predicated on *extricating* ourselves from the logic of evolution, in part because it has selected for morally undesirable features such as xenophobia, racism, and sexism.

---

[16]Some have argued that life in general is so morally valuable that we have a duty to spread life, even microbial life, to otherwise lifeless worlds (Milligan 2016). If this is the case, then perhaps it does not matter what specific capacities these settlers have, as long as they are alive.

Unfortunately, it's just these kinds of morally problematic intuitions which guarantee that many of our fellow humans will view genetically engineered people as abominations, however strong the philosophical arguments against this view are. In the context of genetically engineered settlers, this poses a nasty problem that does have important moral implications, though not the ones those who voice it have in mind. Consider that we would be creating:

1. A new type of society
2. Composed of genetically distinct persons
3. Who are *visibly* different (e.g., little and green) from terrestrial humans.

It seems, at the very least, that we should worry about a new form of racism taking hold in some factions of Earth society, with all the attendant problems we know such attitudes cause. It could even be that the distant future becomes one of open competition, even war, between *them* and *us*. That would be very bad. On the other hand, it's important to keep in mind that this could happen even if the settlement is composed of *unmodified* humans since, over a long enough period of time, any successful settlement will adopt its own cultural trajectory that will be different from that of the homeworld, and likely in ways the mother culture will not approve of (Smith 2016). So, there will be grounds to classify settlers and their descendants as "other", regardless of their biology, and every reason to expect we will use these differences in inappropriate ways, just as we have always done.

This is a significant problem that has not been addressed adequately in the literature, and we cannot solve here. But we suggest that this challenge is best met, not by avoiding the creation of a settlement (at a potentially massive cost to humanity), but rather in the way we normally do, by encouraging cooperation over competition, embracing diversity, etc. And a critical part of that is not overreacting to the creation of human variants, but openly recognizing our shared humanity.[17] To be sure, humans are not very good at this, but that hasn't stopped us from making progress, even if it's frustratingly slow. It would be supremely ironic if, in the name of creating a perfect settlement, we were to embrace the connection between our biology and our moral status that has been used throughout history to justify humanity's worst transgressions.

15.8 Conclusion

There are numerous threats to the existence of humanity (even all life on Earth) which might come to pass if we refuse to leave Earth. This simple fact should act as a clarion call to establish an off-world settlement as soon as practical in order to mitigate this existential risk. Even if that were the only reason to favor off-world settlement (and we think there are more), it is sufficient to justify it.

---

[17] Phillip Zimbardo, the investigator behind the infamous Stanford Prison Study, argues that the psychological distinction between them and us is *the* central feature of all groups that have committed systematic atrocities like the Holocaust (Zimbardo 2007).

We applaud the spirit that motivates many of the criticisms opponents have raised against genetic engineering of settlers. It is a good thing that some among us will always assume the moral high ground, even if they sometimes ignore practical considerations when doing so, since this forces us to assess and reassess the necessity for and implications of the hard choices we must make. On the other hand, we cannot ignore practical realities simply because things would be better if they did not exist. Those advocating the moral high ground thus have a critical role to play, but they should take care to advocate suggestions that are both practical *and* morally acceptable, lest they be ignored by those with the power to make decisions.[18] After all, if humans were to go extinct because we couldn't figure out in time how to create the perfect settlement, then was the "moral high ground" really so high?

We have argued that there are excellent reasons to think genetic engineering will be a practical requirement for future settlements, and that it can be done in ways that are morally acceptable, if not always ideal. Pragmatic concerns are not always paramount, but we cannot postpone a decision until every possible objection is met, especially when the objections are either unworkable in principle or so vague as to be unmeetable in practice. We have only scratched the surface of these issues, and a great deal of careful theoretical work remains to be done, but hopefully we have at least given food for thought to those who are currently opposed to any use of genetic technology in this context. We welcome the inevitable debate.

References

29 CFR § 1625.6 Electronic Code of Federal Regulations, Title 29B, part 1625. Accessed 12/5/19 here: https://www.law.cornell.edu/cfr/text/29/1625.6.

Aiello, L. (2010). Five years of *Homo floriensis*. *American Journal of Physical Anthropology, 142*(2), 167–179.

Andreadis, A. (2013). Making aliens. *Journal of the British Interplanetary Society, 66*, 269–274.

Baumann, J. J. (1999). The ethics of human genetic enhancement: Extending the public policy debate. ProQuest Dissertations Publishing.

Beauchamp, T. L., & Childress, J. F. (2019). *Principles of biomedical ethics* (8th ed.). New York, NY: Oxford University Press.

Borenstein, J. (2011). Shaping our future: The implications of genetic enhancement. *Human Reproduction and Genetic Ethics, 13*(2), 4.

Bostrom, N., & Ćirković, M. M. (2008). *Global catastrophic risks*. New York: Oxford University Press.

Currie, A., & Ó hÉigeartaigh, S. (2018). Working together to face humanity's greatest threats: Introduction to the future of research on catastrophic and existential risk. *Futures, 102*, 1–5.

Fenton, E. (2007). Genetic enhancement a threat to human rights? *Bioethics, 22*(1), 1–7.

Fransolet, D., Roberty, S., & Plumier, J.-C. (2012). Establishment of endosymbiosis: The case of cnidarians and *Symbiodinium*. *Journal of Experimental Marine Biology, 420–421*, 1–7.

Harris, J., & Chan, S. (2008). Enhancement is good for you! Understanding the ethics of genetic enhancement. *Gene Therapy, 15*(5), 338–339.

[18] Indeed, given the basic principle of "ought implies can," the distinction between what is practical and what is morally laudable is to some extent artificial, provided the end being pursued is morally laudable (e.g., securing the future of humanity).

Kareiva, P., & Carranza, V. (2018). Existential risk due to ecosystem collapse: Nature strikes back. *Futures, 102,* 39–50.

Kilner, J. F., Pentz, R. D., & Young, F. E. (Eds.). (1997). *Genetic ethics: Do the ends justify the genes?* New York: Wm. B. Eerdmans.

Kiuru, M., & Crystal, R. G. (2008). Progress and prospects: Gene therapy for performance and appearance enhancement. *Gene Therapy, 15*(5), 329–337.

Manson, N. C., & O'Neill, O. (2007). *Rethinking Informed Consent in Bioethics.* Cambridge, New York: Cambridge University Press.

Marshall Space Flight Center. (2019). Advanced space transportation program. Accessed online 12/8/19 here: https://www.nasa.gov/centers/marshall/news/background/facts/astp.html.

Mehlman, M. J., & Botkin, J. R. (1998). *Access to the genome: The challenge to equality.* Washington, DC: Georgetown University Press.

Milligan, T. (2016). Common origins and the ethics of planetary seeding. *International Journal of Astrobiology, 15*(4), 301–306.

Naser, M. Z. (2019). Extraterrestrial construction materials. *Progress in Materials Science, 105,* 100577.

Nunney, L. (2018) Size matters: Height, cell number and a person's risk of cancer. *Proceedings of the Royal Society B-Biological Sciences, 285*(1889), 20181743.

Parlett, K., & Weston-Scheuber, K.-M. (2005). Consent to treatment for transgender and intersex children. *Deakin Law Review, 9*(2), 375–397.

Reed, B. (1983). Age discrimination of airline pilots: Effects of bonafide occupational qualification. *The Journal of Air Law and Commerce, 48*(2), 383.

Schwartz, J. S. J. (2020). The accessible universe. Present volume.

Smith, K. C. (2016). Cultural evolution and the colonial imperative. In C. Cockell (Ed.), *Dissent, revolution, and liberty beyond earth* (pp. 169–187). New York, NY: Springer.

Smith, K. C., & Abney, K. (2019). Human colonization of other worlds. *Futures, 110,* 1–3.

Szocik, K., & Braddock, M. (2019). Why human enhancement is necessary for successful human deep-space missions. *New Bioethics, 25*(4), 295–317.

Szocik, K., Campa, R., Rappaport, M. B., & Corbally, C. (2019). Changing the paradigm on human enhancements: The special case of modifications to counter bone loss for manned mars missions. *Space Policy, 48,* 68–75.

Walters, L., & Palmer, J. G. (1997). *The ethics of human gene therapy.* New York, NY: Oxford University Press.

Zimbardo, P. (2007). *The Lucifer effect: Understanding how good people turn evil.* New York, NY: Random House.

Chapter 16
Evolving from Earthlings into Martians?

Ted Peters

Abstract The question of whether or not earthlings should colonize Mars is exacerbated by the hostile conditions space travelers will confront: radiation exposure that threatens life; diminished gravity causing loss of bone strength; an alien surface atmosphere requiring cocoon living; and isolation from ongoing Earth history. If planners of a permanent Mars colony elect to create a posthuman species to populate the Red Planet, could CRISPR gene editing speed up adaptive evolution? To prevent interplanetary sin transfer, could genetic engineering pre-program virtue and altruism into the future Martian community? This essay concludes: if a biosphere already exists on Mars, we should treat it as having intrinsic value; but if Mars is currently lifeless, then, despite interplanetary sin transfer, we should take advantage of the opportunity to seed the Red Planet with life borrowed from Earth.

In Ray Bradbury's fictional work, *Martian Chronicles*, colonizers from Earth wandering the Red Planet look for the resident Martians. When they peer into a canal and see their own reflection, they realize they alone are the Martians. "The Martians were there—in the canal—reflected in the water.... The Martians stared back up at them for a long, long silent time from the rippling water...." (Bradbury 2019).

T. Peters (✉)
Center for Theology and the Natural Sciences, Berkeley, California, USA
e-mail: tedfpeters@gmail.com

K. Szocik (ed.), *Human Enhancements for Space Missions*, Space and Society,
https://doi.org/10.1007/978-3-030-42036-9_16

Does simply changing one's address from Earth to Mars immediately make one a Martian? Not quite. Who we are on Earth is largely determined by our historical past, our environmental present, and our anticipated future. We earthlings are contextualized, morphized, and localized by our home planet's *Sitz im Leben.* Charles Darwin's principle of natural selection predicts that if a generation of earthlings survive on the Red Planet they will do so only if they become contextually Martian. Only if they evolve. Only if they adapt. Only if they speciate. To become a Martian is to become a posthuman.[1]

Posthuman in every respect? Or, only some respects? The theologian will be skeptical about excessive promises of total transformation. It is a safe gamble that the pattern of human sin developed on Earth will be borne to every new world we earthlings settle. Perhaps, we need a new theological category: *Interplanetary sin transfer*. Even if earthlings moving to Mars adapt physically, what about their spiritual development? Or lack thereof?

In what follows, we will imagine a speeded-up evolution leading to what might be considered a new species, a posthuman species of Martians. Might we even attempt to direct if not govern this future speciation through CRISPR gene editing?

Our analysis in this essay of the potential to control future evolution through genetic engineering will grant that we can alter the human body; yet we will wonder with skepticism whether we can alter the human soul. To press the matter, we will examine the prospect of enhancing our genetic predisposition to living a life of virtue and to creating on Mars a posthuman colony characterized by altruism and neighbor love. Is it possible, we will ask, to establish a utopian community on Mars? Or, will that future Mars colony only extend what we have on Earth: avarice, greed, competition, war, and environmental desecration? Will we export terrestrial sin to extraterrestrial worlds?

16.1 Can Earthlings Evolve into Martians?

When we imagine a colony of earthlings on Mars, we detonate an explosion of harrowing challenges to be overcome. The Martian environment is so hostile to terrestrial life that our astronauts will either have to create an earthlike cocoon on

[1]The term, *posthuman,* here refers to a successor species to the current human, even if the present generation witnesses the transition from the human to the posthuman in a short span of time. This provocative term has garnered varying meanings in other discourses. In discussions surrounding transhumanism, for example, it refers to the evolution of superintelligence. In postmodern deconstructionist philosophy, this term destabilizes what we've assumed to be human. "Posthumanism is inextricably related to the studies of the differences, referring to the fields of research which developed out of the deconstruction of the 'neutral subject' of Western onto-epistemologies. The deconstruction enacted within the historical and philosophical frame of Postmodernism, by feminist, black, gay and lesbian, postcolonial, and chicana theorists, together with differently abled activists and other outsiders, pointed out the partiality of the construction of the Discourse…of the self-claimed objectivity of hegemonic accounts" Ferrando 2019, 24–25).

the Red Planet or adapt the human species to an alternative environment. Or, most likely, a combination of both.

Just getting to Mars, let alone walking the surface of the Red Planet, will subject us earthlings to increased radiation. Without either high tech protection or adaptation, we will glow momentarily like green embers before blinking out.

Once we have arrived on Mars, we will skip lightly on the surface with only 38% of Earth's gravity. Our eight irons will carry a golf ball further than a number 2 wood at Pebble Beach, which will make the golfers happy.

This lack of gravitational load will change the shape of our bodies. We will rely on different muscles than those we need here on Earth. We will suffer bone loss estimated to be ten times higher than osteoporosis.

Mars' atmosphere is extremely thin; it is less than one percent of Earth's atmosphere. It is composed of about 95% carbon dioxide (CO_2), 3% nitrogen (N_2), and 1.6% argon (Ar), plus some trace gases such as methane and water vapor.[2] From the surface, we will look at the Martian atmosphere through our helmet visors. Or, will we? It is not clear that we will be able to see clearly. Mars' thin atmosphere carries clouds of iron oxide dust. The sky on Mars, rather than blue and bright, is rust colored and dim, somewhere between beige and pink. Everything we look at will seem alien.

Martian dust is finer than anything on Earth. Martian dust will penetrate each and every seal devised by humans from Earth. Nothing will be completely impenetrable except perhaps for a protective suit. We will need to buy Pledge by the case load.

In short, Mars will not provide the grassy plains, snow-capped mountains, and trickling streams sustaining terrestrial life that our home planet has. Biologically speaking, Mars is antipathetic to the life as we have grown up with it.

How should we handle this belligerent threat? By striking first. We strike first either by the technological creation of a safe cocoon or, alternatively, adapting ourselves to live and thrive in that hostile environment. Konrad Szocik, like the Boy Scouts of America, alerts us to *Be Preapred*! "It is worth keeping in mind that living in different—let us call them unnatural places, which are not a part of the environment of evolutionary adaptedness—locations is not problematic per se, if humans are prepared in an appropriate way to live there" (Szocik 2019, 244).

A space suit would provide our colonizers on Mars with a mobile cocoon. But earthlings colonizing Mars would want to do more than take a Sunday afternoon walk. We would live there. And living on the fourth planet from the Sun would wreak havoc on our human biology. Can we change our inherited biology? Can we adapt? Can we deliberately invoke the intentional equivalent of natural selection to enhance certain people for adaptation to Mars?

[2]There is some mystery attached to the Martian atmosphere. It changes seasonally. "The values logged by the SAM instrument for carbon dioxide (CO_2) at Gale were 95% by volume; molecular nitrogen (N_2), 2.6%; argon (Ar), 1.9%; molecular O_2, 0.16%; and carbon monoxide, 0.06%. The constituents were found to mix and circulate in response to seasonal changes in the Martian air pressure. The changes in pressure occur as CO2 freezes over the Martian poles in winter, leading the pressure to fall globally." (Carreau 2019).

Adam Hadhazy suggests that the present generation can take control of future evolutionary adaptation through genetic engineering. "Biologically enhancing people for space travel…would involve altering genes to render would-be astronauts more robust against the ravages of space. The genes could, for instance, make bones superhumanly strong, or ramp up the repairing of DNA strands sundered by radiation" (Hadhazy 2019). We are ready to ask: might the new CRISPR/Cas9 technology for gene editing provide just what we need to govern the creation of a posthuman Martian species? (Gouw 2018).

Evolutionary change over time is expected. Could the present generation of earthlings take control of the reins of evolution? Guide it? Speed it up? Engineer our successor species? Create Martians?

16.2 The Moral Question: Should Earthlings Colonize Mars?

Here is a prior moral question: should earthlings colonize Mars at all? Yes indeed, says Robert Zubrin, who founded the Mars Society in 1998. "Mars can and should be settled with Earth émigrés" (Zubrin 2019, 305). Zubrin is already packing his toothbrush for the Mars trip. Will everyone back home envy his trip and wish him *bon voyage*?

No, we earthlings should not colonize Mars or any other planet, trumpets NASA consultant Linda Billings. Billings alerts us to two dangers lurking beneath colonization ardor. The first danger hiding beneath colonization zeal is that "colonizing other planets and exploiting extraterrestrial resources…is a variant of nationalist ideology…some interpretation of Christian dominion, or dominationist, theology" (Billings 2017, 328–329). Just as European colonization born of manifest destiny reaped injustice on the indigenous peoples of the Americas, Australia, Africa, and Asia, we can only forecast a duplication when earthlings subjugate off-Earth geographies.

The second danger—actually a variant on the first one—is fear of interplanetary sin transfer. If we earthlings have been so destructive to our home planet, we are likely to contaminate other worlds too. "It would be unethical to contaminate a potentially habitable planet for further scientific exploration and immoral to transport a tiny, non-representative, subset of humanity—made up of people who could afford to spend hundreds of thousands to millions of dollars on the trip—to live on Mars" (Billings 2019, 341). We have messed up Earth. Should we mess up another planet too? Why spread the contamination of Earth by human injustice and careless pollution any further in this universe?

Monica Vidaurri would agree with Billings, because terrestrial economic and political injustice is built right into space exploration ideology. "American domination/exceptionalism in space displays an open disregard of all other nations that participate in space, and a disregard for the right that all nations and people reserve for science and exploration. And, of course, colonizing other worlds comes with an astronomical burden to resolve: who will be able to do the colonizing, who is going to set this in motion, and why?" (Vidaurri 2019) In short, the economics of launching space colonization only reinforces structural injustice on our home planet.

Then, according to Vidaurri, there is the matter of the interplanetary transfer of human sin. We on Earth will take our sin with us when we leave Earth. "In this light, ethical exploration and a responsible approach to fair play in space is going to require a serious and uncomfortable assessment surrounding the goals that both public and private sectors have in space, a humbling assessment our technological readiness, and an even more uncomfortable assessment of who the proponents for colonization/settling historically have been and currently are, and why they view colonization as our right" (Vidaurri 2019) History tells us who we are. Traveling to Mars will not in itself cut the ties with our past habits. We will carry our evolved habits with us, replete with the human propensity for competition, avarice, greed, violence, war, and ecological sacrilege.

Whether it is recognized or not, non-theologians in this debate are practicing theology without a license. Perhaps no license is necessary when it comes to grasping the dark side of human nature. Our evolutionary history on Earth has defined who we are; and who we are is observable to anyone with eyes open to see. What Billings and Vidaurri observe is what theologians for centuries have called, *sin* (Peters 1994).

Without using theological terms such as *sin* or *soul*, Ray Bradbury describes the human condition vividly in *The Martian Chronicles*. "We earth men have a talent for ruining big, beautiful things" (Bradbury 2019).

Theologian Noreen Herzfeld draws the connection between non-theological and theological assessments of the human soul. She is even more pessimistic about our terrestrial future. She doubts that *Homo sapiens* on Earth will stave off self-destruction before the rocket to Mars even departs the launch pad. "The mechanisms of evolution that lead to intelligent life and technological development also lead to propensities traditionally labeled as sin. These propensities make it difficult for technological civilizations to survive long enough to escape their home planet" (Herzfeld 2019, 366).

Does the human condition have to be this way? No. Sin may be normal for us, but it is not required by nature. Theologian Reinhold Niebuhr is remembered for declaring that sin is universal; but it is not necessary. "Sin is natural in the sense that it is universal but not in the sense that it is necessary" (Niebuhr 1941, I:242). Here is the implication: earthlings who elect to colonize Mars and become Martians will continue the sinful history begun on Earth. This will happen not by necessity; but it will be predictable.

In sum, we are born into sin and we perpetuate sin throughout our lives, passing it on to future generations. As it stands, we leopards are not likely to give birth to Martian kittens without spots. Unless, of course, we alter our genetic code to erase those spots. Unless, of course, we redesign the human soul. Unless, of course, we turn Martian civilization into utopia.

16.3 Could Gene Editing Turn Sin into Virtue?

A utopian Mars colony would be made up of individuals who exhibit personal virtue along with care for both their neighbors and the common good. *Altruism* is a term frequently used in the scientific community to describe this.[3] Jacob Haqq-Misra uses the term, *deep altruism*, to refer to such a community over time, over a millennium. "Deep altruism can then be defined as the selfless pursuit of informational value for the well-being of others in the distant future" (Haqq-Misra 2019, 145). For the posthuman Martian colony to foster deep altruism, the individual colonizers will have to have undergone a moral transformation. Compared to humans back on Earth, the posthuman utopians will need a renewal of soul, a spiritual revolution, a process of deification. What will it take to accomplish all this? Could gene editing do the trick?

Whether genetic engineering can alter the soul as well as the body depends on just what genes determine and how potent they are. It is easy to attribute more governing power to genes than they deserve. It is easy to overestimate the influence of genetic expression and, thereby, overestimate our capacity for genetic determinism.

"Genes, genes, everywhere," mewls neuroscientist Robert Sapolsky. "Large genetic contributions have even been uncovered for everything from the frequency with which teenagers text to the occurrences of dental phobias…. [But, at most,] genetic influences on behavior often work through very indirect routes" (Sapolsky 2017, 237). The prospect of engineering virtue or altruism into our evolutionary future is dependent on the existence of identifiable genes for virtue and altruism and on our technological capacity for controlling those genes. If Sapolsky is right, such genes may not exist and, if they do, they may be less deterministic than we assume.

Despite the tenuousness of genetic determinism, we still engage in a thought experiment by pressing the question: could we edit our genome by deleting sin genes and replace them with virtue genes? Could the present generation of *Homo sapiens* transform itself or, additionally, give birth to a successor species of posthumans with greater virtue than the current progenitor species?

Our focus here is the connection between genes, virtue, neighbor love, and altruism. Roman Catholic moral theologians, such as Charles Curran, are the most knowledgeable when it comes to virtue. "The classical tradition recognized the three theological virtues of faith, hope, and charity, as well as the cardinal moral virtues of prudence, justice, fortitude, and temperance. From the theological perspective, even these moral virtues involved both the gift of God's grace and the human response" (Curran 2011, 32). These seven virtues are moral ends in themselves, requiring that the virtuous person transcends his or her self to embrace each virtuous trait.

[3]"Selfishness beats altruism within groups. Altruistic groups beat selfish groups. Everything else is commentary" (Wilson and Wilson 2018, 297). "Christian ethicists do not often use the word 'altruism', because the term is not morally helpful" (Pope 2007, 227).

There is more to virtue. Surrendering self-interest on behalf of one or another virtue is to take a giant step toward neighbor love, toward loving the other for the sake of the other, toward altruism if not *agape* love. "The art of living," declares philosopher Ottfried Hőffe, "lies in the good life as the sense of well-being that, admittedly, includes a jolting dosage of interest in the well-being of others as well" (Höffe 2010, 330). In short, to live the life of virtue the individual person must sacrifice self-interest on behalf of the virtue itself and, in addition, on behalf of the well-being of the neighbor. Is this humanly possible? Could we enhance the possibility of virtuous living through genetic engineering?

To approach this question, we must first ask: is the human person changeable? Transformable? Improvable? Yes. We *Homo sapiens* are a work in progress. Our future can be different from the past. In fact, we are beckoned by God to become more than we have been. We are invited even to become the image of God, according to astroethicist and theologian, Jacques Arnould. "So a human individual is not created but formed gradually, through a journey; the individual does not immediately attain the perfect image of God, but represents a divine promise that remains to be achieved" (Arnould 2018, 29).[4] Could we achieve that divine promise by editing our genomes toward virtue enhancement?

At least one philosopher, Mark Walker, answers affirmatively: yes, we can enhance human virtue and even pursue sanctification if not deification through gene editing. "To soul building we must also enhance the biological basis of our humanity [through] genetically engineered virtue" (Walker 2018, 251). Walker sponsors what he calls *The Genetic Virtue Program*, to create caring humans with the so-called Big Five specific complex traits: openness, conscientiousness, extroversion, agreeableness, and neuroticism (Bouchard and McGue 2003, 4). These particular traits require more intelligence than we *Homo sapiens* currently exhibit. So, the first enhancement will be intelligence; and intelligence will in turn lead toward enhanced virtue. Walker fittingly calls his posthuman species, *Homo bigheadus*. Walker hopes these bigheads will continue our altruistic precedent and yield to a still more intelligent species, *Homo biggerheadus*. The present generation of *Homo sapiens* who will make way for a superior successor species would win the laurel for virtuous altruism. Walker's goal is to employ genetic enhancement to "makes us more like God in terms of virtue" (Walker 2018, 267).

Self-transformation through gene editing will make us virtuous, loving, God-like. Right? Well, not if you ask the experts, the theologians.

The difficulty in pursuing virtue through genetic enhancement is that most virtue theorists deny that virtuous living is natural or spontaneous. Rather, virtue requires conscious attention, will power, discipline, and habit formation. Virtue takes time. Virtue takes more than biology. Virtue relies on a process of soul formation guided by the human will and energized by divine grace.

---

[4]Deconstructionist Francesca Ferrando reminds us that the concept of the human is fluid, not fixed. "…radical deconstruction of the human as a fixed notion, emphasizing instead its dynamic and constantly evolving side and celebrating the differences that inhabit the human species itself" (Ferrando 2019, 187).

Roman Catholic bioethicist Lisa Fullam makes this clear. Yes, she would welcome a healthy genetic start, but the running of the race toward the life of virtue would still require a willing self-discipline over time. Moral transformation is a participatory process. "So, could we engineer virtue genetically, at the start of one's life? No, not really. Virtue happens after we inherit our genomes. Virtue involves one's practice toward perfection. Virtue is pursued through schooling the appetites by intellect (including beliefs and commitments) over time. Virtue is the fruit attained after a process of growth in character. Virtue is not a static trait or constellation of traits. We grow into virtue more than we possess it. Indeed, as we make progress in a given virtue, we tend to see more broadly the scope and ramifications that it has for our lives" (Fullam 2018, 321). If what Fullam holds is true, then the skillful editing of astronaut genomes could not in itself determine the level of virtue in a future Mars colony.

Ukranian Orthodox scientist Gayle Woloschak pursues the ultimate virtue, becoming God-like. Even if gene editing could aid in her pursuit of deification, any genetic enhancement would not guarantee Godlikeness for two reasons. First, the opportunity for deification must be egalitarian, open for all persons regardless of their respective genomes. Second, deification requires an act of the human will, a cooperative act that receives God's grace and builds on that grace.

The idea of becoming God-like is the calling of each Christian, and each person needs to be willing to heed the call to deification. Each person must freely consent to living a holy life…this salvation must be available to all people regardless of genetic make-up or environmental influences. Each person is called to the life of growth toward God achieved through discernment, prayer, contemplation, meditation, and virtue. Deification requires not just the grace of God extending to humanity, but it also requires the willingness of the human person to choose this path of righteousness and goodness. It is a full cooperation between God and the person with the human being freely choosing to accept God's will and action in his/her life.

Salvation (and deification as a component of it) must be accessible to each human person. Genetics cannot be a precondition for salvation since it could limit salvation to only selected parts of humanity (Woloschak 2018, 304).

Protestant ethicist Braden Molhoek iterates what appears to be the consensus position. "Genetic engineering has the capacity to enhance the human disposition to moral behavior, but gene editing cannot create virtue because virtues are stable, habituated dispositions, acquired over time" (Molhoek 2018, 279).

Let us pause to summarize our findings on the question of genetic moral enhancement. First, genetic influence is at best limited. Genes predispose; they do not determine (Peters 2003). Even inserting a special virtue gene would not guarantee that a person's resulting behavior would be moral let alone God-like. No number of genetic strings can turn a human person into a moral puppet.

Second, everyday morality along with the passionate pursuit of virtue requires two things not found in the DNA, namely, free will combined with commitment over time. One must choose virtue for it to be virtue. More, one must choose virtue daily, repeatedly, and habitually. Even when God's grace strengthens the human will, the decision to live virtuously and to love one's neighbor requires freely elected human agency.

What does this imply for our future colony on Mars? The level of virtue in that future Mars colony cannot be pre-programmed genetically in advance. If Martian colonialists are to live in an altruistic community characterized by moral excellence and mutual care, it will require willful participation on the part of astral travelers who choose the life of virtue. The road to utopia is not paved by genes.

16.4 How Do Sin and Grace Interact?

The power of the human self to transcend itself when embracing either virtue or altruistic neighbor love is not itself a human power. It is a divine power. When that divine power grasps us within our daily life it is called *grace.*

Grace comes to us when we are passive and when we are active. The divine expression of unconditional love, mercy, and forgiveness, can be received by us only passively.[5] We have no control over the sovereign God, so we can at most accept with gratitude God's grace as a gift. The scholastics alluded to this as *gratia operans* and the Protestant Reformers as *sola gratia.* God's grace is unmerited, unasked for, and uncontrollable.

There is more. There is also an active relationship offered to us by God's grace. "Grace reorders the self and his or her relation to all other objects of love," avers Stephen Pope in the context of a discussion of evolution (Pope 2007, 236). Grace has the power to change, to transform, and to renew.

When God's grace grasps us, it empowers us. Grace overcomes our inherited propensity to serve the self with self-survival, self-merit, and self-aggrandizement. God's grace liberates the self from the self's past, opening a future where it is possible to embrace virtue for virtue's sake or embrace the neighbor for the neighbor's sake.

Reinhold Niebuhr outlines the dialectic between sin and grace. "Grace represents on the one hand the mercy and forgiveness of God by which [God] completes what [we humans] cannot complete and overcomes the sinful elements in all of [our] achievements.... Grace is on the other hand the power of God in man; it represents the accession of resources which [humanity] does not have of [itself], enabling [us] to become what [we] truly ought to be. It is synonymous with the gift of the Holy Spirit" (Niebuhr 1941, II:98–99).

This dialectic constitutes our active sharing in the dynamics of grace, a sharing with divine and human dimensions. In scholastic theology it was known as *gratia cooperans* or synergy, that is, divine grace that cooperates with human free will.

---

[5] Martin Luther champions the passive reception of divine grace. "No human being can be thoroughly humbled until knowing that one's salvation is utterly beyond one's own powers, devices, endeavors, will, and works, and depends entirely on the choice, will, and work of another, namely, of God" (Luther 2016, 178).

Methodists know it as sanctification and the orthodox as deification. Lutherans tend to discourage such synergy, relying almost solely on the passive reception of grace.

The grace empowered person experiences both continuity and discontinuity with his or her previous self. Transformation both negates past sin while fulfilling the beauties of our naturally bequeathed identity. "Grace is related to nature partly as a fulfillment and partly as negation" (Niebuhr 1941, II:245). One implication we may draw for the present discussion is this: a graced life fulfills our humanity. Grace does not make us posthuman; rather, it makes us truly human.

The true human, according to the New Testament, is Jesus Christ, the *eikon tou Theou,* the *imago Dei,* the image of God. Grace empowers us to grow in likeness to Christ, to grow into our true humanity. Sanctification amounts to growth into fulfillment, not replacement. This would apply to altruistic Martians as well as those of us who remain at home on Earth.

16.5 Again: Should Earthlings Terraform or Colonize Mars?

When on the eve of the modern world Europeans colonized the Americas, Australia, Africa, and Asia, the existing inhabitants had to make way for invaders. The colonizers brought a new form of injustice to those previously living on the land. This is not likely to be the case when earthlings travel to Mars. Nobody's currently at home on Mars.

On Mars, presently, there is nobody to be victimized by our interplanetary sin transfer.[6] Only our companion astronauts. Only those making up the new Martian family could become victims of the injustices we have known on Earth.

If the explorers we send to the Red Planet discover a native biosphere, however, the moral situation would change dramatically. Then, the question of intrinsic value would arise. Should we earthlings treat an existing form of Martian life as a moral end rather than merely a means for our own instrumental value? I believe any sign of existing life on Mars would warrant this ethical inquiry (Peters 2018; Race and Randolph 2002).

If we decide that existing Martian life should be treated as having intrinsic value, we will be mandated to embrace the equivalent of virtue. That is, we will find ourselves mandated to protect if not enhance the well-being of that Martian life

[6]Astrobiologist Chris Impey distinguishes European expansionism from Mars colonization. "The historical example of manifest destiny is misleading in the context of space colonization. Countries have grown and gained resources on Earth by seizing territory and displacing or subjugating the original inhabitants. Even in the twenty-first century, the stains of this brutal history persist. Space is a new resource. The people who leave Earth won't be taking land from anyone" (Impey 2019, 107). Still, there is a debate regarding microbial life on Mars. "NASA allegedly found strong evidence of life on Mars way back in July 1976" (Axe 2019).

for its own sake. To care for the welfare of an off-Earth life form will require a sacrifice of our own self-interest on behalf of the interest and perhaps even the flourishing of that biosphere. Can we forecast that earthlings will rise to that level of altruistic behavior? Can we count on God providing sufficient grace to transform terrestrial sin into extraterrestrial virtue?

This leads us back to our earlier question: should we earthlings colonize Mars? If we confirm that Mars is without life, should we take life to an apparently lifeless planet?

NASA's Christopher McKay answers, yes, indeed. McKay's ethical calculus begins with a premise: life is better than non-life. This premise implies two corollaries. First, if life already exists on any off-Earth planet or moon, then we earthlings are morally obligated to treat it with intrinsic value. Second, if life does not already exist on Mars, then we should consider bringing life to that heavenly body. "Here is my astroethical premise," writes McKay, "*the long-term goal for astrobiology is the enhancement of the richness and diversity of life in the universe*" (McKay 2018, 381, McKay's italics). Regardless of the prospect of interplanetary sin transfer, McKay is ready to pack his toothbrush for the trip.

I concur with McKay. If it turns out that Mars is already inhabited by microbial life, we need to consider a possible obligation to treat that life as having intrinsic value. This may require protecting that life from our own exploitation and destruction. It may even require enhancing that life form's future.

The astroethicist enjoys some flexibility here. The principle of intrinsic value, like the principle of dignity borrowed loosely from philosopher Immanuel Kant, goes like this: *treat life as an end in itself and not merely as a means to some further end.* Note the term, *merely*. An off-Earth life form just like the waitress in a restaurant can, under certain circumstances, function as a means as long as he or she is not *merely* a means. The waitress provides a means for placing a hot meal on the table. Yet, in a prior and more fundamental sense, the waitress is a person deserving dignity, deserving treatment as a moral end for her own sake and well-being. By analogy, life discovered on an off-Earth site might become a means for some degree of exploitation while, at the same time, we would respect and protect that life as a moral end as well. In short, just what *intrinsic value* requires of us must be determined on a case by case basis.

If it turns out that Mars is not inhabited by existing life, then the question of terraforming Mars or even colonizing Mars opens up. Fear of rocketing terrestrial injustice to the Red Planet is insufficient grounds for staying home, in my judgment. The Martian colony need not achieve utopian status to warrant its establishment.

16.6 Conclusion

The hurdles to be jumped by Earth émigrés to Mars are skyscraper high: radiation exposure that threatens us with death; diminished gravity causing loss of bone strength; an alien surface atmosphere requiring cocoon living; and isolation from ongoing Earth history. We have asked: if planners of a permanent Mars colony elect to create a posthuman species to populate the Red Planet, could CRISPR gene editing speed up adaptive evolution?

This was followed by a related question dripping with theological stickiness: to prevent interplanetary sin transfer, could genetic engineering pre-program virtue and altruism into the Martian community? We concluded that, even if gene editing could engineer physical adaptation, pre-programming genes could not guarantee in advance that posthuman Martians would be virtuous or altruistic. Only freely elected consent and commitment would make that possible.

We have grappled with these concerns because of the hotly debated issue: should earthlings colonize Mars or not? Those opposing colonization remind us of the injustices resulting from Europe's conquering of new worlds, the economic inequities built into existing space programs, and the havoc the human species has wreaked on the ecology of our home planet, Earth. Even though they avoid theological language, opponents of Mars colonization fear interplanetary sin transfer. They fear that colonizers will disturb the now tranquil Mars with avarice, greed, competition, violence, war, and environmental desecration.

I have offered this conclusion: if a biosphere exists on Mars, then we should treat it as having intrinsic value. But if Mars is currently lifeless, then, despite interplanetary sin transfer, we should take advantage of the opportunity to seed the Red Planet with life for the sake of its future.

References

Arnould, J. (2018). Space exploration: Current thinking on the notion of otherness. *Theology and Science, 16*(1), 54–61. https://doi.org/10.1080/18746700.2018.1413814.

Axe, D. (2019, November 3). NASA scientist: We found life on mars 43 years ago. *The Daily Beast.* https://news.yahoo.com/nasa-scientist-found-life-mars-100255821.html.

Billings, L. (2017). Should humans colonize Other planets? No. *Theology and Science, 15*(3), 321–322. https://doi.org/10.1080/14746700.2017.1335065.

Billings, L. (2019). Should humans colonize mars? No. *Theology and Science, 17*(3), 341–346. https://doi.org/10.1080/14746700.2019.1632524.

Bouchard, T. J., & McGue, M. K. (2003). Genetic and environmental influence on human psychological differences. *Journal of Neurobiology, 54*(1), 4. https://www.semanticscholar.org/paper/Genetic-and-environmental-influences-on-human-Bouchard-Mcgue/5941daec6de7e3f11b7668336b26b03b9249d6b1.

Bradbury, R. (2019). Quotes form Martian Chronicles. https://news.yahoo.com/nasa-scientist-found-life-mars-100255821.html.

Carreau, M. (2019, December 2). Mars oxygen variations baffling scientists. *Aerospace Daily.* https://mail.google.com/mail/u/0/?tab=rm#inbox/FMfcgxwGCHBTCvKFPzsfFDGDVwlrVcXt.

Curran, C. E. (2011). How does Christian ethics use its Unique and distinctive Christian aspects? *Journal of the Society of Christian Ethics, 31*(2) (Fall/Winter), 23–36.

Ferrando, F. (2019). *Philosophical posthumanism.* London: Bloomsbury.

Fullam, L. (2018). Genetically engineered traits versus virtuous living. *Theology and Science, 16*(3), 319–329. https://doi.org/10.1080/14746700.2018.1488522.

Gouw, A. M. (2018). Challenging the therapy/enhancement distinction in CRISPR gene Editing. In D. Boonin (Ed.), *The Palgrave handbook of philosophy and public policy* (pp. 493–508). New York: Macmillan, Palgrave.
Hadhazy, A. (2019, July-August). Homo Sapiens astronautica. *Aerospace America*. https://aerospaceamerica.aiaa.org/features/homo-sapiens-astronauta/?fbclid=IwAR2cj0zI_L2J9pGidJFpHu3cvl8wGpSKNBw7RKUhRor5l-NZGrK_NLqUQO0.
Haqq-Misra, J. (2019). Can deep altruism sustain space settlement? In K. Szocik (Ed.), *The human factor in a mission to mars: An interdisciplinary approach* (pp. 145–156). Heidelberg: Springer.
Herzfeld, N. (2019). Where is everybody? Ferme's paradox, evolution, and sin. *Theology and Science, 17*(3), 366–372. https://doi.org/10.1080/14746700.2019.1632547.
Höffe, O. (2010) *Can Virtue Make us Happy?* Tr. Douglas R. McGauhey and Aaron Bunch. Evanston IL: Northwestern University Press.
Impey, C. (2019). Mars and beyond: The feasability of living in the solar system. In: K. Szocik (Ed.), *The Human factor in a mission to mars: An interdisciplinary approach* (pp. 93–113). Heidelberg: Springer.
Luther, M. (2016). The bondage of the Will (1525). In V. Leppin & K.I. Stjerna, *The annotated Luther study edition*. Minneapolis: Fortress.
McKay, C. (2018). Astroethics and the terraforming of mars. In T. Peters, M. Hewlett, J. M. Moritz, & R. J. Russell (Eds.), *Astrotheology: Science and theology meet extraterrestrial life* (pp. 381–390). Eugene OR: Cascade Books.
Molhoek, B. (2018). Raising the virtuous bar: The underlying issues of genetic moral enhancement. *Theology and Science, 16*(3), 279–287. https://doi.org/10.1080/14746700.2018.1488474.
Niebuhr, R. (1941). *The nature and destiny of man,* Gifford Lectures. 2 Vols. New York: Scribners.
Peters, T. (1994). *Sin: Radical evil in soul and society*. Grand Rapids MI: Wm. B. Eerdmans.
Peters, T. (2003) *Playing god? Genetic determinism and human freedom* (2nd ed.). London and New York: Routledge.
Peters, T. (2018). Does extraterrestrial life have intrinsic value? An exploration in responsibility ethics. *International Journal of Astrobiology, 17*(2), 1–7. https://doi.org/10.1017/S14735504170 0057X; and Margaret S. Race and Richard O. Randolph (2002).
Pope, S. J. (2007). *Human evolution and Christian ethics*. Cambridge, UK: Cambridge University Press.
Race, M.S., & Randolph, R. O. (2002). The need for operating guidelines and a decision making framework applicable to the discovery of non-intelligent extraterrestrial life. *Advances in Space Research, 30*(6), 1583–1591. https://www.seti.org/pdfs/m_race_guidelines.pdf.
Sapolsky, R. (2017). *Behave: The biology of humans at our best and worst*. New York: Penguin Books.
Szocik, K. (2019). Human place in the outer space: Skeptical remarks. In K. Szocik (Ed.), *The human factor in a Mission to mars: An interdisciplinary approach* (pp. 233–252). Heidelberg: Springer.
Vidaurri, M. (2019, October 21). What happens when you leave empty seats at the table? *The Space Review*. https://www.thespacereview.com/article/3817/1.
Walker, M. (2018). Genetic engineering, virtue-first enhancement, and deification in neo-irenaean theodicy. *Theology and Science, 18*(3), 251–272. https://doi.org/10.1080/14746700.2018.1488472.
Wilson, D.S., & Wilson, E. O., cited by Kohn, M. (2018, November 20) The needs of the many. *Nature, 456*, 7220, 296–299.
Woloschak, G. E. (2018). Can we genetically engineer virtue and deification? *Theology and Science, 16*(3), 300–307. https://doi.org/10.1080/14746700.2018.18.1488447.
Zubrin, R. (2019). Why earthlings should colonize mars! *Theology and Science, 17*(3), 305–316. https://doi.org/10.1080/14746700.2019.1632519.

Chapter 17
Human Enhancement and Mars Settlement—Biological Necessity or Science-Fiction? The Special Case of Biomedical Moral Enhancement for Future Space Missions

Konrad Szocik

Abstract This chapter discusses arguments for human enhancement during space missions. I show that objections to human enhancement are usually based on a weak justification. I emphasize that a space environment creates a specific ethical framework which substantially changes the ethical and axiological value of human enhancement when compared with analogous terrestrial applications. I focus my considerations on one particular case study—biomedical moral enhancement for the purpose of future space missions.

17.1 Introduction

Both ideas—human enhancement and Mars settlement—may sound to many like a science-fiction vision of the human future. However, human enhancement, understood as the modification of the human body and mind aimed at improvement of some human functions, is an old and well-established human practice. Many kinds of human enhancement, including diet, physical exercise, education, and others, are usually not considered to raise ethical issues. Therefore, I focus my attention only on such kinds of human enhancement which are controversial and, as such, generate ethical concerns. That kind of controversial human enhancement can be called a radical or substantial enhancement. Radical forms of enhancement may include pharmacological cognitive or moral enhancement, brain-computer interfaces, or aggressive genetic modification and editing. The latter seems to be the most controversial form of enhancement. In contrast to the idea of Mars settlement, human enhancement based on gene editing is already becoming a reality (Lea and Niakan 2019). However, there are good reasons to consider the idea of a human Mars mission or any other space settlement in terms of a realistic future scenario for which there

K. Szocik (✉)
Department of Social Sciences, University of Information Technology and Management in Rzeszow, Sucharskiego 2 Street, 35-225, Rzeszów, Poland
e-mail: kszocik@wsiz.rzeszow.pl

K. Szocik (ed.), *Human Enhancements for Space Missions*, Space and Society,
https://doi.org/10.1007/978-3-030-42036-9_17

may be many implications. This chapter discusses some arguments for the idea of human enhancement for the purposes of a future Mars settlement. The particular idea of moral bioenhancement is discussed here as a case study. The idea of moral bioenhancement was originally discussed by Persson and Savulescu (2008, 2015) as a remedy for new global threats such as climate change or terrorism (Persson and Savulescu 2014). As such, its application differs from possible applications discussed here for the purposes of spaceflights and space colonization. I aim to show that an ethical framework for moral bioenhancement—as in the case of physiological and psychological human enhancements—is different for terrestrial and space applications.

17.2 Does Human Enhancement Have to Be an Ethical Issue?

Human enhancement as such is not necessarily controversial because humans cannot survive without enhancement. This is a basic biological and cultural truth about the human species, reported by German philosophers such as Helmuth Plessner and Arnold Gehlen (1988). Humans are probably the unique animal species which is not adapted to live in an unaltered natural world. But modification of the natural environment is connected with some kinds of modification of humans themselves. As they evolved, humans were obligated to improve their physical performance and cognitive skills mostly to adapt themselves to new environments and new challenges. Such enhancement is not controversial, and it is difficult to find a precise border between acceptable and unacceptable kinds of enhancement. As we know, the issue is changed when new ways of medical enhancement are considered. This is challenging from the biological and anthropological points of view when modification and enhancement are required for humans to survive and to flourish.

One could ask why some people reject the idea of human enhancement by biomedical means when humans have already practiced their own enhancement for thousands of years. We can make that question more precise and ask: what exactly is unacceptable in human enhancement by biomedical means when compared with enhancements by other means even if they lead to the same results? For instance, improvement in intelligence and knowledge may be provided by traditional academic education or by hypothetical cognitive, for instance, genetic or pharmacological enhancement. Why do some humans care for the means if the results are always the same, or even better when bioenhancement is permanent or faster than enhancement by traditional means? If the result of enhancement is the same, it means that some people care about the means (which are obviously different) but not about the purpose of enhancement (which is definitely the same). If this is really the case, the question becomes more complicated: why are the means of achieving a particular target so important? Let us not take into account such evident obstacles addressed to human enhancements by biomedical means as the criteria of safety and efficacy.

My argument goes as follows: an intentional human enhancement is an alteration which is obligatory for human survival. There are good reasons to accept some limitations for enhancement which are aimed at avoiding risk and harm. But when such criteria are met, it is hard to argue against biomedical human enhancement when its beneficial effects can be proven. People who argue against biomedical enhancement usually do so for metaphysical or religious reasons (Trothen and Mercer 2017). They argue against any attempt to modify so-called human nature. They argue that individual human freedom may be limited when a person is enhanced in a radical way. However, such an objection is not convincing for several reasons. For instance, the idea of human nature is challenging and hard to defend beyond philosophy and the religious and theological frameworks. People are prone to use such terms as human nature as a kind of technical mental shortcut, and it may be acceptable as long as they do not assign to it any metaphysical form. For instance, philosophers talk about human nature usually when they discuss some human moral biases and moral intuitions, mostly those which are perceived as a common basis for the human species. Taking care of offspring by parents is a part of human nature, yet there are parents who do not care for their offspring, and even if statistically most parents do, it does not mean that there is something like human nature in the sense of form. We can talk only about a set of biases, functions and abilities which are usually a domain of a human being.

Second, even if we agree that there is such a set of patterns, another challenge appears, of evolution and changeability. If an opponent of radical human enhancement argues against it because of a prohibition on applying any changes into human nature, he is obliged to make his argument clear due to the fact that many human patterns and biases change over time because of exogenous or anthropogenic factors. For instance, let us assume that our opponent is against any kind of radical human enhancement even if it has therapeutic aims such as to prevent disease. But humans do their best to increase their immunity levels. The immunity level is also a factor which changes synchronically and diachronically, respectively, in human history and in different human populations. The same is true about other human factors and functions such as performance, growth, some cognitive capacities and their collaterals like the rate of literacy, and many other examples.

Consider, for instance, some kind of cognitive and intellectual enhancement such as education in general. Education is a broadly accepted means of non-radical human enhancement. However, can we say what is a "natural" human starting point in relation to factors such as cognitive, intellectual performance? Can we say, for instance, that a particular IQ is typical for a human being and, as such, should not be modified? Probably no one should defend such a statement. But the same people who are supporters of human education are often against educating humans by biomedical means. My point is the fact that no one is able to defend the idea of human immutability in the light of evolution—biological and cultural, but also in the light of human development. Humans still evolve by natural and cultural selection. Human intelligence, cognitive capacities, physical performance and many other functions and factors are changing over time even within each individual's lifespan. This is why no one can say that human nature—independently of the meaning of that term—is fixed, not flexible. If people accept human enhancement aimed at risk avoidance and increase of benefits, and human "nature" is flexible, there are no objections toward application of radical, biomedical human enhancements other than the mentioned criteria of safety and efficacy. Defenders of so-called human nature should define what "nature" and "natural state" mean when referred to the human species. There is a high variability of features and functions among humans, and many regularities are context dependent.

Another challenge lies in the fact that almost no human being lives in a so-called natural state, because all human effort is used to cope with and to modify the natural state. Science and technology from their very beginnings are aimed at struggling with natural challenges. Plessner and Gehlen point out that something like a human natural state does not exist. Humans change their own set of functions, factors and capacities over their lifespan. Nothing remains natural in human beings. Opponents of radical human enhancement are obliged to defend their position in a situation where a particularly radical way of human enhancement is safe, and its beneficial effects are proven.

17.3 Weak and Strong Rationale for Human Enhancement for Space

In this section, I show that the border between a so-called weak and strong rationale for human enhancement for space is hard to define in a clear and unequivocal way. I suggest that a weak rationale for human enhancement for space may be easily and clearly separated from a strong rationale. Someone could say that a weak rationale includes only trivial reasons which are not needed for survival, adding that a strong rationale includes reasons necessary for human survival. According to that intuitive logic, an enhancement of human morality or cognition for spaceflight is the former rationale, while preventing disease and modification aimed at increasing human resistance to space radiation or altered gravity is an example of the latter form of rationale. However, that distinction is increasingly challenging when applied to future space missions.

Currently, applied countermeasures are not effective for longer space missions, most of which last more than two years. Such missions will have hazardous physiological and psychological effects which are discussed in detail in other chapters of this volume. One of the main challenges reported during spaceflights is the risk of bone loss. Such countermeasures as physical exercise and a special diet, including vitamin D supplements, are not able to mitigate all the negative effects of the space environment on bone loss (Cristofaro et al. 2019). The same is true of the long-term impact of space radiation.

Consider the following example. Let us assume that mission planners will apply somatic gene editing to the future long-term mission astronauts to increase their resistance to space radiation. This kind of gene editing meets the criteria of preventing disease and may work as a kind of therapy. All kinds of biomedical enhancement which are aimed at protecting human health and life belong to that category. Such enhancement should be accepted and, as such, is not controversial either on Earth or in space. In space, such an enhancement gets a stronger rationale due to a permanently hazardous environment. Let us assume that the enhancement is morally permissible and possibly even morally required.

Consider another example, let us assume that mission planners decide that not only physiological but also psychological and cognitive enhancement should be applied to the future Mars mission astronauts. The criteria of health and life harm avoidance which are met in the example discussed above, do not work here in a direct way. One could rightly argue that the health and life of astronauts do not necessarily need to be threatened when they are not modified in their intelligence, cognitive abilities or moral intuitions. However, when we agree that such enhancements are not necessary to protect their health and life, no one is able to prove that performance and efficacy in cognitive and/or moral issues will not affect human health and life in a space colony. While cognitive or moral enhancements considered on Earth seem relatively trivial when compared with physiological modifications, the space environment changes their trivial ethical status on Earth. I argue that two other criteria such as the performance and well-being of astronauts and the space base/colony, and mission success possess a high value which is equal or almost equal to the value assigned to their health and life. There are good reasons to assume that some minimal level of cognitive capacities and moral attitudes is required to provide for the well-being of astronauts and, in consequence, of an entire population living in a space base/colony.

One enhancement worthy of consideration for the purposes of a future space colony is a biomedical moral enhancement. While we cannot test before the mission is launched to determine if such moral enhancement by biomedical means is required, we also cannot prove that non-biomedical moral enhancement will be sufficient for a space mission. Astronauts should in general be selected and trained in relation to their moral behavior. The unique geophysical factors in Mars such as space radiation and reduced gravity, and some specific psychological issues caused by habitat confinement, isolation and distance from Earth shape a unique moral and behavioral framework with no analogy on Earth. Currently applied non-biomedical methods of moral and behavioral enhancement known on Earth may not work in a Martian environment. There are good reasons to assume that currently tested enhancements will not necessarily work in an environment of deep-space or, at minimum, they will not work in an expected and predictable way. This creates uncertainty for a decision-making process about future space missions.

In such conditions of uncertainty, the three following decision scenarios may be depicted. One is a mission cancelation. Mission planners should avoid any risks for the crew's health and safety, and such uncertainty is a good reason to cancel such a mission—if there is not at least one really strong argument to send humans on a deep-space mission. Another scenario is a decision to postpone such a mission to some point in the future when progress in space technology will eliminate all of the expected risks. There are a couple of desirable technical solutions which are expected to cope with the main challenges in space. They include, among others, artificial gravity, faster spacecraft or hibernation technology. None of them is available today or in the near future, but some may be considered a point of reference and as the minimal technological requirements for launching such a mission. The third option in decision making is a decision to apply a radical human enhancement to astronauts. That solution can ignore premises which are essential for the situations described in the former two cases. In this context, it becomes clear that the concept of human enhancement for space is connected with a rationale for human space missions. If a rationale for sending humans to space is strong, arguments for human enhancement are also strong. But, paradoxically, if a rationale for human mission is relatively weak, a rationale for human enhancement may remain still as strong as in the case of a strong rationale for human space mission.

17.4 Biomedical Moral Enhancement for Long-Term Human Space Missions

In this section, I consider one particular example of human enhancement for space missions—so-called biomedical moral enhancement. Biomedical moral enhancement should be considered a separate kind of enhancement which differs conceptually and axiologically from enhancements aimed at improving physiological parameters. There are some differences which make a moral bioenhancement qualitatively different from enhancements aimed at physiological or psychological enhancements. One of the main differences lies in the fact that human moral decisions and moral behaviors are complex and complicated. In contrast to physiological performance and adaptivity, human morality is not a simple physiological process which is predictable, and which always works in a stable way when the same parameters and criteria are held. People can share the same moral intuitions and they can know the same moral rules, but they behave differently in moral terms independently of their knowledge and moral intuitions. For Vojin Rakić and Milan M. Ćirković such a difference is a substantial kind of difference, and this is one of the reasons why they divide humans into "mere persons" and "post-persons" (Rakić and Ćirković 2016). This is the commonly known specificity and the challenging nature of human morality. The fact that we know moral rules is not a guarantee that we will always follow them. This trivial truth about human moral and behavioral specificity provides rationale for a moral enhancement but, at least apparently, not necessarily for biomedical moral enhancement. However, this fact may offer rationale also for moral bioenhancement when we assume that we are obligated to make our world better in a moral sense, and we already know that humans find challenging the mentioned correlation between their beliefs and intuitions and behaviors. As Rakić and Ćirković (2016) point out, we should not be afraid of potential harms caused by morally bioenhanced persons toward non-enhanced persons because enhanced persons, by definition, should always possess a higher level of morality. As such, they are not able to behave in a morally wrong way or, if they do, their behaviors interpreted as wrong or hazardous for non-enhanced persons will be justified by their higher morality. Such biomedically morally enhanced persons may be beyond good and evil according to our current non-enhanced standards of morality. If our current moral intuitions are created by "standard"—non-enhanced persons—and by natural selection working on such persons, the hypothetical new morality of biomedically morally modified persons should be justified in an analogous way.

One could argue that a moral bioenhancement should be applied for future long-term space missions as a preventive tool. It means that independent of real needs and hazards, it is always better to send enhanced astronauts. This rule may be referred to both moral and physiological enhancements. That way of reasoning possibly is more justified in regard to a moral bioenhancement than in regard to other kinds of enhancement—it is always better to send astronauts who will possess a desirable set of moral intuitions and biases. It is expected in relation to a moral bioenhancement that those moral intuitions and biases will be enhanced or, at least, that an enhanced individual will always behave according to a particular moral intuition, bias or belief. There are strong reasons to take for granted a need for such moral bioenhancement for space. However, paradoxically, objections to application of such enhancement may be stronger than in the case of other enhancements. One of the basic objections is the fact that we do not know what kind of moral intuitions, biases and beliefs should be enhanced. While this task is relatively easy at the theoretical level in regard to human enhancement for anti-radiation or anti-microgravity protection, the challenge arises when we ask for enhancement of human morality. There are some good candidates for enhancement such as empathy, altruism or a sense of justice, just to mention a few. However, independent of a targeted intuition, bias and belief, the basic challenge connected with a moral bioenhancement lies in the fact that, in contrast to physiological enhancement, no one is able to predict moral decisions and behaviors of members of a deep-space mission. It is also difficult to assume that a particular intuition or belief is the most appropriate target of moral enhancement. Nicholas Agar argues that a moral bioenhancement is risky for, among others, the following reason: morally enhanced individuals may be biased to always follow a particular intuition and belief without taking into account a context which an average non-enhanced person usually takes into account. Such context is one of the reasons why the above-mentioned moral inconsistency between knowing moral prescriptions and an attempt at their application works so often in the human species. Agar talks about a need for keeping "the right balance between moral reasoning and moral emotion" (Agar 2015). A morally enhanced person may lose such a balance toward an attempt to always apply a particular moral belief such as empathy or altruism, and to follow only moral reasoning or moral emotion. Consequently, we cannot be sure if a biomedically morally enhanced person is a person who will always follow the so-called right morality, or if they may be hyperactively biased toward a particular belief and intuition. This is a kind of speculative uncertainty which is similar to a philosophical reflection on a future superintelligence. Superintelligence is considered an entity which, on the one side, can cause a moral threat to humans but, on the other side, which does not need to be interested in human life (Szocik et al. 2020). Agar adds that another challenge with a moral bioenhancement lies in the fact that there is a diversity of moral theories and opinions about a so-called right morality (Agar 2015). This challenge is increasing in regard to a space mission environment which is new and unpredictable for humans. This is why we are not able to identify before the mission is launched which set of moral rules and beliefs may be the most appropriate for a mission to Mars. I assume that such a moral set may differ depending on mission targets. This moral diversity and context dependence make that issue more challenging. If a moral bioenhancement is irreversible, people enhanced for one kind of mission may be not able to adapt to a new situation when the mission targets and specificity will be changed. For instance, a small group of astronauts sent to space on a reconnaissance mission may be bioenhanced toward particular moral beliefs and intuitions. Then, at some point, in the future, the same astronauts may live in new ecological and societal conditions in which their moral bioenhancement may be inappropriate and even harmful. The permanent challenge is still the above-mentioned unpredictability of Martian ecology which, possibly, may be reduced in further missions when some behavioral data completed during previous missions will be available.

The idea of moral bioenhancement is challenging for epistemic reasons. This challenge is discussed by, among others, Parker Crutchfield, who identifies a couple of epistemic problems. One of them is the so-called reflection problem. He points out that an individual with modified moral beliefs will lack a justification for those beliefs. As such, he may be not able to follow those new beliefs. Consequently, modified belief does not possess the power to motivate to a desirable behavior (Crutchfield 2016, p. 392). Another problem, the moral hallucination problem, means that an individual may know that his current moral belief is a result of biomedical intervention. Therefore, he may feel himself to be manipulated (Crutchfield 2016, p. 393). Crutchfield concludes that because of these epistemic challenges, an implementation of moral bioenhancement may succeed only when such a program will remain covert.

The idea of moral bioenhancement in the way that it was originally presented by Persson and Savulescu may be criticized on many points. While a critical assessment of that idea is not the purpose of this chapter, it is worth taking into account some of the objections which are discussed by critics and discussing them in relation to the space environment. The foundational idea of moral bioenhancement states that there is an evolutionary mismatch between our old psychological patterns and the current moral ecology including the lethal power of science and technology. The key idea states that our morality is not adapted to the current environmental challenges. But this assumption is far from being obvious. Michael Hauskeller argues that there are good reasons to assume that our morality evolved from targeting kinship ties to targeting large groups (Hauskeller 2016, p. 154). While moral patterns which have been evolved in the Holocene epoch are not necessary evolutionarily so deeply rooted as the patterns evolved in the Pleistocene, there are good reasons to take for granted that moral progress really happens (Buchanan and Powell 2018).

However, it is worth noting that while arguments of critics of the idea of moral bioenhancement may be right, they usually are not applicable to the context of space missions. Some critics discuss objections such as a need for informed consent which is especially important for the idea of moral bioenhancement. An informed consent and voluntary moral bioenhancement do not have only ethical value connected with human freedom. The issue is more complicated in relation to morality and, in contrast to the physiological enhancements, may affect the effectivity and success of moral bioenhancements. Inmaculada De Melo-Martin and Arleen Salles consider a scenario of a voluntary moral bioenhancement—let us take for granted that the voluntary nature of that procedure is assumed as the optimal and desirable option. They argue that a particular individual who considers an opportunity to be modified morally should be self-aware of his moral deficits and should want to be better in moral terms. Consequently, such a person—if we want to protect the voluntary nature of moral bioenhancement—should possess some relatively high level of personal morality which makes it possible to detect their deficits and to desire moral progress (De Melo-Martin and Salles 2015, pp. 226). This remark is considered by many authors as a kind of a *vicious circle* but only when the model of a voluntary bioenhancement is being considered. If a moral bioenhancement should be applied to the kind of people who need a betterment of morality more than others, there are some reasons to assume that they may not necessarily feel themselves to be obligated to improve their morality. Consequently, a compulsory moral enhancement is considered a reasonable option which not only is able to solve that problem but is also able to solve the above-mentioned epistemic problems associated with an awareness of being externally manipulated. However, that objection cannot be applied to the context of space missions for a couple of reasons. First, astronauts selected for the long-term space missions will possess a high moral compass and appropriate behavioral biases. If they need a moral bioenhancement, they should understand its rationale and therefore they should want it voluntarily. Second, the epistemic objections do not apply to the space missions either. Astronauts will understand the hazardous and challenging nature of the space environment and they will know that in such an environment, their moral intuitions, biases and beliefs—even if right—may be or should be supported and enhanced by all possible means. Third, there are good reasons to apply a moral bioenhancement for space missions as a compulsory procedure which is a part of the pre-launch selection and training process. In this context, the moral package has the same status as the human body, psyche and all human biological systems. If mission planners accept an enhancement of some biological organs and processes to prevent humans against space radiation and microgravity, the same should be true about the protection of human morality. The question arises of whether such moral bioenhancement could remain covert—having in mind, all of the virtual benefits mentioned above by Crutchfield. If being not informed increases the efficacy of enhancement, probably mission planners should not inform astronauts who will be morally bioenhanced. But in such a scenario, one of the ethically challenging questions remains of the risk of possible interference with psychological integrity (Shaw 2019).

I argue that the space environment and the context of space missions are an exceptional environment which shapes a unique ethical framework. Consequently, I assume that some of the philosophical considerations which are of high importance for ethics and philosophy, and do not necessarily matter for space missions. One of such philosophical issues addressing the idea of moral bioenhancement is the value of motivation and moral reasoning contrasted to behavioral outputs. Bioenhancement may target only behaviors but, at least theoretically, human motives and moral reasoning may remain untouched (Hauskeller 2017, p. 371). This is of high importance from the anthropological, philosophical, ethical and just "human" points of view. However, it may be assumed that as far as a mission to Mars is considered an exclusivist rather than inclusivist mission, it should not matter why space travelers will behave in a morally desirable way. But the challenge may arise when a mission to Mars will become a colonization mission. Then, a basic question about human moral "nature" may appear including evaluation of human behaviors in terms of moral good and evil. Here, the concepts of freedom and responsibility are crucial because humans may be responsible only if they could act differently.

There are good reasons to assume that some of the objections traditionally discussed in regard to human enhancement related to the physical properties are not applied to moral bioenhancement. One of such objections is the risk of irreversibility. This is a strong objection due to the fact that a modification which is irreversible may make a preadaptation to a previous environment impossible. As such, chances for survival and/or daily standards of individual life are decreased. But this is not necessarily the case of moral bioenhancement. Let us consider the following scenario. The future deep-space mission astronauts may be modified toward a higher level of empathy, altruism and sense of justice—main properties discussed by Persson and Savulescu. Let us assume—for the sake of argument—that the same properties may be desirable or even required for the purpose of the long-term space missions or space settlement. If morally bioenhanced astronauts come back to Earth at some point in the future, their enhanced moral predispositions may fit well with earthly conditions of life. As far as we can predict, the future climate change will make life on Earth harder and harder. In this specific sense, conditions of life on Earth will not be too much less challenging than conditions of life on Mars or in another space refuge—of course, in the sense of a selective pressure on a specific kind of moral values and ethical norms in a hazardous environment. Human life on Earth in the near future may become again more like hunter-gatherer communities of the Pleistocene in which an unstable climate made long-term agriculture impossible (Gowdy 2020). Space refuge, at least in its preliminary phase, may be more like the unstable Pleistocene human communities than like the large and stable Holocene societies. Consequently, an artificial moral adaptation to conditions in a space refuge may be beneficial and useful for astronauts who come back to Pleistocene-like Earth communities. However, there are good reasons to assume that independently of the specificity of future human life on Earth, every kind of irreversible moral bioenhancement has a chance to be profitable in a different environment—as long as human behavioral dynamics will be expressed in terms of altruism, collaboration and justice. And last but not least, as Marshall (2014) points out, morally bioenhanced people may always contribute to all projects and efforts in particular communities. I may say that it is always desirable for society to have people who are better in a moral sense because they virtually always may contribute beneficially to society.

17.5 Conclusions

Biomedical moral enhancement is an idea of growing attention in recent years. As long as we assume that some part of current problems is caused by human moral decisions and behaviors, the idea of modification of human morality finds its specific rationale. As I showed in this chapter, the idea of moral bioenhancement for space missions is rooted in another rationale. Humans are not perceived as the cause of problems but as victims of the hazardous space environment. Consequently, a prevention policy should take into account not only modification of the human body but also some of the moral intuitions, biases and beliefs. This idea is especially controversial mostly due to the fact that a moral bioenhancement assumes interference with the human mind and possibly human personality—something private and intimate for everyone. While this idea today sounds more like science-fiction than reality, this is a challenge for scientists of how to affect precisely human morality. In this chapter, I discussed only an idea and its possible usefulness for purposes of long-term space missions. When criteria of safety are met, morally bioenhanced astronauts seem to be better candidates for such an effort than non-enhanced personnel. One point which remains unclear is a targeted moral intuition/bias or belief. One of the possible scenarios discussed by Persson and Savulescu includes enhancing human altruism, empathy or sense of justice. But it is not clear if some of these moral predispositions are especially desirable for space missions. Another scenario includes inhibition of emotions which are usually considered causes of morally wrong behaviors.

Acknowledgements I am grateful to Chris Impey for his useful comments.

References

Agar, N. (2015). Moral bioenhancement is dangerous. *Journal of Medical Ethics, 41,* 343–345.

Buchanan, A., & Powell, R. (2018). *The evolution of moral progress*. A Biocultural Theory: Oxford University Press.

Crutchfield, P. (2016). The epistemology of moral bioenhancement. *Bioethics, 30*(6), 389–396.

Cristofaro, F., Pani, G., Pascucci, B., Mariani, A., Balsamo, M., Donati, A., et al. (2019). The NATO project: Nanoparticle based countermeasures for microgravity-induced osteoporosis. *Scientific Reports, 9,* 17141. https://doi.org/10.1038/s41598-019-53481-y.

De Melo-Martin, I., & Salles, A. (2015). Moral bioenhancement: Much ado about nothing? *Bioethics, 29*(4), 223–232.

Gehlen, A. (1988). *Man, his nature and place in the world*. New York: Columbia University Press.

Gowdy, J. (2020). Our hunter-gatherer future: Climate change, agriculture and uncivilization. *Futures, 115*, 102488.

Hauskeller, M. (2016). The art of misunderstanding critics. The case of Ingmar Persson and Julian Savulescu's defense of moral bioenhancement. *Cambridge Quarterly of Healthcare Ethics, 25*, 153–161.

Hauskeller, M. (2017). Is it desirable to be able to do the undesirable? Moral bioenhancement and the Little Alex problem. *Cambridge Quarterly of Healthcare Ethics, 26*, 365–376.

Lea, R. L., & Niakan, K. K. (2019). Human germline genome editing. *Nature Cell Biology, 21*, 1479–1489.

Marshall, F. (2014). Would moral bioenhancement lead to an inegalitarian society? *The American Journal of Bioethics, 14*(4), 29–30.

Persson, I., & Savulescu, J. (2008). The perils of cognitive enhancement and the urgent imperative to enhance the moral character of humanity. *Journal of Applied Philosophy, 25*, 162–176.

Persson, I., & Savulescu, J. (2014). *Unfit for the future: The need for moral enhancement*. Oxford: Oxford University Press.

Persson, I., & Savulescu, J. (2015). The art of misunderstanding moral bioenhancement: Two cases. *Cambridge Quarterly of Healthcare Ethics, 24*(1), 48–57.

Rakić, V., & Ćirković, M. M. (2016). Confronting existential risks with voluntary moral bioenhancement. *Journal of Evolution and Technology, 26*(2), 48–59.

Shaw, E. (2019). Counterproductive criminal rehabilitation: Dealing with the double-edged sword of moral bioenhancement via cognitive enhancement. *International Journal of Law and Psychiatry, 65*, 101378.

Szocik, K., Tkacz, B., & Gulczyński, P. (2020). The revelation of superintelligence. *AI & Society*. https://doi.org/10.1007/s00146-020-00947-7

Trothen, T. J., & C. Mercer (Eds.). (2017). *Religion and human enhancement. Death, values, and morality*. Cham: Palgrave Macmillan.

Chapter 18
Anthropocentrism and the Roots of Resistance to Both Human Bioenhancement and Space Colonization

Milan M. Ćirković

Abstract Both human bioenhancement and human activities in space are two great projects of our species for this millennium. Both are often motivated by exalted and noble purposes, the most important being the avoidance of the "ultimate harm" or a similar extremely undesirable outcome. Both projects are often met with fierce and even fanatical opposition. Motivations for opposition are heterogeneous—at least when surveyed superficially. This chapter puts forward the hypothesis that the ultimate source of this often-hysterical opposition is anthropocentrism: the tendency to separate humans, as they are today, from their natural and evolutionary embedding and assign high essential value to their current make-up. There is a strong anthropocentric cartel with vested interest in human exceptionalism, geocentrism, and other forms of parochial, self-indulgent, small-scale, and solipsist thinking. The cartel is the main source of resistance and opposition to both human bioenhancement and colonization of the universe. The anthropocentric stance against colonization of the universe is incoherent and implies magical, rather than scientific, way of thinking. The argument for delayed human exploration and colonization of the universe is, like any other delay-type argument, just a cop-out for avoiding responsibility which resolves nothing of substance.

18.1 Introduction: Cartels and Regressives

A cartel is usually defined a group of independent actors with heterogeneous goals, whose one common goal is to improve their profits and/or market position by reducing their mutual competition. A special—or not that special at all—kind of cartel are drug or other criminal cartels in which, for instance, crime families gather in order to monopolize black market of prohibited substances, or in order to efficiently control a territory whose government is weak or corrupt. Historical examples of such cartels are the Italian-Jewish-American Mafia (especially after the famous meeting in Atlantic City in 1929, which was attended by Al Capone, Benjamin

M. M. Ćirković (✉)
Astronomical Observatory of Belgrade, Volgina 7, 11000 Belgrade, Serbia
e-mail: mcirkovic@aob.rs

K. Szocik (ed.), *Human Enhancements for Space Missions*, Space and Society,
https://doi.org/10.1007/978-3-030-42036-9_18

"Bugsy" Siegel, Dutch Schultz, representatives of the Gambino clan, and others; cf. Bernstein 2002); Mexican drug cartels like the Gulf Cartel or the Guadalajara Cartel; or the now dissolved Columbian Cali Cartel. The fight against these extreme forms of organized crime is so difficult exactly because of the strong support of their "clients" as long as the cartels enjoy monopolistic position in either a geographical region or a social stratum. Some totalitarian and authoritarian regimes, such as the fascist Italy or Spain under Franco endeavored to construct a total cartel system comprising all or most trades of the economy.

Cartelization exists, however, in the domain of ideas as well—and often has at last as ominous meanings and implications. From sophists in the time of Socrates, to the present-day social-constructivists on the academic extreme left, thinkers of otherwise heterogeneous bents and interests have joined forces in order to promote particular idea, meme, or the way of thinking, not on the merit of intrinsic value, but on the quasi-Hegelian notion of converting "quantity into quality" and the vernacular "strength in numbers." In the rest of this paper, I wish to briefly defend the following theses:

- There is a strong anthropocentric cartel with vested interest in human exceptionalism, geocentrism, and other forms of parochial, self-indulgent, small-scale, and solipsist thinking.
- The cartel is the main source of resistance and opposition to both human bioenhancement and colonization of the universe, as the main human project for this millennium.
- The anthropocentric stance against colonization of the universe is incoherent and implies magical, rather than scientific, way of thinking.
- The argument for delayed human exploration and colonization of the universe is, like any other delay-type argument, just a cop-out for avoiding responsibility which resolves nothing of substance.

In the following Sects. 18.2–18.5, I shall consider each of these theses in some detail, before summarizing the threat to human progress and flourishing in the concluding section.[1]

18.2 Anthropocentric Cartel and Its Discontents

Anthropocentrism can best be understood as assigning special position to humans in either spatiotemporal, or some other parameter space of relevance in any particular field or assigning special teleological importance to *Homo sapiens*. Naïve versions of anthropocentrism were related to our spatiotemporal position in the universe, or our position in the terrestrial biosphere—and those versions have been efficiently dispelled by the Copernican and Darwinian revolutions. This process continues to

---

[1] Some further elaborations, with the emphasis on the space colonization imperative, are in Ćirković (2014, 2019).

apply to the twentieth-century follow-ups of the latter, in particular, the physical-cosmological and molecular-biological revolutions in science.

Copernicus and his followers showed that there is no astronomical—or indeed any physical-science-related—reason to assign special position to humans, our planet, our planetary system, or even, in the modern extension, our Galaxy. Darwinism was, in a sense, an extension of Copernicanism, not along the spatial scale, but along the dimensions of historical origin and complexity. Cognitive sciences extended this into the domain of mental phenomena, successfully linking them to many of our inherited traits, and the ongoing revolution in the field of artificial intelligence threatens to relegate our allegedly supreme and exalted cognitive powers to a rather modest place where they belong in a wider scheme of things (still best elaborated by Mazlish 1967); the discovery of an extraterrestrial intelligence—predominantly likely to be much more advanced than ourselves—will have the same effect (Ćirković 2017, and references therein). Other applied sciences and engineering have in various ways offered great contributions to the project of Copernicanism, and consequently have wrought havoc on our superiority complexes in domains ranging from social sciences like economics or linguistics, all the way toward arts and humanities, where we encounter modern digital arts, techno music, and many other forms of pop-cultures in which the role of human creator is at least decentered.

In light of all this, it is prima facie strange that the anti-Copernican tendencies are not only still strong, but according to all signs have experienced quite a revival in recent years. One might naively think that rejection of Copernicanism is a steep price, unlikely to be paid by anybody except a few religious zealots stuck in the Middle Ages. Unfortunately, this is far from being the truth and, if anything, anti-Copernican activities of the anthropocentric cartel have grown stronger since about the turn of the century. It is obviously a cartel according to the definition above, since it gathers wildly heterogeneous groups, individuals, and ways of thinking, with the common denominator of either vested interests in anthropocentric institutions permeating our society, or ideological blindness for reality underpinning the successes of the scientific method. Wide and dense anti-Copernican front encompass people ranging from opponents of animal rights and other defenders of anthropocentric legal orthodoxies, to various either religious conservative or social-constructivist "warriors on science" and their various allies, from the Discovery Institute, to anti-vaccination lobbies to NaturalNews, Greenpeace, Union of Concerned Scientists, and other promoters of anti-GMO hysteria, to self-proclaimed "progressive humanists" incapable of dealing with the rational facts of science on a psychological level (including indubitably enlightened people like Hannah Arendt or Michael Frayn[2]), to radical futurists believing we need ideological commitment to anthropocentrism in order to ensure the perceived desired future of humanity.

[2]Arendt [1967] (2007), Frayn (2006).

Fighters against perceived "scientism", or "scientific imperialism", or something nebulously dubbed the "hegemony of science", as well as others bothered by the alleged "coldness", or "insensitivity", or even "inhumanity" of modern science, a la Francis Fukuyama or Mary Midgley, hold hands *both* with anti-environmentalists who do not recognize *Genesis* 1:28–30 as the harmful Bronze Age superstition it really is, and with extreme new-age environmentalists worshipping Gaia as—no surprise there!—the center of the universe. Concerned guardians of the "humanistic canon," be it further specified as modern or postmodern, allegedly worried about the tough position of assorted social sciences and humanities in the current university curricula (ostensibly under attack from the Most Evil Dark Forces of science and engineering), join forces in the anti-Copernican camp with assorted media and religious pundits like Leon Kass portraying science and scientists in a uniform Dr. Victor Frankenstein's mold. When Kass (2007) preaches that

> [s]cientific ideas and discoveries about living nature and man, perfectly welcome and harmless in themselves, are being enlisted to do battle against our traditional religious and moral teachings, and even our self-understanding as creatures with freedom and dignity.

He incites the basest conservative instincts (our traditional religious/moral teachings are under attack! *Moral panic! Moral panic!*), but also performs an insidious demagogic maneuver by alleging that our self-understanding is *of necessity* correct and good. Which is grounded in neither neuroscience nor ethics nor historical facts and instances. And to all these instances, one should add legions of their less sophisticated counterparts in much of the developing world, often blending local superstitions and ill-conceived "postcolonial" irrational nonsense into the antiscientific mix (e.g., the "explaining", in large parts of sub-Saharan Africa, of illnesses, AIDS included, as being caused by black magic) and preying on poor educational and public outreach standards still prevailing in many places. In spite of much effort by various environmental groups, in the twenty-first century, a mass murderer of animals, including our closest mammalian and even primate relatives, is still celebrated as a "capable hunter," while nobody would attach that label to, for instance, members of Al-Qaeda or the Norwegian far-right terrorist Anders Behring Breivik, who was convicted for killing 77 people, including 69 children, in 2011.

It would not be an overstatement to claim that anti-Copernicanism in one form or another dominates 99% of public life and thought on this planet—which still serenely revolves around the Sun, an insignificant speck on the periphery of the Milky Way, with our Galaxy itself being only a smudge of light at the outskirts of the Local Supercluster. Evolution goes on, irrespectively of our parochial nonsense, throughout the 99.99999...% of the world; the fact that even formally distinguished intellectuals can completely ignore this simple fact in their thinking is flabbergasting.

Thus, the job of the Copernican revolution is still quite an actual, timely, and risky concern. While the inquisition which condemned Galileo seems unlikely to receive any open support today, I submit that this is more due to their old-fashioned garments and politically incorrect language than any true dissonance of ideas. After all, the underlying concern stays the same: worry about the perceived "well-being of humanity" and its institutions being threatened by "cold" and "soulless" science and technology, as well as the specter of "scientific imperialism" and similar linguistic concoctions devoid of meaning and reason, but quite appealing to *worse* angels of our nature (to paraphrase Lincoln and Pinker 2011).

The focus of the odium has shifted from astronomy in Galileo's time to evolutionary biology, computer science, and environmental science today, but the underlying reality remains the same: below a thin skin of modernity often threatens a surprisingly medieval anthropocentrism. As summarized by the great computer scientist Edsger Dijkstra, "Science is hated because its mastery requires too much hard work, and, by the same token, its practitioners, the scientists, are hated because of their power they derive from it."

Even in science itself, the Copernican revolution often looks like an unfinished business, and indeed many scientists aid and abet the tide of anti-Copernicanism in various ways: by condoning various anthropocentric social and political mores, especially in animal ethics and environmental science; by reintroducing teleological elements into science even where it is clearly unnecessary; by accepting some of the postmodern nonsense about the social construction of physical reality; by seeking "deep" reasons beneath obvious coincidences; by subscribing to conspiracy theories outside one's own discipline; by condoning the abuse of science by politicians, clergymen, and self-proclaimed "humanistic scholars" whenever it suits their own ideological prejudices; and so on. For a particularly egregious recent example, see Comfort (2019).

Like any drug cartel, the anthropocentric cartel offers the pretence of legitimate business dealings: it offers its clients merchandise at a price in a voluntary deal. Except that, just in any other drug deal, the offer is deep down dishonest and the deal is not really voluntary.

18.3 The Unbearable Lightness of Essentialism

The main opposition to the human bioenhancement comes from bioconservatives like Leon Kass who abhor "playing God" and possibly jeopardizing some mystical "human nature" or "human essence"; the other wing, not really that difficult, but tactically less exposed at this juncture consists of people on the self-proclaimed humanist left such as Leon Wieseltier. In the bioethics debate, this side is well-represented by, e.g., Agar (2010), Sparrow (2013). This form of essentialism presents a continuity with other forms of essentialism going back to Aristotle and the Aristotelian misconstruals popular in the context of medieval philosophical thought. A modern rehashing of this scholastic dogmatism has been formulated by Fukuyama (2002), who argues for a mystical "factor X," admittedly unknown and unclear, but endangered by our biotechnological meddling and our transhumanist ambitions. In Fukuyama's view, we may be able to enhance humans in both physical and intellectual sense, while losing the "factor X" in the process, and thus being necessarily at the net loss. He goes further and claims that the transhumanist idea of taking control of our own evolution is "the most dangerous idea"—with the implication that (pseudo)random chance is somehow *safe*. This is obviously in conflict with the thrust of the evolutionary thinking about the origin of humans (e.g., Diamond 1992; Sober 1993; Povinelli 2004).

Why limit oneself to just abstract moral or spiritual dimensions? A part of the localization of essentialism is just a well-disguised geocentrism, which manifests itself in the rejection of space travel (at least insofar it involves humans) and, in more pathological forms, even the rejection of the entire field of astrobiology (Malazita 2017). Why should humans remain on Earth for all times, unless one does believe, with Aristotle, that it is humans' *unique proper place*? In stark contrast to Tsiolkovsky's dictum (often incorrectly cited to refer specifically to Earth) that "[a] planet is the cradle of mind, but one cannot live in a cradle forever," anthropocentrism is a doctrine of eternal infantilism, arbitrary limits, and ultimately closed mindedness.

As Owe (2019) cogently writes:

> It is easier for us to believe that the solutions to the current global problems are external to us, but the external instruments we possess, such as technology and policy, cannot handle these issues unless they are controlled by morally responsible people. The same could be said for major projects along the multiplanetary trajectory, such as terraforming. I would argue that these psychological shortcomings relative to the social and natural environment we have transformed so radically by scientific technology, constitute an internal existential risk. If moral bioenhancement may help us overcome the existential risks the entirety of terrestrial life now faces, then I argue it is something we should seriously consider.

Here, we see an emerging synergy of the great projects of human enhancement and space colonization, as refracted through the same ethical lens. Radical transformative technologies pose new and unexpected problems, which we need to face, and not bury our heads in the sand or treat them as business as usual.

18.4 Grand Irony: Anthropocentrism and the Obsolescence of Humans

As we have seen, the resistance to human colonization of space comes in two flavors, the "theoretical" worries about the Evil Words (they are talking about "colonization", ergo *Moral panic! Moral panic!*) and other postmodern fashionable nonsense, and the "pragmatic" wing which highlights the practical costs and risks of crewed spaceflight. While one can easily identify some rational and well-meaning critics in the latter group (e.g., Weinberg 2013; Szocik 2019b), we can clearly see that this type of criticism is either very temporary, or merges smoothly with a flavor of the anthropocentric cartel rhetoric. The most merciful construal is the "robots are better" argument: we have nothing against space travel as such, we admit its values, but we need to admit and accept that robotic probes and spacecraft are much better, cheaper, and less risky than crewed spaceflight. That is, while there is an intrinsic value to be found in space, in terms of both knowledge and economic resources, we should not send people after it, only robots.

The biggest irony of the "robots are better" argument is obvious to all but the most extreme anthropocentrics still slouching out of the woodwork. If robots are indeed better—in all relevant senses—at space exploration than humans, than robots are arguably better *at everything else* than humans. Not just in playing chess and go, or cleaning carpets, or answering phone calls, or driving trucks, or sex. Space exploration, taken generally and seriously, not just the pitiful baby steps we have made so far, is more complex and involved than any of these human activities. Now, a key question looms here: *Maybe robots are better at writing "philosophical" criticisms of human space exploration as well?* In a Monty-Pythonesque way, it obtains something of a Gödelian flavor.

Space exploration is indubitably both science and technology, especially when we take it in medium- and longer-term sense; and one might argue that it has an important artistic aspect as well (e.g., Malina 1991; Sagan 1994; Ono 2010). So, people who are arguing that humans should relinquish space exploration in favor of robots seem to be arguing that humans could, at least in principle, relinquish any other hitherto human activity. And since robots are already immensely better than humans in so many different activities, from operating heavy machinery to driving in heavy traffic to playing chess and go to operating customer service call centers, after just a blink of an eye in historical and evolutionary terms, there is every reason to think they will be superior in other fields as well before long. So, maybe Karel Čapek—together with some of the modern transhumanist thinkers like Kurzweil or Moravec—was entirely correct that humans should give the way to the next, superior stage in evolution.

Thus, the opponents of human spaceflight who take the "robots are better" road argue that humans are effectively obsolete. This is obviously in contradiction with the anthropocentric attitude of the critics. The wiggle room there is that they assume robotic probes will always and necessary be simple things similar to the *Mariner* and *Pioneer* spacecraft; this is incoherent, since such simple automata could never be competitive with human explorers.[3] In order to be equivalent and superior to humans *as explorers*, future spacefaring robots need to be sophisticated AI systems, capable of independent decision-making, conceiving new research programs in situ, or modifying the ongoing programs in light of contingent circumstances, etc. In light of ever-increasing communication delays between the probe and mission control, their autonomy will be of paramount importance.

In light of all this, it turns out that opponents of human space exploration and colonization will need to resort to some kind of repression, similar to that shown in Sir Ridley Scott's magnificent *Blade Runner* (1982) movie in order to sustain the anthropocentric dogma. Recall that in the movie androids are only allowed in space colonies ("off-world") and prohibited on Earth on the pain of state-sponsored summary execution. Presumably, the prohibition is instituted exactly because androids are both physically and intellectually superior to humans, which is also the reason they have an

---

[3]The incoherency is usually swept under the rug by various rhetorical tricks and appeals to emotions.

artificially shortened lifetime. That form of discrimination is, of course, both immoral (as argued by the movie) *and* stupid, likely to be unsustainable in the long run. It is also hinted—for instance, in the famous monologue of the leader of rebel androids, Roy Batty, to the protagonist, Deckard, on that rainy roof—that at least a part of the androids' superiority *exactly comes from their exposure to experiences in space*, which are favorably contrasted with the narrow horizon of the masses of people remaining on Earth. Do we really wish for a *Blade Runner*-like future of humanity?[4]

18.5 Passing the Buck

The argument for *delaying* space exploration/colonization by humans is occasionally proposed as a "realistic" middle ground between proponents of the cosmic future of humanity and those who would burn them at stake as heretics/witches/Dr. Frankensteins. The argument roughly goes like this: *right now*, it is not a good time for human exploration and colonization, but at some point in the future the time will be right. Recent good representatives in this category are Weinberg (2013), Schwartz (2019), or Szocik (2019a, b); these should be separated from fundamentalist and more toxic criticisms, such as that of Westfahl (1997), Slobodian (2015), Kriss (2017), Torres (2018), or Vidaurri (2019). Here, I wish to argue three points related to the argument for delay: first, that it is in part tautological and vague; second, that it contains conveniently hidden assumptions of rather questionable nature; and third, that it is morally suspicious in the same manner and at least in the same amount as the analogous argument for the delay in climate change actions.

First, since there is no known "natural" timescale for any particular occurrence in human cultural evolution it is very difficult to objectively assess this argument. How big delay is "sufficient"? Or "optimal"? Or "desirable"? Proponents refrain from going open on this central issue. On one hand, the fact that I am writing this now and not working directly on space science and technology could be construed as my support to *some degree* of delay. On the opposite pole, someone who would claim that human space activities should be postponed for a billion years should not be taken seriously in the first place. Usually, we are seeing some vague construals such as "it is a task for the distant future" (Weinberg 2013) or empty platitudes such as "once we have figured out how to make life on Earth work in an environmentally and politically sustainable way" (Williams 2010).

Second, the major questionable assumption of the argument for delay is that we are indeed free to delay it according to our will and intentionality. In other words, the assumption is that there is no well-defined (albeit unknown) window of opportunity for at least the onset of the human colonization of space. In view of the recent

[4]Actually, many in the anthropocentric cartel would prefer futures even *worse* than the *Blade Runner*, where it is implied that at least *some* humans live off-world and participate in the grand exploration and colonization feats. Cf. Klee (2017), Torres (2018) for views that humans could and should not leave Earth at all.

work in astrobiology, for instance, the absence of such window looks highly unlikely (e.g., Chopra and Lineweaver 2016). Everything in evolution, including biological and cultural evolution, seems to occur in windows of opportunity and it is quite possible that, once we blew the first chance—or the first couple of chances—there would not be further ones forthcoming. In particular, it is exactly those extreme global catastrophic risks (one of the major motivations for human space exploration and settlement) which might prevent further possibilities by what bioethicists Ingmar Persson and Julian Savulescu call "the ultimate harm" (Persson and Savulescu 2012). Even a gradual process, such as climate change still is, creates dangerous feed-backs in slowing down economic activity which, ironically, may reduce future emissions and offset some of the natural feed-back cycles (Woodard et al. 2019; Caldeira and Brown 2019), without significant reduction in price of further action.

Third, I submit there is no structural difference between the argument for delaying of space colonization and the infamous argument for delaying climate action (e.g., Lomborg 2001). In both cases, there is a gasping presumption that we shall *inevitably* be in some generalized sense better off in the future than we are now: it is usually said that particular climate change mitigation measures which are expensive today will become much cheaper in the future, so we should undertake them 20 years from now, say. Or 50 years. Or a hundred. Clearly, irresponsibility and naivete of such a view defy belief. Adverse consequences of climate change are what threatens—to a large degree—global economic prosperity today, which is a precondition for the decrease in price tomorrow. While the question which model of temporal discounting best represents reality still lacks clear answer, nobody seriously doubts the necessity of such discounting in the future studies. Neither are catastrophic consequences of the delayed climate change action a secret anymore (Mengel et al. 2018).

Why should it differ for human space exploration and colonization? Insofar as space colonization is regarded as decreasing the risk of human extinction and increasing the chance of preserving human values, the parallel holds. While it is conceivable that, *barring crises and catastrophes*, direct costs of human space colonization projects will decrease in the future, one is still entitled to the same skepticism toward this economic reasoning as in the case of climate change above. Some of the skeptics are quite aware of that (e.g., Szocik 2019b). For all the noble talk about ecological awareness, social justice, and other shibboleths, the cartel is just a bunch of early humans afraid to go on two legs, afraid to go into a dark cave, afraid to start a fire, afraid of a storm, or of a gust of wind, or of a tiger or a bear. And here we come upon the central cowardice of anti-space skepticism: (mis)using this primal fear to justify procrastinations and delays to do what is both useful and right.

18.6 Instead of Conclusions: Say No to the *Blade Runner*

We are living in perhaps the riskiest epoch of *Homo sapiens*'s existence: in addition to classical global catastrophic hazards such as asteroidal/cometary impacts or supervolcanic eruptions, there is a host of new and unprecedented risks stemming from misuse of powerful technologies, like nuclear power, biotechnology, nanotechnology, and artificial intelligence (e.g., Posner 2005; Bostrom and Ćirković 2008). Studying and mitigation of these risks is a global priority in both moral and policy-related terms (Bostrom 2013). Both human bioenhancement and human colonization of space—starting, of necessity, with colonization of the Solar System—are ways to address both the research and practical aspects of this priority. Perhaps there are additional multiple ways of achieving the same goal; it is, however, incumbent on those who are opposed to these two compatible strategies to show that there are those additional ways and that they are feasible. However, a rather clear pathway—and not very novel or radical—is to extend (enhanced) human presence beyond Earth and beyond those hazards and catastrophes which are, by their physical nature, limited to the Earth system. All other benefits come on the top of that and as a bonus.

While human colonization of the Solar System and, ultimately, wider universe sounds utopian in many circles, a proponent should not dissipate energy in combating the moniker. Instead, it should be endorsed, for the wise reason the great designer, architect, and visionary Richard Buckminster Fuller has put in the title of one of his books, originally published in 1963: it is *either utopia or oblivion* (Fuller 2008). Of course, even earlier, H. G. Wells warned that the history of humanity is "a race between education and catastrophe" (Wells 1920). And education is hardly winning.

To the best of our present-day reasoning about existential risks, immunity to such risks (with, perhaps, one or two exceptions) follows from a successful human colonization of the Solar System; let us therefore concentrate on this particular form of colonization of the universe, while noticing that, of course, there is a meaningful sense in which subsequent expansion into the Galaxy is much better still. It is clear why this is so: risks generically associated with the Earth system (asteroid impacts, supervolcanoes, global climate change) will be transcended by expansion of the spatial domain filled with humans and their values. The risks associated with our cosmic environment (giant Solar flares, close supernovae/GRBs, encounters with dense molecular clouds) will be overcome through increased understanding of their astrophysical mechanisms and the construction of protective measures using the same technologies necessary for undertaking the colonization in the first place. The measures to be undertaken to mitigate more extreme cosmic risks are tightly connected—as far as we can see today—with human control and management of the resources of the Solar System. For instance, this applies to constructing swarm shieldings for Earth and other local habitats in a case of predicted close supernova or a Galactic GRB (Ćirković and Vukotić 2016). Similar reasoning, however, is valid for other existential and global catastrophic risks. Risks associated with ecological imbalance and resource depletion will be transcended in an obvious way—by transcending the boundaries of the unique terrestrial ecosystem and vastly increasing the resource base (Zubrin 1999; Cockell 2007; de Grass Tyson 2012).

Even the two risks which are likely to stay with us in some form, cataclysmic warfare and technologically-enhanced totalitarianism, seem less dangerous in the extended domain of colonized space than on Earth. While warfare will remain possible unless the outcome of social and political evolution is a singleton, it will be less likely to destroy *all* multiple ecosystems of an expanding humanity; whereas on Earth, one nuclear or supervolcanic winter or a created superpathogen would be enough. And the diversity of habitats will necessarily increase the diversity of modes of thinking, something by definition antithetical to all totalitarian projects; in addition, the latency imposed by large distances in space might be a further obstacle to enforcement by any centralized authority.[5]

Conclusions about ethical value of space exploration and colonization hold irrespectively of one's moral preferences; in particular, this is true irrespective of whether one trusts consequentialist/utilitarian intuitions in estimating our "cosmic endowment" (Bostrom 2003) in terms of value in information, (post)human lives, or any other way. Quite to the contrary, this is a perfect example of an issue capable of uniting people of various philosophical, ideological, and ethical bents in the same manner as it can unite different other seemingly opposed strands of the present-day cultural fabric. In one sense, the answer to questions often posed about the purposelessness of modern-day humans—e.g., in critical reactions to Pinker's (2018) book—is exactly safeguarding of human species and human accomplishment against risk from existential catastrophes, or the "ultimate harm" of Persson and Savulescu (2012).

Even if no such global cataclysm occurs, humans could become obsolescent through evolutionary means exactly *because of our relinquishment of space exploration and colonization*. That people talk less about slow dysgenic pressures does not make them any less real. In the long run, there is no difference between isolated planet Earth and an isolated island like Rapa Nui: *everything* which has happened in any isolated local system like Rapa Nui *will necessarily* happen on Earth in the fullness of time. Of course, one could negate the temporal qualification by appeal to the Solar evolution and the accompanying increase in luminosity; but even in the most pessimistic accounts, there is still about 1 Gyr, before the Earth becomes uninhabitable, which is more than enough time to play out all options of human history (or tragedy, if we willingly accept the Rapa Nui approach and neglect the wider universe). Of course, the very same projects attacked by the cartel, human bioenhancement and space colonization, could mitigate these dysgenic processes; it is, therefore, quite natural that they are developed in parallel (cf. Szocik and Braddock 2019).

Not that there is any real trend in favor of space colonization either in our navel-gazing civilization; as warned by Nunn et al. (2014), we are already in a highly disturbing situation in which the annual cost of obesity in the USA is about *12 times larger* (!!!) than the annual cost of the national space research and exploration programs. That the anti-space sentiment has become completely normative is obvious to such extent that even a self-professed believer in science and progress like Steven Pinker calls space colonization a "nonstarter," proposed by "naifs" [Pinker 2018, p. 390]. In light of this reality, all this frantic anti-space opposition looks more and more like flogging of a dead horse. The work of the anthropocentric cartel, as exemplified by some of the works quoted above, has been remarkably successful thus

[5] However, see Cockell (2008).

far; not that much by directly obstructing our progress in science and technology, but in inventing and advertising various justifying narratives which divert both attention and resources from our global priorities in Bostrom's sense (Zubrin 2011; Autri and Skran 2019; Ćirković 2019). And there is absolutely no warrant that all this distraction and diversion will not turn into something much worse, similar to what has happened with many strands of anti-nuclear movement or deep ecology/ecoterrorist groups.

One should be crystal-clear about one thing: contemporary anthropocentrism is a deep and insidious form of anti-humanism. Problems, difficulties, calamities, even catastrophes and existential risks facing humanity are always welcomed by the cartel, usually as a motivation for smug, self-congratulatory, and self-indulgent messages of pessimism. It is either the need of "humbleness before the divine," or "the tragic view of the human nature," or even "respect for untamed wilderness of other worlds," or some similar hollow ideological mantra of the day. The obvious naivete of these and of some of the actors parroting them should not blind anyone in thinking they are not dangerous in the extreme. On the contrary, by covering up the detrimental loss of adventuring and exploratory spirit, as well as the expansion drive, by superficially sophisticated arguments, the academic wing of the anthropocentric cartel inflicts immeasurable damage to the welfare and the very existence of future generations of humanity.

Acknowledgements Foremost thanks go to the Editor, Konrad Szocik, for his diligent work on promoting serious thought about cosmic futures of humanity, as well as to Vojin Rakić for sharing his overarching vision of modern, future-oriented bioethics which has been a constant and enormous inspiration. The author also wishes to express gratitude to Richard Cathcart, Andrea Owe, Branislav Vukotić, Ksenija Petrović, Zoran Živković, Jelena Dimitrijević, Anders Sandberg, George Dvorsky, John Smart, Slobodan Perović, Karl Schroeder, Aleksandar Obradović, Milan Stojanović, S. Jay Olson, Goran Milovanović, Srdja Janković, and the late Damian Veal for stimulating discussions on the subject matter of this essay. The author has received financial support from the Ministry of Education, Science and Technological Development of the Republic of Serbia through the projects #ON176021 and #ON179048.

References

Agar, N. (2010). *Humanity's end: Why we should reject radical enhancement*. Cambridge: MIT Press.

Arendt, H. [1963] (2007). The conquest of space and the stature of man. *New Atlantis*, Fall issue, 43–55. https://www.thenewatlantis.com/publications/the-conquest-of-space-and-the-stature-of-man. Last accessed August 15, 2019.

Autri, G., & Skran, D. (2019). Humanity Can't afford to keep space pristine. *Foreign Policy*. https://foreignpolicy.com/2019/10/30/nasa-space-exploration-mars-city-musk-humanity-keep-space-pristine/. Last accessed November 15, 2019.

Bernstein, L. (2002). *The greatest menace: Organized crime in Cold War America*. Boston: UMass Press.

Bostrom, N. (2003). Astronomical waste: The opportunity cost of delayed technological development. *Utilitas, 5,* 308–314.

Bostrom, N. (2013). Existential risk prevention as global priority. *Global Policy, 4,* 15–31.

Bostrom, N., & Ćirković, M. M. (Eds.). (2008). *Global catastrophic risks*. Oxford: Oxford University Press.

Caldeira, K., & Brown, P. T. (2019). Reduced emissions through climate damage to the economy. *Proceedings of the National Academy of Sciences, 116,* 714–716.

Chopra, A., & Lineweaver, C. H. (2016). The case for a Gaian bottleneck: The biology of habitability. *Astrobiology, 16,* 7–22.

Ćirković, M. M. (2014). The greatest gamble in history. *Collapse, 8,* 385–415.

Ćirković, M. M. (2017). Enhancing a person, enhancing a civilization: A research Program at the intersection of bioethics, future studies, and astrobiology. *Cambridge Quarterly of Healthcare Ethics, 26,* 459–468.

Ćirković, M. M. (2019). Space colonization remains the only long-term option for humanity: A reply to Torres. *Futures, 105,* 166–173.

Ćirković, M. M., & Vukotić, B. (2016). Long-term prospects: Mitigation of supernova and gamma-ray burst threat to intelligent beings. *Acta Astronautica, 129,* 438–446.

Cockell, C. S. (2007). *Space on earth: Saving our world by seeking others*. London: Macmillan.

Cockell, C. S. (2008). An essay on extraterrestrial liberty. *Journal of the British Interplanetary Society, 61,* 255–275.

Comfort, N. (2019). How science has shifted our sense of identity. *Nature, 574,* 167–170.

de Grass Tyson, N. (2012). *Space chronicles: Facing the ultimate frontier*. New York: Norton.

Diamond, J. (1992). *The third chimpanzee*. New York: HarperCollins.

Frayn, M. (2006). *The human touch: Our part in the creation of a universe*. London: Faber and Faber.

Fukuyama, F. (2002). *Our posthuman future: Consequences of the biotechnology revolution*. New York: Farrar, Straus & Giroux.

Fuller, B. R. (2008). *Utopia or oblivion: The prospects for humanity*. Zürich: Lars Müller.

Kass, L. R. (2007). Keeping life human: Science, religion and the soul. Wriston Lecture, Manhattan Institute for Policy Research. https://www.manhattan-institute.org/html/2007-wriston-lecture-keeping-life-human-science-religion-and-soul-8894.html. Last accessed 1 November, 2019.

Klee, R. (2017). Human expunction. *International Journal of Astrobiology, 16,* 379–388.

Kriss, S. (2017, March). Think twice about escaping earth to an exoplanet. *The Atlantic*. https://www.theatlantic.com/science/archive/2017/03/space-travel-wont-save-you-from-capitalism/518853/. Last accessed May 27, 2019.

Lomborg, B. (2001). *The skeptical environmentalist: Measuring the real state of the world*. Cambridge: Cambridge University Press.

Malazita, J. W. (2017). Astrobiology's cosmopolitics and the search for an origin myth for the Anthropocene. *Biological Theory, 13,* 113–120.

Malina, R. F. (1991). In defense of space art: The role of the artist in space exploration. *International Astronomical Union Colloquium, 112,* 145–152.

Mazlish, B. (1967). The fourth discontinuity. *Technology and Culture, 8,* 1–15.

Mengel, M., Nauels, A., Rogelj, J., & Carl-Friedrich Schleussner, C.-F. (2018). Committed sea-level rise under the Paris Agreement and the legacy of delayed mitigation action. *Nature Communications, 9,* 601 (10pp).

Nunn, A. V. W., Guy, G. W., & Bell, J. D. (2014). The intelligence paradox; will ET get the metabolic syndrome? Lessons from and for Earth. *Nutrition & Metabolism, 11,* 34 (13pp).

Ono, A. (2010). Art: Art as a psychological support for the outer space habitat. *Lunar Settlements,* 197–214.

Owe, A. (2019). *Environmental ethics in outer space: A macrostrategic space journey through cosmism, posthumanism and moral enhancement*. Master's Thesis in Development, Environment and Cultural Change, University of Oslo. https://www.duo.uio.no/bitstream/handle/10852/69331/1/Owe-Andrea-15-05-2019-Environmental-Ethics-in-Outer-Space-MAthesis-.pdf.

Persson, I., & Savulescu, J. (2012). *Unfit for the future: The need for moral enhancement*. Oxford: Oxford University Press.

Pinker, S. (2011). *The better angels of our nature: Why violence has declined*. New York: Penguin Books.
Pinker, S. (2018). *Enlightenment now: The case for reason, science, humanism, and progress*. New York: Penguin Books.
Posner, R. A. (2005). *Catastrophe: Risk and response*. Oxford: Oxford University Press.
Povinelli, D. J. (2004). Behind the ape's appearance: Escaping anthropocentrism in the study of other minds. *Daedalus, 133,* 29–41.
Sagan, C. (1994). *Pale blue dot: A vision of the human future in space*. New York: Random House.
Schwartz, J. S. (2019). Space settlement: What's the rush? *Futures, 110,* 56–59.
Slobodian, R. E. (2015). Selling space colonization and immortality: A psychosocial, anthropological critique of the rush to colonize Mars. *Acta Astronautica, 113,* 89–104.
Sober, E. (1993). *Philosophy of biology*. Boulder: Westview Press.
Sparrow, R. J. (2013). The perils of post-persons. *Journal of Medical Ethics, 39,* 80–81.
Szocik, K. (2019a). Should and could humans go to mars? Yes, but not now and not in the near future. *Futures, 111,* [illegible].
Szocik, K. (2019b). Human place in the outer space: Skeptical remarks. In *The human factor in a mission to mars* (pp. 233–252). Cham: Springer.
Szocik, K., & Braddock, M. (2019). Why human enhancement is necessary for successful human deep-space missions. *The New Bioethics, 25,* 295–317.
Torres, P. (2018). Space colonization and suffering risks: Reassessing the 'maxipok rule.' *Futures, 100,* 74–85.
Vidaurri, M. (2019, October 21). What happens when you leave empty seats at the table? *The Space Review*. https://www.thespacereview.com/article/3817/1.
Weinberg, S. (2013). Response: Against manned space flight programs. *Space Policy, 29,* 229–230.
Wells, H. G. (1920). *The outline of history: Being a plain history of life and mankind*. New York: Macmillan.
Westfahl, G. (1997). The case against space. *Science Fiction Studies, 24,* 193–206.
Williams, L. (2010). Irrational dreams of space colonization. *Peace Review, 22,* 4–8.
Woodard, D. L., Davis, S. J., & Randerson, J. T. (2019). Economic carbon cycle feedbacks may offset additional warming from natural feedbacks. *Proceedings of the National Academy of Sciences, 116,* 759–764.
Zubrin, R. (1999). *Entering space: Creating a spacefaring civilization*. New York: Jeremy P. Tarcher/Putnam.
Zubrin, R. (2011). *Merchants of despair: Radical environmentalists, criminal pseudo-scientists, and the fatal cult of antihumanism*. New York: Encounter Books.

Chapter 19
Religion as Human Enhancer: Prospects for Deep Spatial Travel

Lluis Oviedo

Abstract Plans for long-time space missions have to deal with many factors and need to prepare crews able to cope with very difficult conditions. The present paper examines the case for religion as a possible assistance for human enhancement in such harsh circumstances. To that end, analysis of what religion can really enhance is required, based on historical record and experience, and what would be expected to be enhanced in a spatial mission, which could match with religious deliverance. However, even if the argument can be advanced, some serious problems arise and invite caution, at the time that the case for religion can be made probably from a different angle.

Religion has been perceived in several historical forms as an institution aimed at enhancing specific human traits. This is quite obvious in traditions whose scope was to prepare and train their followers to free body and soul from what was perceived as their worse tendencies and conditions. 'Virtue' has been pursued as a main goal, including an ability to dominate or control passions, but even elementary needs, like nourishment, sleeping, or fatigue. Ascetic ideals have been paramount to different religious traditions, aimed at forming very resistant characters, able to fight in the struggle against temptation and evil, or simply to overcome passions and reach a deep state of calm and indifference before life vagaries and changing emotions.

From the described point of view, religious traditions are not strange to ideas of 'human enhancement', but the opposite is true: possibly what is now emerging is the concurrence between old and new means to enhance human nature, a perception that increases when the enhancement looks for moral disposition, to render all us 'better persons'. The contrast arises now between sacred and secular approaches to enhancement; or between the traditional ways to dominate and tame body with its passions and needs, and new technologies applied to improve our faculties and endow us with greater resistance and abilities. The contrast between the old and the new could become somewhat misleading, since religious forms evolve and new needs

L. Oviedo (✉)
Antonianum University, Rome, Italy
e-mail: loviedo@antonianum.eu

K. Szocik (ed.), *Human Enhancements for Space Missions*, Space and Society,
https://doi.org/10.1007/978-3-030-42036-9_19

arise demanding new interventions and focused assistance. Religious forms appear in many settings as ‘coping strategies’ and assisting in ‘human flourishing’, an idea that could imply many things, including some ideal at excellence—moral or intellectual (Van Eyghen 2019). This is possibly not the same as ‘human enhancement’ in its more extreme and futuristic technical expressions, but just in a very broad sense. What would signify the difference is the contrast between means guided by religious motivations, or assisted by transcendent impulses; and those that resort to technical devices or to genetic and neurological intervention, which rely mostly on scientific advances, and less on a traditional wisdom and learned practice. However, it would be wrong to represent both approaches as just antithetic or concurring: why not conceiving them as collaborative and coordinate them in order to really improve human condition? Does the religious enhancement programme exclude necessarily the secular technical one?

Current research in the study of religion, its functions and performances in advanced societies or the conditions characteristic of secular cultures point to some aspects that render religion still meaningful or at least useful for some population segments and for some cases beyond or beside usual secular management. Religion would work in those cases as an instance managing whatever other social systems do not achieve or cannot manage by their own specialized means. Where uncertainty and indeterminacy are the rule—as could be the case in extreme situations or where too many factors can go wrong—religious faith and its means to cope become more needed and applicable. From this point of view, religious functions would be always required as a necessary supplement when we cannot eliminate completely any contingency or uncertainty: when technical solutions are no longer available, then religious means become convenient, just in case.

Focusing more on the central topic of this paper, at least two main issues need to be reviewed when trying to discern religion’s capacity to enhance human crews in a long-term space mission. The first one concerns which aspects or traits could—in due case—enhance religious faith and practice in that extreme context, since that function is clearly limited or applies just to a sub-set. The second issue is about how such enhancement could work in that situation and for that long time. To answer these issues, it is necessary to start with a general review about what are actually the possibilities religions provide to enhance specific dimensions—not all—in human mind and behaviour, for later trying to apply these abilities or performances to the expected conditions in deep space travel, once such conditions and the required needs are better described and assumed inside a systemic and multi-dimensional programme that includes human aspects, since those sent so far in deep space are humans and not robots or chimpanzees.

19.1 What Can Religion Enhance?

The question that motivates this first step in our research is not rhetoric, but is supported by long traditions and experiences, reaching to many centuries of practices trying to control one's own body and its impulses, as part of a programme aimed at providing 'salvation' through a more conscious dominion of what could be perceived as negative and opposed to the ways to reach sanctity or quietness, depending upon tradition's own goals. Such programmes have given place to elaborated techniques that have pursued a purification or limitation of passions or emotions deemed as dangerous or evil, as means to reach perfection or to achieve some higher level of illumination and internal peace. The German sociologist Max Weber described at the beginning of twentieth century those strategies as an expression revealing 'religious rationalization', intending a programme to better adjust the available means to the goals conceived in most evolved religious traditions (Weber 1920). Indeed, that view entails an evolutionary schema that follows a plan to survive and adapt to new conditions.

All those programmes—ascetical and mystical—can be seen as something out of touch with our current culture, something belonging to a different world and time, but nevertheless the idea at its ground continues to be meaningful: religious beliefs and practices often can assist in strengthening human character, and in this way in preventing its worst developments and tendencies. This same dynamic applies to the concept of 'religious coping', a well-researched programme with many applications and fields based on the ability of religion to restrain bad effects or to limit the reach of negative events and experiences. Then, in line with the strengthening idea, religions often help to educate and form personal traits that render people more prone to behave in positive and constructive ways, avoiding selfish and toxic attitudes.

Here we are talking about religions as 'salvation schemas', as a distinctive class, distinct from other perceptions that understand religion as a fuzzy spiritual experience closer to the aesthetic one. Salvation is not just about feeling a beautiful bliss or internal peace, it is about being assisted to get rid from the worst that haunts human nature, something which can be extended to our relationships—always threatened by evil interests—and even to social bodies or communities as a whole, which can be subject to healing and transforming grace to free them from the most disintegrative or entropic tendencies. Theologians would remind us that salvation is a much a broader concept, and its main aim is to restore broken human relationships with God. In that sense, it would be reductive to render it just in instrumental terms; this idea makes sense and becomes effective only if salvation is related to God's mystery of love.

If religion is seen as a salvation instance, or a structure aimed at freeing from evil and every negativity, then, one of its main functions will be to assist humans to resist that evil and to improve in their virtue or in their capacity to do good and to live a fulfilling life. Now, how does that noble programme find operative expression in real life and in the present time? To what extent what salvation religions proclaim becomes efficient means to enforce better living standards and to overcome negative limits and hindrances? To answer these questions, we need to look the available literature and recent research showing how specific religious forms help to improve mental health (for systematic reviews and metanalyses on a huge body of published literature: AbdAleati et al. 2016; Agli et al. 2015; de Bernardin Gonçalves et al. 2017); to cope with stress (Park 2005; Folkman 2011); or to deal with solitude and other negative affects (Whitehead and Bergeman 2012). The idea is similar in all the described cases: religion and spirituality moderates negative psychological effects providing greater balance, motivations or strength to overcome or to elaborate those adverse experiences and to avoid their worse consequences. Even if that result is limited and not too robust when compared with control groups, the broad consensus is that religion and spirituality work at that level and become a reliable means to cope with adverse situations and to resist temptations to avoid negative experiences.

A different orientation when observing religion as a human enhancing disposition looks more to positive building character or the ability to render persons better, stronger and more confident or 'in control' of their lives and circumstances. This is an old story that finds several echoes in our days: asceticism has been paramount to all Post-Axial religions, and its aim has been always to control one's own body, reactions and emotions, to achieve a state of perfection or closer to the ideal that each religion has fixed as the most desirable state. To what extent traditional asceticism can be assimilated to a form of human enhancement is a matter of contention, since the goals and the means could appear as too distant and their meaning very different. However, asceticism as a form of gaining greater control over one's own body and capacities appears quite close to what is looked at new technical means. After all, forms of intense training include several exercises that clearly reflect what in former times was seen as ascetic practices. This is still more clear regarding specific diets, body regulation and psychological exercises to render our body and mind ready for the task or competence that is ad hoc programmed.

Obviously, the question now is not to what extent the traditional techniques were more or less efficient, but whether they can compete today with the new secular ones, which are much more rational, as conceived to attain the aimed goals for higher performing in several activities, both physical and mental. Indeed, several voices see in this trend a clear secularization of traditional religious inspired asceticism, becoming today secular means to achieve better health, strength or greater performance in sports or other activities that require higher concentration and better physical or mental shape (Hoverd 2005; Peeters et al. 2011; Petersen 2019).

A related factor in the religious human enhancement project points to implementing social skills or a more convivial attitude, needed to live in communities and to collaborate in shared enterprises that require higher coordination and the downplaying of selfish behaviour. This has been a very discussed issue in the recent wave of the scientific study of religion, and some consensus recognizes that such ability belongs only to most evolved religious forms, and still more to those that insist on compassion and other's concern as part of their religious precepts (Galen 2012). In any case, if religious faith and practice manages to implement a greater interest and attention towards others and renders believers better at sharing and collaborating, then we can assume those virtues as a real enhancement, at least in the sense usually attributed to the concept of 'moral enhancement'.

Summarizing the reported data, several religious traditions, assuming a format of 'salvation religions', provide interesting and useful motivations and impulses to improve human conditions. That function can be conceived, first, as coping strategies to confront stress, psychological pain, and negative or exhausting events; and second, as a stimulus for virtue through asceticism and restraint, which allows the development of superior capacities or a training for higher spiritual achievements. In this last capacity, should be included a force to engage in more prosocial attitudes, in line with a programme of 'moral enhancement'. Religion, in other words, becomes a transforming force for humans and groups, through means that cannot be completely naturalized—otherwise they will stop being religious—and can be applied to different settings. This last point is discussed by those pleading for a 'religious naturalism' (Leidenhag 2018); however, it is still to be shown that such versions are able to perform religious forms as do supernatural or transcendentalist ones (Leidenhag 2018). From a historical perspective, such transforming force has helped many populations under high stress to survive and adapt, to avoid desperation and to motivate new strength to move on and to explore new possibilities. It would be much harder for those groups without that resource, even if this counter-factual statement becomes very hard to test.

19.2 What Needs to Be Enhanced in a Deep Space Mission?

Some studies have already attempted to deal with human enhancement in spatial missions (Szocik and Wójtowicz 2019; Szocik et al. 2019). Their focus is placed on physical aspects, like how to deal with persistent radiation, absence of gravity or chronicle fatigue. More studies have reported on the psychological circumstances which would mostly influence the life and living conditions in a long-time mission away from the Earth orbital area, especially in a planned mission to Mars (Kanas 2011, 2014; Goemaere et al. 2016). When we get a better view on those circumstances, then it would be easier to design strategies to cope with them or to train crews to adapt to such harshness and precarious conditions. Even if the current research is more hypothetical, some scenarios can be foreseen quite accurately and invite more engagement and foresight when preparing a mission to a distant objective able to deal with most of the details and circumstances that determine such a programme.

Obviously, the most urgent needs regard physical resistance and strength; this is more apparent when the extreme conditions and dangers that threaten the astronauts—radiation, absence of gravity—are considered. Then, other factors appear as important in such a mission when its success has to be ensured. Preparing a long-standing space travel would require caring for the cognitive and emotional factors that could influence the success or the failure of such a mission; this involves building cognitive abilities to deal with problems and to look for fitting solutions in extreme and highly difficult circumstances. However, in my opinion, and after reading reports describing stressors and potential harshness, the features that need to be improved, besides the best cognitive and emotional levels, could be the following:

- A strong sense of meaning and purpose that can motivate a mission with unpredictable outcomes and many hardships and dangers. This provision can be made in many ways, since that mission could be highlighted and explained in several manners: resorting to human progress and expansion, to scientific enquiry, or to a spirit of conquest and self-affirmation. However, at a deeper level, and when risks increase in an incontrollable way, the question arises about more resistant and far-reaching meaning systems, which can account for high contingency and even for failure, and being supported by strong hope, perhaps beyond actual disasters and technical means.
- A psychological ability to stand adversity and to [illegible] unexpected and damaging situations, without feeling defeated and without too many resources, or unwilling to pursue and try again. Since failures and unexpected bad events usually trigger states of depression, and those would endanger a mission and even the general health of entire crews, all the means that could help to cope with such situations and to hold a positive and hopeful feeling should be promoted and nourished.
- Special virtues for that extremely difficult environment, which could include temperance to control of nourishment and other physical needs; prudence in taking decisions or in controlling one's own will; patience before adversity or negative inputs; and fortitude or endurance in those long times with many challenges and trials. It is not an easy task to educate or train those virtues playing a positive role, but they should be part of every programme trying to improve the conditions for a crew in a long-time space travel.
- Social conditions, including empathy, compassion, collaborative disposition and altruistic attitudes are also important. Such dispositions are essential in a mission in which team-work and collaborative abilities, even self-sacrifice, are expected as a clue for survival in extreme situations. The available studies show how such dispositions depend in great measure upon personality and deeply rooted inclinations. It is not easy to train or implement that disposition, and possibly more needs to be done to 'enhance' moral attitudes, like those here described, just through technical means.

The general lesson to be learned after this short review is that these human attitudes are important and needed when planning a long-time space mission, but on the other hand, they appear as hard to implement by sheer technical means, as has been the dream of 'human enhancement' programmes. The big issue is to what extent—if such attitudes or features need to be implemented—we must look somewhere else, beyond psychological or hard training technics, in order to assist in such programme and to provide the right supplement required to cope and to improve limited human capacities when they will be pushed to their limits.

19.3 Open Questions on Religion as Human Enhancer in Spatial Missions

It is now relatively easy to match the first and the second sections in the present analysis, and to link the religious enhancers—coping, virtue building and prosocial motivation—with the perceived needs in that long-foreseen travel: meaning; endurance before adversity; special virtues; and prosocial attitudes. Religion appears as a candidate to cover most of these psychological requirements, provided that it will work in a smooth way, without severe secondary effects, and can deliver what in theory is promised or grounded on historical practice and records.

The last statement opens a first question and invites a more nuanced view, since the available evidence does not ensure an efficient application in most cases of religious resources to every coping eventuality. The second issue is more difficult still; indeed, any attempt to resort to religious means in an instrumental way appears as doomed to failure and deeply flawed, except that astronauts would be motivated to regard the space mission itself as a religion, and hence, religion might be instrumentalized for space missions. The third question to deal with concerns the way religion can compete with current and foreseen technical means of human enhancement, some still in the way or being expected to develop in the next few years. Rightly positing and addressing these questions is very important if we want to make sense of religion in that special and uncertain context.

Regarding the first issue, about the efficiency and applicability of religion as human enhancer in the context of a long-term spatial mission, scepticism can be more than justified when the available literature in religious coping and religion as a source of human flourishing is closely examined. Indeed, the positive effect religious beliefs and practices promises is reduced to a moderate—and less robust—effect, when taking into account control groups that do not resort to religious means to cope and to grow. The general impression is that religion triggers more good than harm for those holding it, but that it becomes hard to measure and ascertain to what extent that positive influence can be extended and be applied to different contexts, far beyond those that have been tested until now. It is out of discussion the therapeutic effect that religious beliefs and practices can exert, but at the same time, in other cases, religions can trigger undesired negative and even pathological attitudes, and it depends upon many factors that it is not always possible to control or manipulate. Religion's essence is to resist being manipulated or used for other scopes than its own goal: salvation through reference to the divine. Then, a second concern regards the asceticism present in several religious traditions and liable to the enhancement of virtuous characters and greater endurance. Those practices appear often as a training that is just motivated by religious ideas, but that could be easily disentangled from that context to become just secular practices aimed at empowering and endowing subjects with some virtues or abilities. It happens when selected people are prepared for some mission or professional activity that requires special resistance, attention or endurance. Religious faith would clearly add more motivation in those cases—like a sort of 'vocation'—but would not be that unavoidable element.

The second perplexity has already been indicated: religion defuses instrumental use or application by its own nature, since its aim is to connect with the divine as the highest principle, and hence it should avoid serving other scopes or interests. This issue is far reaching, since modern times—and even before—people have tried to understand and legitimate religion by its benefits or functions, both for individuals and for societies alike, rendering religion an instrument for sheer immanent goals, like moral motivation or social integration. However, such a strategy leads in the limit to paradoxical blind ends, since religion—at least the most evolved forms—becomes useful when it is perceived as avoidable or not necessary for the social order or to build a meaningful personal existence. The case to apply religious means to cope with hard situations or to improve character and endurance would appear as deeply flawed when it is not kept at the same time the last meaning or goal of religious faith, one that completely transcends any human expectation or limited interest. However, a way out from this dilemma is to distinguish between the primordial goal of a religious faith, and its positive consequences for community and personal life: looking for these positive effects does not mean to downplay the main scope of a religion.

A further difficulty arises from the plurality of religious choices available. It could be possibly sorted out if all a mission's selected crew already belongs to the same religious tradition. Indeed, if only some crew members would profess a religious faith, it could become a source of conflict and tension, except in cases where religious beliefs invite inclusivism and openness. In any case, such a selected group would be not trained into religious practices just for the spatial travel and its scope, since it would be already available among a population of religious subjects; at most they would need a particular supplement, as those who feel called to become monks need and go through a special formation time. However, this choice poses further questions about selection polices and possible discrimination criteria.

The third issue arising in the attempt to resort to religion as human enhancer has been insinuated: it is far from clear to what extent such a proposal could be fitting into a broad schema of foreseeable practices for human enhancement, nowadays gaining big predicament in the media and even in academic circles, often under the label of 'transhumanist' programmes. Indeed, those programmes promise much more for much lower cost, without hard training or accomplishing repetitive and boring rituals with all their temporal cost. When new technical means would be applied, just some genetic editing, neuro-friendly devices, or just some last-generation drugs or synthetic hormones and neurotransmitters, would be enough to provide all that is needed for an effective human enhancement, even the moral sort. But we cannot do this now. For this reason, it seems that using religion is still preferable. In other words, religious enhancers could supply some desired effects while we have not yet achieved the technical level or mastery to deliver those expected results. Perhaps at some time, we could get rid of those other 'primitive' means, since we get now much more efficient and updated ones. In my opinion, possibly a better answer is that both means or enhancing systems can collaborate and contribute to the success of that dangerous and open-ended space mission, not competing but collaborating with each other.

19.4 Concluding Remarks

This chapter has made the case for religion as a human enhancing system able to improve psychological and even physical features in an astronaut crew that would have to undertake a long and far from our planet mission. Since religion is about tackling contingency or indetermination that goes too far or is very hard to manage by other means, religious faith and practice seems to be a good contribution when planning such an uncertain and risky mission. What is less clear is that religious means can serve for just specific needs in enhancing features that would be very helpful in such harsh circumstances. Religious faith, at least its higher evolved expressions, is like a 'package deal': you cannot take just some elements or pieces at your choice and convenience, and leave aside others; it is about 'taking or leaving' the entire pack, it is about transforming one's own life and going into a salvation schema that resorts to transcending means, or what Christians traditionally call 'grace' as a helping force working through a full transformation or what is called a 'conversion'. In that sense, the only thing we can predict is that possibly religious faith lived in a balanced and deep way by some members can become an asset when selecting that crew, and could translate into a better, more meaningful and hopeful attitude, something that would mean a clear difference in that difficult and unpredictable conditions.

References

AbdAleati, N. S., Zaharim, N. M., & Mydin, Y. O. (2016). Religiousness and mental health: Systematic review study. *Journal of Religion and Health, 55*(6), 1929. https://doi.org/10.1007/s10943-014-9896-1.

Agli, O., Bailly, N., & Ferrand, C. (2015). Spirituality and religion in older adults with dementia: A systematic review. *International Psychogeriatrics, 27*(5), 715–725. https://doi.org/10.1017/S1041610214001665.

de Bernardin Gonçalves, J. P., Lucchetti, G., Menezes, P. R., & Vallada, H. (2017, October 19). Complementary religious and spiritual interventions in physical health and quality of life: A systematic review of randomized controlled clinical trials. *Plos One*. https://doi.org/10.1371/journal.pone.0186539.

Folkman, S. (2011). *The Oxford handbook of stress, health, and coping*. Oxford, New York.

Galen, L. W. (2012). Does religious belief promote prosociality? A critical examination. *Psychological Bulletin, 138*, 876–906.

Goemaere, S., Vansteenkiste, M., & Van Petegem, S. (2016). Gaining deeper insight into the psychological challenges of human space flight: The role of motivational dynamics. *Acta Astronautica, 121*, 130–143.

Hoverd, W. J. (2005). *Working out my salvation: The contemporary gym and the promise of "self" transformation*. Oxford: Meyer & Meyer Sport.

Kanas, N. (2011). From Earth's orbit to the outer planets and beyond: Psychological issues in space. *Acta Astronautica, 68*, 576–581.

Kanas, N. (2014). Psychosocial issues during an expedition to mars. In *The new martians: A scientific novel*. Heidelberg, Dordrecht: Springer. https://doi.org/10.1007/978-3-319-00975-9_2.

Leidenhag, M. (2018, May 28). Religious naturalism: The current debate. *Philosophy Compass*. https://doi.org/10.1111/phc3.12510.

Park, C. L. (2005). Religion as a meaning-making framework in coping with life stress. *Journal of Social Issues, 61*(4), 707–729.

Peeters, E., Van Molle, L., & Wils, K. (Eds.). (2011). *Beyond pleasure: cultures of modern asceticism.* New York, Oxford: Berghahn Books.

Petersen, A. K. (2019). Training and secession suspended between individuality and collectivity. *Numen, 66*(5–6), 465–498.

Szocik, K., & Wójtowicz, T. (2019). Human enhancement in space missions: From moral controversy to technological duty. *Technology in Society, 59*, 101156. https://doi.org/10.1016/j.techsoc.2019.101156.

Szocik, K., Campa, R., Rappaport, M. B., & Corbally, C. (2019). Changing the paradigm on human enhancements: The special case of modifications to counter bone loss for manned mars missions. *Space Policy, 48*, 68e75. https://doi.org/10.1016/j.spacepol.2019.02.001.

Van Eyghen, H. (2019). Is supernatural belief unreliable formed? *International Journal for Philosophy of Religion, 85*(2), 125–148. https://doi.org/10.1007/s11153-018-9671-4.

Weber, M. (1920). [illegible] *[illegible] Religionssoziologie I* (pp. [illegible]). Tübingen: Mohr.

Whitehead, B. R., & Bergeman, C. S. (2012). Coping with daily stress: Differential role of spiritual experience on daily positive and negative affect. *The Journals of Gerontology: Series B, 67*(4), 456–459. https://doi.org/10.1093/geronb/gbr136

Index

K. Szocik (ed.), *Human Enhancements for Space Missions*, Space and Society,
https://doi.org/10.1007/978-3-030-42036-9

The manufacturer's authorised representative in the EU is Springer
Nature Customer Service Centre GmbH, Europaplatz 3, 69115 Heidelberg,
Germany. If you have any concerns regarding our products, please
contact ProductSafety@springernature.com

Printed and bound by CPI Group (UK) Ltd, Croydon, CR0 4YY
30/09/2026
02225503-0006